Ultrathin Two-Dimensional Semiconductors for Novel Electronic Applications

Ultrathin Two-Dimensional Semiconductors for Novel Electronic Applications

Mohammad Karbalaei Akbari
Serge Zhuiykov

CRC Press
Taylor & Francis Group
Boca Raton London New York

CRC Press is an imprint of the
Taylor & Francis Group, an **informa** business

First edition published 2021
by CRC Press
6000 Broken Sound Parkway NW, Suite 300, Boca Raton, FL 33487-2742

and by CRC Press
2 Park Square, Milton Park, Abingdon, Oxon, OX14 4RN

© 2021 Taylor & Francis Group, LLC
CRC Press is an imprint of Taylor & Francis Group, LLC

Reasonable efforts have been made to publish reliable data and information, but the author and publisher cannot assume responsibility for the validity of all materials or the consequences of their use. The authors and publishers have attempted to trace the copyright holders of all material reproduced in this publication and apologize to copyright holders if permission to publish in this form has not been obtained. If any copyright material has not been acknowledged please write and let us know so we may rectify in any future reprint.

Except as permitted under U.S. Copyright Law, no part of this book may be reprinted, reproduced, transmitted, or utilized in any form by any electronic, mechanical, or other means, now known or hereafter invented, including photocopying, microfilming, and recording, or in any information storage or retrieval system, without written permission from the publishers.

For permission to photocopy or use material electronically from this work, access www.copyright.com or contact the Copyright Clearance Center, Inc. (CCC), 222 Rosewood Drive, Danvers, MA 01923, 978-750-8400. For works that are not available on CCC please contact mpkbookspermissions@tandf.co.uk

Trademark notice: Product or corporate names may be trademarks or registered trademarks, and are used only for identification and explanation without intent to infringe.

Library of Congress Cataloging-in-Publication Data

Names: Akbari, Mohammad Karbalaei, author. | Zhuiykov, Serge, author.
Title: Ultrathin two-dimensional semiconductors for novel electronic
   applications / by Mohammad Karbalaei Akbari and Serge Zhuiykov.
Description: First edition. | Boca Raton, FL: CRC Press, 2020. | Includes
   bibliographical references. | Summary: "Offering perspective on
   scientific and engineering aspects of 2D oxide semiconductors, this book
   discusses how to successfully engineer 2D metal oxides for practical
   applications. The work provides comprehensive data about wafer-scale
   deposition of 2D oxide semiconductors, from scientific discussions up to
   the planning of experiments and reliability testing of fabricated
   samples. It discusses wafer-scale ALD of 2D metal oxides and
   investigates ALD techniques. It covers sub-nanometer surface
   functionalization and device fabrication and post-treatment. It
   addresses a range of scientific and practical applications of 2D oxide
   materials for electronic and optoelectronic devices"-- Provided by
   publisher.
Identifiers: LCCN 2020006027 (print) | LCCN 2020006028 (ebook) | ISBN
   9780367275112 (hardback) | ISBN 9780429316784 (ebook)
Subjects: LCSH: Semiconductor films. | Thin film devices.
Classification: LCC TK7871.15.F5 A39 2020 (print) | LCC TK7871.15.F5
   (ebook) | DDC 621.3815/2--dc23
LC record available at https://lccn.loc.gov/2020006027
LC ebook record available at https://lccn.loc.gov/2020006028

ISBN: 9780367275112 (hbk)
ISBN: 9780429316784 (ebk)

Typeset in Times
by Deanta Global Publishing Services, Chennai, India

To my parents, Fatti and Hossein.

*For all of your dedicated kindness, inspiring love,
and wonderful colors added to my life.*

Mohammad Karbalaei Akbari

**This book is dedicated to my beloved mother
Mrs. Alla Zhuiykova**

Посвящаю эту книгу моему самому родному и любимому человеку. Моей маме. Где бы я не находился и что бы я не делал, я всегда чувствовал ее любовь, заботу и поддержку. Пусть эта книга будет маленькой крупицей той огромной благодарности, что мне хотелось бы ей подарить.

Сережа, Сеул, Южная Корея, 2020

Contents

Preface... xiii
Acknowledgments.. xv
Authors.. xvii
Introduction.. xix

Chapter 1 Chemical Vapor Deposition of Two-Dimensional Semiconductors..... 1

  1.1 Overview 1
  1.2 The Key Parameters of CVD Growth of 2D Materials............. 4
    1.2.1 Precursor 5
    1.2.2 Temperature.. 5
    1.2.3 Pressure 6
    1.2.4 Substrate 6
    1.2.5 Other Technological Parameters 8
  1.3 The CVD Growth of 2D Materials 9
    1.3.1 Grain Size of 2D Materials 9
    1.3.2 Layer Number of 2D Materials 10
    1.3.3 Orientation... 12
    1.3.4 Morphology.. 12
    1.3.5 Phase... 12
    1.3.6 Doping... 13
    1.3.7 Quality and Defects... 14
  1.4 The Wafer-Scale CVD Growth of Continuous 2D Materials..... 15
    1.4.1 Continuous CVD Growth of 2D TMDC Films
           on Rigid Substrates.. 16
      1.4.1.1 MOCVD 16
      1.4.1.2 Conventional CVD Techniques 16
      1.4.1.3 Other Thin-Film Deposition Method 18
    1.4.2 Continuous CVD Growth of 2D TMDC Films
           on Flexible Substrates 18
  1.5 The CVD Growth of Heterostructured 2D Materials 21
    1.5.1 CVD Growth of Vertical 2D Heterostructures 22
      1.5.1.1 CVD Growth of Metal/Semiconductor
               Vertical 2D Heterostructures...................... 22
      1.5.1.2 CVD-Grown Semiconductor/
               Semiconductor Vertical 2D
               Heterostructures... 22
      1.5.1.3 CVD Growth of Semiconductor/
               Insulator Vertical 2D Heterostructures....... 24
      1.5.1.4 CVD Growth of Metal/Insulator
               Vertical 2D Heterostructures...................... 25
    1.5.2 CVD Growth of Lateral 2D Heterostructures............. 26

vii

viii   Contents

		1.5.2.1	CVD Growth of Semiconductor/ Semiconductor Lateral 2D Heterostructures	26
		1.5.2.2	CVD Growth of Metal/Insulator Lateral 2D Heterostructures	28
	1.6	Future Outlook of CVD of 2D Materials		28
	References			33

Chapter 2   Atomic Layer Deposition of Two-Dimensional Semiconductors 43

2.1	Overview of ALD Technique of 2D Materials		43
2.2	ALD Parameters		47
	2.2.1	ALD Window	47
	2.2.2	ALD Precursors	48
2.3	ALD of 2D Metal Chalcogenide Films		49
	2.3.1	ALD of 2D MoS_2	50
	2.3.2	ALD of 2D WS_2	54
	2.3.3	ALD of 2D WSe_2	56
	2.3.4	ALD of Layered SnS and SnS_2 Films	56
	2.3.5	ALD of Layered Metal Dichalcogenide Heterostructures	59
2.4	ALD of 2D Metal Oxide Films		60
	2.4.1	ALD of 2D MoO_3	63
	2.4.2	ALD of 2D WO_3	65
	2.4.3	ALD of 2D TiO_2	67
	2.4.4	ALD of 2D Al_2O_3	67
2.5	Conclusion		71
References			73

Chapter 3   Self-Limiting Two-Dimensional Surface Oxides of Liquid Metals 79

3.1	Introduction			79
3.2	The Fundamental Properties of Liquid Metals			80
	3.2.1	Monophasic and Biphasic Liquid Metals and Alloys		80
	3.2.2	Characteristics of the Liquid Metal Interface		82
3.3	Synthesis and Applications of 2D Surface Oxides of Liquid Metals			83
	3.3.1	Screen Printing of 2D Semiconductors		84
		3.3.1.1	Screen Printing of Ga_2O_3 and GaS 2D Films	84
		3.3.1.2	Wafer-Scale Screen Printing of GaS 2D Films	85
	3.3.2	Reactive Environment for Synthesis of 2D Semiconductor Films		88

Contents

ix

	3.3.3	Liquid Metal Media for Green Synthesis of Ultrathin Flux Membranes 92
	3.3.4	The vdW Exfoliation and Printing Methods of 2D Semiconductors 95

               3.3.4.1   Semiconducting SnO Monolayers 95

               3.3.4.2   Semiconducting 2D Gallium Phosphate Nanosheets 97

               3.3.4.3   Semiconducting GaN and InN Nanosheets 99

               3.3.4.4 SnO/In_2O_3 2D van der Waals Heterostructure 101

        3.3.5   Sonochemical-Assisted Functionalization of Surface Oxide of Galinstan 102

Conclusions 107

References 107

Chapter 4   Hetero-Interfaces in 2D-Based Semiconductor Devices 111

   4.1   Introduction to 2D Hetero-Interfaces 111

   4.2   Properties of 2D Hetero-Interfaces 113

        4.2.1   Band Alignment 113

        4.2.2   Charge Transport in Hetero-Interfaces 116

        4.2.3   Generation of Interlayer Excitons 118

   4.3   The Atomic Structures at Hetero-Interfaces 119

        4.3.1   Homogenous Junctions 119

               4.3.1.1   Structural Properties at Hetero-Interfaces 119

               4.3.1.2   Doping and Passivation 122

               4.3.1.3   Strain and Dielectric Modulation 122

        4.3.2   Heterogeneous Junctions 124

               4.3.2.1   Semiconductor/Semiconductor 2D Hetero-Interfaces 124

               4.3.2.2   Metal/Semiconductor (MS) Heterogeneous Junctions 125

        4.3.3   Electrical Contact at 2D Semiconductors 127

               4.3.3.1   The Geometry of Interfaces 128

               4.3.3.2   The Charge-Injection Mechanism 130

   4.4   Devices Based on Heterostructured 2D Films 131

        4.4.1   Electronic Devices Based on Heterostructured 2D Films 131

        4.4.2   Magnetic Devices Based on 2D Heterostructured Materials 133

        4.4.3   Spintronic and Valleytronic 2D Heterostructured Devices 137

Conclusion 138

References 139

x Contents

Chapter 5 Photonic and Plasmonic Devices Based on Two-Dimensional
Semiconductors 145

  5.1 Introduction 145
  5.2 Plasmonic 2D Nanostructures 146
    5.2.1 Principles 146
  5.3 Plasmonic Phenomena in Graphene 148
    5.3.1 Structural Design of 2D Graphene for Plasmonic
      Tuning 149
    5.3.2 Plasmonic Tuning of Hybrid 2D Graphene-Based
      Devices 152
  5.4 Plasmonic Phenomenon in 2D Materials beyond the
    Graphene 154
    5.4.1 Plasmon Tuning by Doping of 2D Materials
      beyond the Graphene 154
      5.4.1.1 Doping of 2D MoS_2 154
      5.4.1.2 Surface Functionalization of 2D
        Surface Oxide of Galinstan Alloy 156
  5.5 Fabrication of Nanostructured Arrays for Hybrid
    Plasmonic Devices 158
  5.6 Application of Plasmonic Devices Based on Ultrathin 2D
    Materials 161
    5.6.1 Plasmonic Two-Dimensional $Au\text{-}WO_3\text{-}TiO_2$
      Heterojunction 161
    5.6.2 Plasmonic-Assisted 2D Photodetector Devices 163
  Conclusions 165
  References 166

Chapter 6 Memristive Devices Based on Ultrathin 2D Materials 171

  6.1 Introduction to Memristor Characteristics 171
  6.2 Solid States Electronics of Resistive Switching 172
    6.2.1 The Missing Memristor 172
    6.2.2 Memristive Materials and Their RS Mechanisms 174
      6.2.2.1 Anion-Based Memristors and Their RS
        Mechanisms 174
      6.2.2.2 Cation-Based Memristors and Their
        RS Mechanisms 179
  6.3 Fabrication of Memristor Devices Based on 2D Materials 182
    6.3.1 Device Structure 182
      6.3.1.1 Vertical Memristors 182
      6.3.1.2 Lateral Memristors 183
      6.3.1.3 Tip-Based Memristors 183
    6.3.2 Atomic-Layered 2D Materials for Memristor
      Devices 184
    6.3.3 Fabrication Techniques of 2D Materials for
      Memristors 184

Contents

	6.3.4	Electrodes for 2D-Based Memristor Devices 185
6.4		Electrical-Pulse-Triggered Memristor Devices Based on Ultrathin 2D Materials 186
	6.4.1	Complementary Resistive Switching (CRS) in TaO_x, HfO_x, and TiO_x Devices 187
	6.4.2	CRS in Heterostructured Memristor Devices 191
	6.4.3	RS Phenomenon in TiO_2 Thin Films 194
		6.4.3.1 Filamentary RS in TiO_2 Ultrathin Film......194
		6.4.3.2 Complementary RS in In-Doped TiO_2 Ultrathin Film.. 196
		6.4.3.3 Bipolar RS in Pt/TiO_2/Ti/Pt 2D Memristors.. 199
	6.4.4	RS in Exfoliated 2D Perovskite Single Crystal........ 201
	6.4.5	Memristor Devices Based on 2D MoS_2 204
		6.4.5.1 Memristor Device Based on Oxidized MoS_2 Nanosheets..................................... 205
		6.4.5.2 Memristor Device Based on CVD MoS_2 2D Film.. 206
		6.4.5.3 2D MoS_2 Memristor for Efficient-Energy Radio Frequency 208
	6.4.6	Memristor Devices Based on 2D MoS_2, $MoSe_2$, WS_2, and WSe_2.. 210
	6.4.7	Memristor Devices Based on 2D Insulating h-BN 212
6.5		Optoelectronic Memristor Devices Based on Ultrathin 2D Materials 215
	6.5.1	Nonvolatile Optical Resistive Memories versus Optical Sensors.. 215
	6.5.2	Case Study of Optical Memristor Devices............... 216
		6.5.2.1 Optical Memories Based on 2D MoS_2...... 216
		6.5.2.2 Heterostructured 2D-Based Nonvolatile Optical Memory Devices 217
	Conclusion.. 221
	References 222

Chapter 7 Artificial Synaptic Devices Based on Two-Dimensional Semiconductors 229

7.1		Introduction to Artificial Synaptic Functionalities................ 229
	7.1.1	Challenges of Data Processing................................. 229
	7.1.2	Biological Synapse versus the Artificial Synapse 230
		7.1.2.1 Biological Synapse.................................... 230
		7.1.2.2 Artificial Synapse 233
7.2		Synaptic Electronics Based on 2D Materials 236
	7.2.1	Design and Structure of Devices.............................. 236
	7.2.2	Electronic/Ionic Artificial Synapses 241
		7.2.2.1 Polycrystalline 2D MoS_2 Synaptic Device 241

xii Contents

| | 7.2.2.2 | Electro-Iono-Photoactive 2D MoS_2 Synaptic Device | 242 |

7.2.2.2 Electro-Iono-Photoactive 2D MoS_2 Synaptic Device 242

7.2.2.3 Ionic Transport in 2D Perovskite Synaptic Devices 245

7.2.2.4 In-Ion Doped Ultrathin TiO_2 Optical Synaptic Devices 247

7.2.3 Electrochemical Metallization/Conductive Bridge (ECM/CB) Artificial Synapses....................249

7.2.4 Redox/Valence Changing Artificial Synapses 252

7.2.4.1 Proton Intercalated Quasi-2D α-MoO_2 Artificial Synaptic Transistor 252

7.2.4.2 Li-Ions Intercalated Quasi-2D α-MoO_2 Artificial Synaptic Transistor 255

7.2.5 Phase-Change Synapses 257

7.2.6 Thermochemical/Joule Heating Synapses 259

7.2.7 2D Heterostructured-Based Synapses.......................260

7.2.8 Optical-Based Synapses 263

7.2.8.1 Optical Synaptic Devices versus Photodetectors 263

7.2.8.2 Optical RRAM Synaptic Devices 264

7.3 Conclusions...268

References 269

Chapter 8 Sensorimotor Devices Based on Two-Dimensional Semiconductor Materials 275

8.1 Introduction to Bioinspired Sensorimotor Devices................275

8.2 Optoelectronic Devices 276

8.2.1 Sensorimotor Device Based on Ultrathin TiO_2 Film ... 278

8.2.1.1 Working Mechanism of Sensorimotors....278

8.2.1.2 Ion Intercalation for Visible Light Sensitivity 278

8.2.1.3 The Sensorimotor System.........................282

8.2.2 Optic-Neural Synaptic Device Based on h-BN/ WSe_2 2D Heterostructure 286

8.2.3 Ultrathin ZnO Photodetectors for Stretchable Sensorimotors...287

8.3 Ultrathin Oxide Films Nociceptor Devices...........................290

8.3.1 Artificial Nociceptors Based on 2D SiO_2 Diffusive Memristors 290

8.3.2 Thermal Nociceptor Based on 2D SiO_2 Diffusive Memristors 295

8.3.3 Artificial Nociceptors Based on Ultrathin HfO_2 Memristors 296

Conclusion..299

References 300

Index..309

Preface

The undeniable role of electronic devices in society is pushing scientists, engineers and, at a higher level, decision makers to employ new strategies to respond to the requirements of communities. The significance of this challenge would be more obvious if we recognize that the semiconductor industry is now a strategical battlefield for multinational corporations; it also a game-changer in the competition among superpowers. It is estimated that municipal management of modern cities, communication technology and transportation industry will integrate the artificial intelligence and electronic systems in their new strategical designs. There are still a number of unseen applications of semiconductor-based devices that are vital for quality of life. Thus, access to the new generation of efficient semiconductor materials has always been a main concern of semiconductor scientists and engineers.

Atomically thin two-dimensional (2D) semiconductors are the important class of nanostructured materials that offer a variety of distinctive physicochemical properties. The remarkable surface properties of 2D semiconductors originate from their defect-free surface structure. The highly polarizable ions cause large nonlinear and non-uniform distribution of charges within their structure, leading to electrostatic screening in the range of 0–100 nm. Exceptional local surface and interfacial properties are generated, resulting in emerging specific energy states near and on the surface of ultrathin 2D semiconductors. The unconventional semiconducting properties of 2D materials originate from the different hybridization states of orbitals on the surface of 2D structures. Because ionic components in 2D materials can adopt different oxidation states and various binding configurations, a large number of 2D metal structures can be characterized. Subsequently, various electronic properties ranging from metallic to insulating behavior for the 2D semiconductors at different stoichiometry states can be detected. Due to these advantages, a wide range of applications have been introduced and many other promising developments are expected to be realized in the future.

Ultrathin nanostructured 2D films are an integral part of several advanced technologies. Some of their applications are well established in electronic devices such as field effect transistors. However, there are other applications that have been recently introduced or are still under development, for example, the engineering applications of 2D surface oxide of liquid alloys, plasmonic devices, 2D heterostructures, and memory and sensory applications. Other systems utilize the ultrathin nature of 2D films to explore several new phenomena or to improve device performance. There are few technologies which empower engineers to successfully achieve the wafer-scale deposition of ultrathin 2D films. Thus, one purpose of this book is to bring together the practical examples of wafer-scale deposition of 2D semiconductors. Atomic layer deposition and chemical vapor deposition are the main vapor phase growth techniques which are discussed in the ensuing chapters.

Apart from the synthesis techniques, there are several requirements for device fabrication which considerably affect the performance of 2D oxide-based devices. Heterostructures and hetero-interfaces are the integral components of electronic devices. In reality, the properties of 2D films are defined based on their interaction with substrates or other hetero-interfaces. Thus, one of the main aims of this book is to address characteristics of 2D semiconductor-based heterostructures and the charge transfer phenomena at 2D hetero-interfaces.

Today, new technologies require transparent semiconductor films for a wide range of electronic and optoelectronic applications. The 2D surface oxide of liquid metals and alloys is one of the most natural 2D structures. The idea of employing these natural 2D oxides in electronic devices can open up the world of transparent devices. However, there are still several obstacles, such as the challenge of developing and tailoring 2D surface oxides of liquid alloys into practical devices. The other challenge is attributed to the high bandgap values of these 2D oxides restricting their visible light absorption.

This book is organized as follows: Chapter 3 is allocated to the engineering applications of this family of 2D films. Several technical aspects are discussed, giving firsthand information about device fabrication to readers. The other chapters address several novel electronic applications of 2D semiconductors. The Plasmonic and optoelectronic devices, as one of the first practical application of 2D films, have been addressed in Chapter 5. There are several novel trends in 2D semiconductors which are introduced in Chapters 6, 7, and 8. These novel applications include the memristive devices, opto-memories and artificial synapses. The last chapter specifically focuses on the application of sensorimotor and nociceptor devices based on 2D semiconductor films.

This book principally is of interest to researchers, scientists and production engineers who are focused on the design of new functional optoelectronic and electronic 2D semiconductor-based devices. As a valuable training source, students of materials engineering, electrical and electronic engineering, physics, and chemistry can gain fundamental knowledge about the basic performance of 2D semiconductor-based devices. Numerous references are cited in the book, all of which will be of interest for those readers who require more information on various aspects of 2D semiconductors and electronic devices.

Acknowledgments

Notwithstanding the authors' personal dedication, hard work, and contribution to this work, it is common that most research is done through direct and/or indirect collaborations. In this regard, the authors would like to express their greatest gratitude to the following people who helped to move this project from an idea to the published product.

Dr. Mohammad K. Akbari would like to express the highest level of appreciation to Prof. Christophe Detavernier, Dr. Ranjith Karuparambil Ramachandran, and Dr. Eduardo Solano from the CoCoon Laboratory and the Center of Conformal Coating of Nanomaterials at Department of Solid-State Sciences of Ghent University, Belgium, for their unlimited contributions during his PhD research study, and because of their practical help and unlimited support.

Specifically, Prof. Serge Zhuiykov would like to express his gratitude to his lifelong friend Prof. Janusz Nowotny from the Centre for Solar Energy Technologies, University of Western Sydney, Australia. He also wishes to thank another lifelong friend and collaborator, Prof. Kourosh Kalantar-Zadeh, University of New South Wales, Australia, as one of the world-leading experts in the research of a new class of nano-materials: two-dimensional (2D) semiconductors.

In addition, both authors want to acknowledge support from the Ghent University Global Campus (GUGC) management, South Korea, and especially from V.P. Prof. Taejun Han and V.P. Dilruk De Silva. Since GUGC is a branch of the Ghent University, Belgium, in South Korea, the authors would like to acknowledge the support given by Ghent University, Belgium, for all GUGC initiatives. In this regard, help from Prof. Dr. Ir. Guido Van Huylenbroeck, Academic Director of Internationalization is invaluable. Additionally, the authors are grateful to many other colleagues from the Ghent University, Belgium, including but not limited to Prof. Dr. Ir. Christophe Walgraeve.

Appreciation is also extended to the authors' fellow GUGC colleagues: Prof. Dr. Ir. Stephen Depuydt, Prof. Dr. Ir. Philippe M. Heynderickx, Prof. Dr. Francis Verpoort, and the authors' Chinese collaborators Prof. Hong Liang Lu of the Fudan University, Prof. Chenyang Xue, and Prof. Hongyan Xu of the North University of China, Taiyuan, and Prof. Jie Hu of the Taiyuan University of Technology.

Prof. Zhuiykov would like to extend special thanks to Dr. Lachlan Hyde of the Swinburn University of Technology, Melbourne, Australia, as well as to three special friends: the energetic Mr. Alex Marich, ALNIGI Electronics Pty. Ltd., Australia, Dr. Radislav A. Potyrailo, Principal Scientist, GE Research, United States, and Dr. Vlad Maksutov, Australia. Their support and encouragement have proven invaluable. Prof. Zhuiykov also acknowledges the help of his great PhD student Zihan Wei. Her suggestions were greatly appreciated.

At this juncture, Prof. Zhuiykov would especially like to thank Mr. Eugene Kats, Senior Project Manager, EcoLight Up Pty. Ltd., Australia. He is thankful for friends who are always prepared to give their time and support: Prof. Andrei Kolmakov, Project Leader, Center for Nanoscience & Technology, NIST, United States, D.Sc.

Pavel Shuk, Principal Technologist, Rosemount Analytical, United States, Ms. Carol Roberts, Mr. David Kennedy, Mr. Andrew Ahmetov, Mrs. Olga Ahmetov, Mr. Alex Shelov, Dr. Sergei Rybalko, Mrs. Olga Rybalko, Mr. Boris Rotshtein, Australia, and Prof. Alfred Anthony Christy, Norway. Prof. Zhuiykov's fellow classmates: Mr. Igor Podoprigora, Dr. Alex Orlov, Mr. Slava Muljar, and Mr. Uri Muljar.

Moreover, Prof. Zhuiykov's very special tribute and appreciation to those who, unfortunately, are no longer with us today: his grandmother Maria Falevich, his father Ivan Zhuiykov, Mr. Alex Solienko, and Mrs. Irina Zvarich.

Certainly, this book is a thank you to Prof. Zhuiykov's father, Ivan Zhuiykov, who didn't have power and money but who gave him an example of how to live life with your own mind and to be strong and charismatic, and to his beloved mother, Alla Zhuiykova, who taught him love and kindness. This is a thank you for her unconditional support his entire life. Very special thanks go to his best friend, wife and partner, Tatiana, who is, as always, the perfect combination of prudence, support, and patience. Prof. Zhuiykov is indebted to her and his three children, Maxim, Slava, and Michael for their encouragement and support throughout his career.

Finally, anyone who has written a book understands how important is to have invaluable help from the publishing team. In this regard, constant encouragement from Allison Shatkin, Senior Publisher, Engineering, as well as Gabrielle Vernachio, Editorial Assistant at Taylor & Francis/CRC Press are greatly appreciated. The publishing team summarized the obtained materials and helped to create the finished product.

Dr. ir. Mohammad Karbalaei Akbari and *Prof. Dr. Serge Zhuiykov*

Authors

Dr. ir. Mohammad Karbalaei Akbari is a postdoctoral fellow in the faculty of Bioscience Engineering, Department of Green Chemistry at Ghent University and a teaching and research assistant at the Center for Environmental and Energy Research at Ghent University Global Campus (GUGC). He has practical experience in R&D sectors in the fields of nanostructured materials. He was also a recipient of several financial incentives and awards in these fields. His interests and experiences cover several fields of materials engineering, including the atomic layer deposition (ALD) of semiconductors, liquid metals and alloys, nanocomposites, light interaction with semiconductor materials, optical memristors, artificial synapses, sensorimotor systems, and optoelectronic devices.

Prof. Dr. Serge Zhuiykov is a senior full professor with the Faculty of Bioscience Engineering, Department of Green Chemistry and Technology of Ghent University Global Campus (GUGC). He is also Director of the Center for Environmental and Energy Research at GUGC. He has more than 28 years of academic experience from universities in Australia, Japan, South Korea, and Europe. As an expert, Prof. Zhuiykov was the official Head of Australian delegation at the International Standards Organizations (ISO) TC-21/SC-8 Technical Committee from 2002–2015. He is also the recipient of 2007, 2011, and 2013 Australian Academy of Science and 2010 Australian Government Endeavour Executive Awards for his work on advanced functional nanocrystals and their applications. Recently, he was selected as Distinguished Expert of the very prestigious *100 Talents* program of Shanxi Province, China. His research interests include development and fabrication of new advanced semiconductor nanomaterials and their heterostructures, 2D nano-crystals for various applications such as photovoltaics, opto-electronics, chemical sensors, and environmental applications. He is the author or coauthor of more than 270 scientific publications including 15 international patents, 7 book chapters, and 3 monographs (*Electrochemistry of Zirconia Gas Sensors*, CRC Press, 2007; and *Nanostructured Semiconductor Oxides for the Next Generation of Electronics and Functional Devices: Properties and Applications*; and *Nanostructured Semiconductors*, Woodhead Publishing, 2013).

Introduction

The latest advances in nanoscience and nanotechnologies during the past decade enabled unprecedented and remarkable development of the scientific and technological landscapes. Following the award of the Nobel Prize in Physics for graphene in 2010, advanced two-dimensional (2D) nanomaterials beyond graphene have emerged. These materials have developed at an increasing rate, accompanied not only by exponential growth of the scientific publications dedicated to 2D nanomaterials but also by establishing new scientific journals, such as *2D Materials, Advanced Science, Applied Materials Today, Advanced Intelligent Systems, Small,* and so on. In fact, during the past 10 years, 2D nanomaterials have clearly established themselves as valuable alternatives to graphene.

As was recently reported by the *World Economic Forum* (https://weforum.org/agenda/2016/01/the-fourth-industrial-revolution-what-it-means-and-how-to-respond/), the next industrial revolution is currently taking place. There are many emerging technological breakthroughs that have been reported, based on functionalized 2D nanomaterials and their sandwich nanostructures. In this regard, it should be stressed not only that 2D nanomaterials have rapidly been developed but also that their fabrication and functionalization techniques have advanced. For instance, the latest improvements in the atomic layer deposition (ALD) technique enabled the development of recipes for different semiconductor oxides to be fabricated with the thickness of just one fundamental layer, ensuring precise control of the deposition process on the Angstrom level. In most of these cases, the thickness of semiconductor oxide is less than 1.0 nm. More importantly, these recipes warranted deposition of the conformal, defects-free, thin films, and various 2D heterostructures on the wafer scale. Notably, the emerging technological breakthroughs based on 2D nanomaterials allowed the rapid growth in novel electronic fields such as artificial intelligence (AI), nanorobotics, bio-inspired nanostructures and devices, the Internet of Things, autonomous vehicles, wearable electronics, quantum computing, and so on, promising to radically modify our future and the ways in which we conceive our society.

A different yet also very promising trend could be observed in chemistry of nanomaterials. For example, the surface functionalization of various 2D nanostructures by different biological objects at nanoscale has added another opportunity for tailoring of nanostructured properties and enabled "nanomaterial-on-demand" strategies for complex 3D nano-architectures. One of the examples of such an approach is modification and tailoring of the metal organic frameworks (MOFs). All of this has brought materials science, organic and inorganic chemistry, and physics much closer to each other. What is more intriguing is that during the past few years, completely new approaches have emerged, based on bio-inspired nano-objects such as sensorimotor devices, nociceptors, artificial eyes, and others that were innovatively assembled into new structures and instruments and that have demonstrated completely new functionalities.

Furthermore, both funding and research in 2D-related fields have been growing exponentially worldwide over the past few years. While the AI global market is booming, the number of papers published on topics relevant to "intelligent systems" is rising and is expected to continue to increase at a rapid rate. In keeping with the needs of the intelligent systems and advanced 2D nanomaterials communities and considerable interest in the latest developments in the fields, the authors are pleased to offer a new high-quality overview of the latest approaches and recently reported developments. This book attempts to provide state-of-the-art approaches for fabrication and further functionalization of 2D nanomaterials for high-quality scientific and engineering research on artificial intelligence systems that recognize, process, and respond to stimuli/instructions and learn from experience. Specifically, the book covers interdisciplinary topics including, but not limited to, fabrication techniques for 2D nanomaterials, methods of their functionalization, various artificial, and bio-inspired smart/responsive systems and devices.

1 Chemical Vapor Deposition of Two-Dimensional Semiconductors

1.1 OVERVIEW

Since the introduction of low-dimensionality topics into scientific communities in the 1980s, there have been rapid developments in the field of low-dimensional (LD) materials. Compared with bulk microstructures, known as three-dimensional (3D) materials, LD materials cover a broad range of structures with nanoscale dimensionality [1]. In fact, quantum confinement is a unique property that originated from the nanoscale dimensionality of LD materials [2]. The precise tunability of materials property is provided by the atomic-level control of dimension. Generally, LD materials are divided into two-dimensional (2D), one-dimensional (1D), and zero-dimensional (0D) groups. Among these, 2D materials are in an extraordinary position compared with 0D and 1D materials because of their layered structures. This feature leads to several noticeable differences among 2D materials and other types of LD materials. From a synthesis point of view, 1D and 0D materials are commonly synthesized by bottom-up approaches, while 2D materials can be prepared either by top-down exfoliation techniques from their host material or be synthesized by bottom-up methods. The outstanding properties of 2D materials originate from the interactions between the layered structures. The unique specifications of 2D materials provide excellent platforms to study fundamental physics and chemistry of materials. For instance, graphene, the monolayer of graphite, a well-known representative of 2D nanomaterials, has demonstrated extraordinary charge carrier mobility, high electrical and thermal conductance, mechanical strength and a large surface area. Furthermore, fascinating and unexpected physical phenomena including the lack of bandgap and the quantum Hall effect were observed during the investigation of properties of graphene. However, the lack of bandgap does not allow us to employ graphene as a 2D semiconductor material [3]. These scientific observations have led to the increase of research and development (R&D) activities toward finding other types of 2D materials similar to graphene, including transition metal dichalcogenides (TMDCs), hexagonal-boron nitride (h-BN), black phosphene, and transition metal oxides. An important feature of 2D materials is that they exhibit a wide spectrum of electronic characteristics, covering a broad range of properties from metals to semimetals and from semiconductors to insulators. This wide range of functionalities originates from the broad energy bandgaps of 2D materials, as demonstrated in Figure 1.1 [4]. The gray strip shows the diversity of a measured bandgap for an individual 2D material. The variation in the bandgap values stems from the number of layers, doping and the surface functionalization of 2D films [4]. This wide range of bandgap values, even for an individual 2D film, confirms the capability of 2D materials for application in optoelectronic and photonic devices. Some of these well-known applications are thermal imaging (to detect wavelengths longer than 1200 nm), fiber optics communication (employing wavelengths in the 1200–1550 nm range), photovoltaics (which requires semiconductors that absorb in the 700–1000 nm range) and displays and light-emitting diodes (requiring semiconductors that emit photons in the 390–700 nm range) [5, 6]. In addition to the wide range of electronic properties, the high-surface area, presence or the lack of dangling bonds, the free nature of surface state, distinguished spin-orbit coupling and quantum spin Hall effects are found in these groups of nanostructured materials. These properties put 2D materials in a unique position to explore the novel physical and chemical properties and to develop outstanding electronic devices [7–12].

Either during the synthesis process or fabrication stage, 2D materials are designed to be positioned together with various geometrical features, mostly vertical or lateral

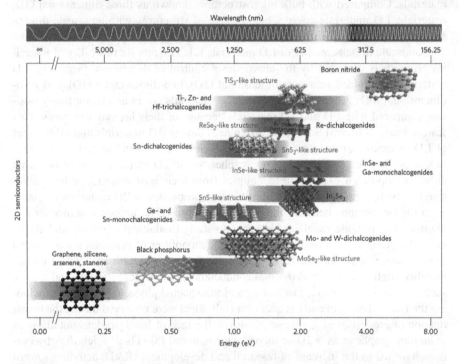

FIGURE 1.1 The comparative graphical scheme of the bandgap values for different 2D semiconductor materials with their corresponding crystalline structure. The gray horizontal bars indicate the range of bandgap values that can be spanned by changing the number of layers of the individual 2D films, straining or alloying. This wide bandgap range for even an individual 2D film shows the capability of 2D films for application in optoelectronics and photonics. Reprinted with permission from Castellanos-Gomez, A. (2016). Why all the fuss about 2D semiconductors? *Nature Photonics, 10*, 202–204.

configurations. The interface between 2D films can be atomically sharp, providing a perfect platform to study the interaction between 2D layers at the atomic scale. The hetero-interfaces of 2D materials can be witness to several distinguished physical phenomena arising from the charge transfer between 2D layers. Thus, the hetero-structure engineering of 2D materials facilitates the fabrication and development of novel devices with unusual functionalities. Taking into account that the unique features of heterostructured 2D films allow them to be employed in many potential applications including heterostructured electronics, optoelectronics, biosensing, environmental sensors, catalysis, wearable and flexible electronics, memristors and synaptic devices [13–16].

Indeed, 2D materials have generated a great deal of attention in scientific communities. Considerable number of 2D materials with diverse and tunable properties are available now. The properties of 2D materials are closely connected to their physical and chemical structures, dimensions, number of layers, morphologies, orientations, structural phases, doping defects, amorphicity or crystallinity and the grain boundaries in their structure [17]. For instance, the number of fundamental layers in 2D materials has a crucial impact on the electronic structure [18]. Furthermore, the numbers of layers determine the performance of electronic and optoelectronic 2D-based instruments [19]. These 2D materials can either be defect-free (single-crystal) or contain some level of impurities and grain boundaries. The single-crystalline structure facilitates high charge mobility and low interface scattering as well as excellent electronic performance. Therefore, properties of 2D materials are highly affected by their geometrical characteristics. The synthesis and preparation techniques of 2D materials determine most of their functionalities and also their applications in electronic devices. To date, several techniques have been generally adopted to synthesize 2D materials including mechanical exfoliation, liquid exfoliation and vapor phase growth [19, 20]. Some of these techniques are still viable for preparation of 2D materials for research. For example, mechanical exfoliation is capable of producing high-quality 2D materials with ideal properties for fundamental studies. However, these methods have failed to fulfill the requirements for large-scale synthesis of 2D materials. At the same time, the liquid phase exfoliation of 2D materials is a suitable approach for low-cost, large-scale production of 2D materials. However, uniformity of size, quality and control of the characteristics of 2D nanostructures are of concern for high-precise electronic, optoelectronic, memristive and synaptic devices. To exploit the distinguished properties of 2D materials, it is critical to have a uniform film with ultra-precise thickness. Additionally, the lack of thickness uniformity, even in the range of one fundamental layer, can ultimately change the resistive characteristics of memristor devices or unfavorably affect the energy consumption of an artificial synapse [21]. The thickness uniformity is also vital for the performance of transistor instruments based on 2D materials [22]. Thus, in order to extend applications of 2D materials from the research domain into an industrial scale, it is important to develop fabrication techniques with outstanding capabilities for large-scale fabrication of ultrathin 2D materials. Furthermore, the electronic interactions at hetero-interfaces of 2D materials provide a unique as well as an unpredictable platform for the development of an almost infinite number of distinguished properties. Thus, it is also essential to develop the fabrication method with the capability of scalable production of heterostructured 2D materials. In the case of the quality of hetero-interfaces, the sharpness and structural atomic arrangement at the hetero-interfaces between 2D films are critical factors, determining the final properties of 2D heterostructured nanomaterials.

In this regard, the vapor phase–based direct growth of the 2D layers is the most reliable and applicable synthesis and preparation approach to develop high-quality films with ultraprecise dimensional specifications [23]. These facts lead the researcher and engineers to gain a deep understanding of the growth mechanism of 2D materials with controlled quality and structure. In this view, chemical vapor deposition (CVD) offers a scalable and controllable approach for the growth of high-quality large-area 2D films [24, 25]. There are several important specifications that put the CVD method in a strong position as a reliable technique for the controllable deposition of 2D materials. The CVD is based on the reaction of gaseous materials in a vapor state or on the surface of a substrate following by the formation of solid products on the surface [26, 27]. It is possible to precisely control the number of layers of 2D films, their lateral sizes, morphology, crystalline structure, the orientation and even the amount of doping elements and structural defects by manipulation of growth parameters of CVD process [28]. The control of CVD technique was also found as a helpful approach to fabricate heterostructured 2D materials with novel properties distinguished from the singular 2D films. Furthermore, the developed novel CVD techniques make it possible to grow large-area high-quality 2D films. The technological development of CVD technique accompanied by the rapid scientific progress in the synthesis of novel 2D materials can improve the capability of CVD technique for the large-scale and high-quality synthesis of 2D films [29, 30].

The present chapter focuses on the fundamental aspects of nucleation and growth of 2D materials as well as the wafer-scale synthesis and growth mechanisms. The topics specifically focus on the CVD of 2D TMDCs, since the CVD is one of the most recommended techniques for deposition of transition metal dichalcogenide–based 2D semiconductors. A comprehensive platform is provided for readers on the recent technological progress in controllable synthesis and growth of 2D materials. Furthermore, several key challenges and also opportunities in fabrication and deposition of 2D based semiconductor materials are discussed. Finally, the development of heterostructured 2D nanofilms with CVD technique is also discussed.

1.2 THE KEY PARAMETERS OF CVD GROWTH OF 2D MATERIALS

The properties of 2D materials are highly dependent on geometrical features, morphology, structural phases and interfacial properties. Thus, these properties could be modulated by the careful tuning of the CVD growth parameters for 2D materials. Therefore, it is very important to understand the general mechanisms of CVD growth. In this view, it would be possible to realize how the CVD parameters including temperature, pressure, substrate and precursors can affect the mass and heat transfer, interface reactions, and consequently, determine the growth mechanisms of 2D materials. In the following sections, the key parameters for the CVD growth of 2D films are discussed.

1.2.1 PRECURSOR

Precursors are chemical materials serving as reactants during the CVD process. The conversion of precursors into the desired products are involved by three different reactions, including thermal decomposition, chemical synthesis and chemical transport reactions. The precursors could be either gaseous or solid. The gaseous precursors are convenient for the CVD since the flow rate of precursors can be controlled easily and accurately over the wide range of concentrations and pressures. It allows to precisely control the structural morphology and the size of 2D films. One of the most common examples is the CVD of graphene by CH_4 and H_2. Furthermore, the precise doping of CVD film is conveniently possible by introducing doping gases. For example, N and P atoms can be doped into graphene structure during CVD process by using NH_3 and PH_3 gases, respectively [31, 32].

The other groups of CVD precursors are solid sources that are mostly used for deposition and growth of TMDCs. In this typical CVD process the transition metal oxides (MoO_3, WO_3) or chlorides ($MoCl_5$) are commonly used as the metal sources while the solid sulfur and selenium are consumed as the source of sulfide and selenides. Since the vapor pressure of solid precursors is highly dependent to the temperature of chamber, the CVD of 2D materials with solid precursors is highly challenging and the controllability of TDMC growth is difficult. In this regards, other gaseous precursors are introduced to improve the growth controllability and uniformity of TDMCs films by CVD method [33].

1.2.2 TEMPERATURE

Generally speaking, the temperature in CVD system affects several different parameters including the flow rate of carrier gas, the rate of chemical reactions of precursors and the growth rate of CVD film. Consequently, uniformity, layer number and film composition can be determined by the control of reactor temperature. Generally, high-quality films can be deposited at the relatively higher temperatures, but only at the price of high-energy consumption. The concentration gradient is another drawback of using high temperature CVD process. High temperature during CVD process causes concentration gradient near to the deposited film in CVD chamber. The growth of TMDCs is critically sensitive to CVD temperature since it directly affects the saturation pressure of gasified solids, and consequently, affects the growth rate of CVD film. Furthermore, the mass transport of precursor spices is controlled by the CVD temperature, which tangibly affects the growth rate of films. For example, the higher temperature results in the higher gas concentration of sulfur and selenium during deposition of TMDCs films. Higher gas precursors concentration can facilitate the controllable chemical reactions. On the other hand, the insufficient CVD temperature results in the limited mass transport and insufficient growth rate. During the nucleation stage, at the vapor-solid interface the chamber temperature directly affects the solidification and growth mechanism. A high CVD temperature leads to a thermodynamic-activated deposition mechanism while a low CVD temperature usually leads to the kinetic growth process. Furthermore, it was found that the CVD temperature directly affects the number of fundamental layers of the deposited 2D films suggesting that higher temperature facilitates the faster growth rate [34].

1.2.3 PRESSURE

Pressure has direct effect on the gas flow in the CVD chamber and it can be variable from few atmospheres to several millimeters. At low pressure the volume flow and the velocity of precursor gas are much higher than those of the high-pressure chambers, which is based on the ideal gas relation (PV = nRT). It is well known that the low concentration and high velocity of the mass feed provide the best conditions for a controllable CVD process. A low-pressure CVD approach is usually more reliable for the wafer-scale growth of continues TMDC films [35, 36]. In addition, it was also realized that the partial temperature of each individual precursor could affect the uniform layer-by-layer growth of TMDC films [35, 36]. For example, during the growth of MoS_2 film, the second fundamental layer can only start to grow at the grain boundaries of the first fundamental layer at the low pressure. However, at a higher CVD temperature, the growth of second layer starts randomly over the first layer and the final structure would be the mixture of monolayer and multilayered films [35, 36].

1.2.4 SUBSTRATE

The substrate has several fundamental roles on the deposition process of 2D materials and its role is not merely restricted to the conventional role of substrate as a basement for deposition of 2D films. Inert Si/SiO_2 film, mica and polyimides are commonly used for deposition of 2D TMDCs for electronic and optoelectronic applications [37–39]. Furthermore, the deposition of 2D films on the metallic substrate such as gold, tungsten and catalytic active nickel, copper and Ag films are reported [40–42]. The role of substrate is vital since the microstructure and crystalline lattice of 2D films can be substantially affected by the crystalline structure of substrate. For instance, when an Au foil was used as the substrate, the nucleation and domain size of TDMCs films was affected by the crystalline orientation of the substrate film. The preferential nucleation and growth of 2D TMDC films on the specific faces is technically originated from the facet-dependent binding energy between the metallic substrate and following deposited TDMC film. The substrate orientation also affects and controls the morphology of as-grown CVD samples. The successful growth of continues MoS_2 film with domain size of 20 μm was achieved by using a vertically oriented substrate in CVD chamber. Some substrates are intentionally used to promote the nucleation of CVD film such as perylene-3,4,9,10-tetracarboxylic acid tetrapotassium salt (PTAS), perylene-3,4,9,10-tetracarboxylic dianhydride (PTCDA), and reduced graphene oxide (rGO) [43]. Moreover, the type of seeding promoters can also be different. For example, the alkali metal halides and some chlorides can react with metal oxide precursors with high melting temperature and form volatile intermediates. This technique helps to increase the domain sizes of TMDCs and broad the growth window of CVD process [44, 45].

In the case of metal oxides, substrate effects are also very important. From a structural point of view, the crystalline bulk of oxide materials can be considered as

CVD of 2D Semiconductors

the sequential order of alternating crystalline layers or as the combination of metal oxygen polyhedral coordination blocks, which are connected via shared corners. In 2D structures, the limitations caused by the structural features of bulk materials are removed, alleviated or altered. The epitaxial effects of substrate on the growth of 2D oxide films alone, the interaction between the oxide ultrathin films and their substrates experience both charge distribution and mass transfers [46]. Compared with a bulk oxide, the chemical bonding, the electronic structure and its levels, and interfacial interactions are fundamentally changed in 2D oxide nanofilms [47–49], which each individual alteration alone can unpredictably affect the properties of ultrathin nanostructures. For the practical targets, the deposited monolayer oxides are usually supported by a solid substrate [50, 51]. The automatically clean surface of inert metals is the most convenient substrate for deposition of 2D oxide films. The rigid support for 2D oxide layer can be gained by using noble metals of group Ib (Cu, Ag, Au) as substrate. The resulting interface would be an abrupt but not necessarily chemically inert one, keeping in mind, however, that the interplay and interaction between the metal atoms of substrate and the metal oxide components would be effective in determining the geometrical features of deposited 2D oxide films and overwhelmingly affect the resulting properties of 2D oxide nanostructures [52]. The conducting characteristic of metal substrates is beneficial for the charge transfer from oxide monolayer to metal substrate for subsequent experimental measurements of electrical properties of 2D oxide materials [53]. The conductance of metal substrates also gives the opportunity to characterize the structural and chemical features of 2D oxide films by electron-based probe techniques, scanning tunneling microscopy (STM) and electron diffraction techniques.

The interface of 2D oxide with Au could be considered as the practical example for interaction between 2D oxide films and the metal substrates. Once the oxide film was grown, the mass transfer between the oxide film and the metal substrate usually can be ignored, since the Au surface is inherently inert. However, the interfacial structure should be considered for the growth of the first monolayer. Thus, the bonding energies between Au and oxygen (O) or a metal (M) component of 2D oxide film play a significant role in determining the chemical bonding structures at the 2D oxide–Au interface [54]. Considering the weak adsorption of oxygen on Au surface and facile decomposition of Au oxide, the preferential bonding between Au and metal atoms of oxide film is predicted, mostly resulting in the development of Au–M–O interfaces [55–57]. The level and the quality of interaction between 2D oxide films and metal substrate can tangibly affect the structural growth of the developed oxide thin films. A strong interaction between metal substrate and oxide thin films facilitates the adaptive growth of 2D oxide films from the crystalline structure of metal substrate [58–60]. So far, a variety of 2D oxide films have been grown on Au substrates including TiO_x [60], VO_x [61], CoO [62], MoO_3 [63], MgO [64], ZnO [65] and WO_x [66]. The bulk Au has a face centered cubic (fcc) crystalline structure, while different crystalline planes of Au can provide different crystalline orientations. As an example, the unconstructed Au (111) consists of hexagonal lattices, while a reconstructed Au (111) plane shows a complex structure [67, 68]. It was observed that various deposited oxide films on the Au substrate lifted the herringbone reconstruction, which is caused by the strong interaction between Au (111) facet

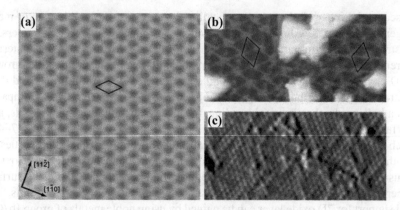

FIGURE 1.2 The STM image of (a) the honeycomb Ti$_2$O$_3$ structure, and (b) pinwheel TiO monolayer grown on Au (111), and (c) atomically resolved STM image of pinwheel structures on Au (111). Reproduced with permission from Wu, C., Marshall, M. S. J., & Castell, M. R. (2011). Surface structures of ultrathin TiO$_x$ films on Au(111). *Journal of Physical Chemistry C, 115*(17), 8643–8652.

and the ultrathin oxide film. Nevertheless, by the formation of oxide-Au interface, there is no longer an energetic advantage to adapt from the top layer of Au films. Consequently, the original hexagonal lattice form of Au (111) faces acts as a template for the growth of ultrathin oxide films [60]. In the case of TiO$_x$ film grown on the Au (111) substrate, the coexistence of honeycomb and pinwheel structures with the different stoichiometry was reported, as displayed in Figure 1.2 [69]. When the Ti atoms occupy the threefold hollow sites of Au lattice and O atoms position at the bridge sites of Ti atoms, a honeycomb structure with stoichiometry of Ti$_2$O$_3$ is formed, shown in Figure 1.2 (a). In the other mechanism, the superposing of a metal/O lattice over the Au (111) surface forms *Moiré* patterns resulting in the growth of different appearance of pinwheel structures with stoichiometry of TiO, as demonstrated in Figure 1.2 (b) and (c), respectively [69]. The hexagonal structures are the most commonly observed *Moiré* patterns of ultrathin 2D oxide films grown on Au (111) substrates. In addition to TiO, the hexagonal *Moiré* patterns were characterized in FeO [70], CoO [71] and ZnO [72] ultrathin films developed on Au (111) substrate.

1.2.5 Other Technological Parameters

The required energy for breaking the chemical bonds in CVD process can be supplied by conventional heaters, plasma light or lasers. Among them, plasma-enhanced CVD is a powerful technique enabled the deposition of several different type of 2D films, which are very difficult or even impossible to be deposited by the regular CVD [73]. Plasma is ionized gas with negatively charged electrons, positively charged ions and neutral component. The temperature of lightweight electron species is thousands of Kelvin, thus can induce unusual phenomenon which are not possible to occur at ordinary temperatures such as dissociation of precursor molecules. For example, the CVD temperature of graphene 2D film was decreased from 1000°C in conventional CVD down to 400°C by the PECVD method [74, 75]. Furthermore, the strong

CVD of 2D Semiconductors 9

interaction between the ionic species and substrate tangibly increases the density of deposited film and improves the quality of the CVD film by removing contaminants. In this process the control of plasma parameters can tangibly modify the properties of CVD film. The induced coupled plasma chemical vapor deposition (ICP-CVD) is another technique, one that permits the deposition of films even at lower temperature than the PECVD technique [76]. There is a great potential for deposition of high-quality 2D films, especially the deposition of transparent films on polymeric or glass substrates. The role of these advanced methods is to assist the regular CVD technique to facilitate the growth of various types of novel 2D materials.

1.3 THE CVD GROWTH OF 2D MATERIALS

The typical growth of 2D TMDCs consists of the thermal evaporation of metal oxide precursors to their own sub oxides. The process is followed by the reaction of products to the other reducer vapor in chamber. In case of 2D TDMCs, the reducer vapor can be sulfur or selenium. The nucleation and growth of 2D materials in CVD process are affected by the various parameters of CVD process. The nucleation and growth mechanisms in 2D materials can be categorized based on the morphological features of deposited 2D films. Up to date, several different nucleation models are observed including the layer-by-layer (LBL) growth, layer-over-layer (LOL) morphology, dendritic growth [77] and screw-dislocation driven (SDD) growth [78]. The growth model can be determined by the various factors and the level of super-saturation plays the main critical role among them.

The initiation stage of growth of 2D materials in LBL model requires a high super-saturation condition, since high barrier energy should be overcome to let a new layer starts to grow over the old one. In LBL growth model, the precise control of CVD parameters and precursor concentration may lead to the controlled growth conditions in which the number of fundamental layers can be determined [79]. In the SDD model, the new layer starts to nucleate and grow at the center of bottom one, and the growth rate in upper layer is always higher than the bottom layer. However, the growth of upper layer is restricted to the border line of bottom layer. This model results in the pyramid-like growth of 2D materials. This always happens when the CVD process is operated under the low saturation condition and thus provides a potential capability to precisely control the number of 2D layers or even control the stacking rotation of 2D layers [80–82]. The relations between the CVD parameters and the growth model of 2D materials are discussed below.

1.3.1 GRAIN SIZE OF 2D MATERIALS

The presence or the lack of grain boundaries in 2D materials is always a critical issue during the synthesis and then evaluation of the 2D films properties. It is always a general concept among the researchers that a polycrystalline granular structure with grain boundaries may deteriorate the electronic and optoelectronic properties of 2D films. It is confirmed that the charge carrier mobility in the polycrystalline film is deteriorated due the carrier scattering at the grain boundaries [83, 84]. In case of optical photodetectors, a large-area single crystal of 2D material provides more active area for photodetection. Furthermore, it directly affects the high sensitivity

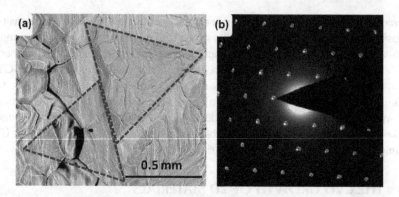

FIGURE 1.3 (a) The optical image of micron-sized single-crystal WS$_2$ domains film grown on the Au foil, accompanied by (b) the selected area electron diffraction (SAED) pattern of 12 individual parts of WS$_2$ films produced by transmission electron microscope (TEM). The SAED image confirms the single-crystalline structure of WS$_2$ film. Reprinted and reproduced with permission from Gao, Y., Liu, Z. B., Sun, D. M., Huang, L., Ma, L. P., Yin, L. C., ... Ren, W. (2015). Large-area synthesis of high-quality and uniform monolayer WS$_2$ on reusable Au foils. *Nature Communications*, 6, 8569.

and short response time of photodetector. Although the large areas single-crystal 2D films are preferable structure for the application of 2D films in high-performance optoelectronic, thermal transfer and wearable and flexible devices, some other instruments may use the profits of polygranular structure. For example, the mechanisms of resistive switching of memristor units can be affected by grain boundaries in 2D films or even can fundamentally alter the energy consumption of synaptic devices. There are several parameters, which determine the gain size of 2D films, among them the maintenance of the reactivity of CVD components for continues deposition process is fundamentally important. The main approach is the decrease of the energy barrier for the CVD reactions. The typical example is the self-limited catalytic-assisted growth of 2D WS$_2$ single-crystal film over Au foils [83]. In this method, the S$_2$ dissociates into two S atoms over the Au surface and then decreases the energy barrier for the sulfurization of the WO$_3$ precursors. It finally resulted in the growth of high-quality single-crystal WS$_2$ domains with size up to millimeter level (Figure 1.3). In another research the 2D single-crystal WS$_2$ films with 135 μm length were grown on the c-plane of sapphire by maintaining the activity of the growth process [84]. However, still the dimensions of single-crystal CVD-grown TMDC semiconductor films are far smaller than the CVD-grown nonsemiconductor graphene. To achieve the large-area single crystal of 2D TMDCs researchers still need to explore the mechanism of grain boundary formation, then reduce the number of nucleation sites and at the same time increase the CVD growth rate. Furthermore, the stable and continuum condition for long period CVD growth should be provided.

1.3.2 Layer Number of 2D Materials

One of the most interesting characteristics of 2D materials is attributed to their tunable properties by changing the number of their fundamental layers. For most of the

TMDCs the bandgap decreases as the number of layers increases from a fundamental layer to several few layers. Furthermore, the transition from indirect to direct bandgap was observed when the number of fundamental layers decreased to a monolayer for MoS_2, WSe_2, WS_2 and $MoSe_2$ films [85–88]. These properties are highly valuable for electronic and optoelectronic applications in light-emitting diode, laser diode and photovoltaic devices. The number of layers of 2D films can be controlled by different approaches. The CVD temperature can tangibly affect the number of developed 2D TMDCs films. For instance, the CVD growth of 2D films at 750°C, 825°C and 900°C, respectively resulted in the growth of monolayer, bilayer and three layer MoS_2 films with distinguished morphological features (Figure 1.4a, b, c) [89]. It was also shown that the type of substrate and precursor could also change the number of CVD-grown layer. Moreover, it is also possible to control the stacking order of TDMC films. As an example, the MoS_2 bilayer crystal showed two individual

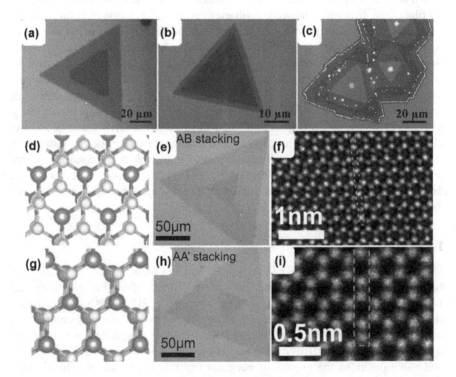

FIGURE 1.4 The control of the number of 2D layers during CVD growth. (a–c) Optical images of (a) 1-2 layer, (b) 1-3 layer, and (c) 1-4 layer $MoSe_2$. (d) The atomic configuration, (e) the optical microscopy image, and (f) the STEM image of AB stacked bilayer MoS_2 film. (g) Atomic configuration, (h) optical image, and (i) the STEM image of AA′ stacked bilayer MoS_2. (a–c) Reprinted with permission from He, Y. M., Sobhani, A., Lei, S., Zhang, Z., Gong, Y., Jin, Z., ... Ajayan, P. (2016). Layer engineering of 2D semiconductor junctions. *Advanced Materials*, 28(25), 5126–5132. (d–i) Reprinted with permission from Xia, M., Yin, K., Capellini, G., Capellini, G., Niu, G., Gong, Y., ... Xie, Y. H. (2015). Spectroscopic signatures of AA′ and AB stacking of chemical vapor deposited bilayer MoS_2. *ACS Nano*, 9(12), 12246–12254.

stacking order, i.e., AB order with 0° twist angle between the adjacent layers and AA′ order with 60° twist angle. The different stacking orders are shown by the optical (Figure 1.4e, h) and high angle annular dark field scanning TEM microscope images (Figure 1.4f, i) [89]. The property is interesting for the applications, which are dependent on the stacking order of materials, for examples the 3R phase TMDCs with ABC type stacking and noncentrosymmetric structure are employed in spintronic and valleytronic applications.

1.3.3 Orientation

The orientation control of deposited nanostructures is highly important to obtain appropriate alignment of 2D films, as was discussed previously about the substrate effects on growth of 2D films. The angle between the substrate and the first CVD-grown layer is the fundamental bases of understanding of nucleation and growth of 2D films. The disoriented lattice growth may cause the formation of grain boundaries and undesired defects during the nucleation of adjacent domain of 2D films. The nonoriented 2D domains would have negative effects on the electronic and optoelectronic properties of instruments. Therefore, it is quite meaningful to control the orientation and facilitate the growth of well-oriented TDMC domains to prepare wafer-scaled single-crystalline 2D TMDCs films for the fabrication of large-scale arrays with the lowest number of grain boundaries [90]. The orientation control is a serious challenge in CVD growth of 2D film, so that the further development of 2D TDMC based units is significantly dependent on the uniformity of wafer-scale CVD based devices. The substrate indeed plays the key role in the control of lattice orientation of the TMDC layer. The magnitude of van der Waals interaction and the degree of lattice matching between the substrate and deposited 2D films determine the orientation preference of the CVD-grown sample.

1.3.4 Morphology

Morphology control of the CVD-grown 2D films is highly related to nucleation and growth of the 2D layers. Because of the threefold lattice symmetry, 2D TMDCs grow as angular crystals. The morphology of 2D TMDCs covers different categories including dendritic, triangles, three-pointed stars, hexagonal and six pointed stars [91–93]. The formation of irregular morphologies was attributed to the high edge diffusion barrier. When the atoms cannot diffuse to their preferable positions the growth of irregular crystals is predicated. The changes in concentration of the elemental component of 2D films also affect the morphology and the edge structure of 2D films. Generally, the shape and the edge of 2D TMDC films are largely dependent on the precursor concentration, substrate and flow rate of the gases.

1.3.5 Phase

Phase control is another critical aspect during the CVD growth of 2D films. The structural phase of TMDCs can ultimately change the electronic properties of 2D

CVD of 2D Semiconductors 13

films. For example, the 1T phase with octahedral orientation of metal atoms is a metallic phase in TMDCs, while the 2H phase with trigonal prismatic coordination of the metal atoms is a semiconducting phase in the same material [94]. The semiconductor to metal transition can be related to the stacking sequence of 1T phase, thus it is simple approach to alter the electronic properties of 2D TMDC films without adding, removing and doping of atoms into 2D film structures.

The direct CVD growth of metastable phases is highly challenging target, since the thermodynamic conditions are in the favor of feasible growth of single stable phases. Previously, it was observed that the step-by-step movement of atomic elements in MoS_2 film caused the phase transformation from 2H to 1T phase during the CVD process of MoS_2 film [95]. It was also reported that the post treatment of CVD-grown films by electron beam irradiation, plasma treatment, alkali metal intercalation and even mechanical forces can help to have metastable phases in 2D films [95].

1.3.6 DOPING

The elemental doping provides a feasible platform for tuning the properties of 2D materials. Specifically, both structure and electronic properties of 2D TMDCs materials can be modified by the elemental doping. The similarity of atomic structure in various types of TMDCs gives this opportunity to prepare ternary TMDCs with metal or chalcogen dopants. Considering the structural characteristics of TMDCs, it is highly challenging to dope metal atoms than to dope chalcogen atoms in CVD growth. It is because the metal atoms are stacked between adjacent chalcogen atoms in the structure of TMDC layers. Nevertheless, it is possible to introduce a dopant precursor into a CVD chamber during the film deposition process and then control the concentration and position of dopant atoms in TMDCs. A common example is the CVD growth of $MoS_{2x}Se_{2(1-x)}$ 2D film by simultaneous introduction of S and Se precursors into the CV chamber [96]. However, when the less reactive chalcogenide atoms (for example, Te) are present in CVD chamber, the regular doping method does not work efficiently. In an innovative approach, the $WSe_{2(1-x)}Te_{2x}$ single-crystal ternary films were grown by chemical vapor transport followed by mechanical exfoliation to prepare monolayer $WSe_{2(1-x)}Te_{2x}$ film (0 < x < 1) [97].

It was confirmed that the structural phase of 2D $WSe_{2(1-x)}Te_{2x}$ film can be precisely tuned by playing with the percentage of Te element (Figure 1.5) [97]. The 2H structure was detected when the concentration is 0 < x < 0.4. The mixed 2H and 1Td phases were characterized when the x = 0.5 and 0.6. Higher concentration of doping elements (x = 0.7–1.0) results in the formation of 1Td structure [97]. The Raman spectroscopy was found as a reliable technique to evaluate the concentration of doping elements by measuring the peak shift and sharpness of the two characteristic E_{2g} and A_{1g} peaks in the Raman spectra (Figure 1.5b and c). In another approach, the post treatment of as-deposited CVD-grown film helps to obtain ternary TMDCs. For instance, as-grown monolayer MoS_2 film was targeted by Ar plasma to remove the certain sulfur atoms. Then, sulfur vacancies were occupied by the selenium atoms from the diselenodiphenyl source. The final structure was the crystalline $MoS_{2(1-x)}Se_{2x}$ 2D film [98].

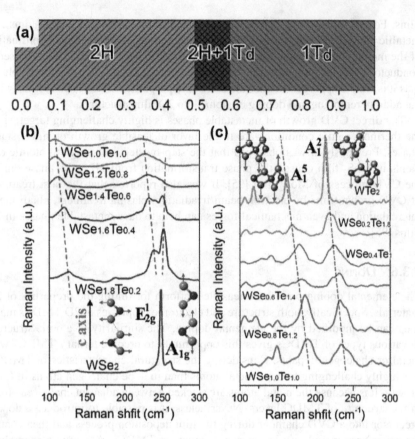

FIGURE 1.5 The doping process of 2D TMDCs films. (a) The graph shows the dependence of concentration of doping element (x) on the created phase in 2D WSe$_{2(1-x)}$Te$_{2x}$ film. (b) The Raman spectra of monolayer 2h WSe$_{2(1-x)}$Te$_{2x}$ film when the $0 < x < 0.6$. (c) The Raman spectra of monolayer 1T WSe$_{2(1-x)}$Te$_{2x}$ 2D films when the $0.6 < x < 1$. Reprinted and reproduced with permission from Yu, P., Lin, J. H., Sun, L. F., Le, Q. L., Yu, X., GVao, G., ... & Liu, F. (2017). Metal-semiconductor phase-transition in WSe$_{2(1-x)}$Te$_{2x}$ monolayer. *Advanced Materials*, 29, 1603991.

1.3.7 Quality and Defects

The presence of structural defects in as-grown CVD film is inevitable and these charged impurities, defects, and traps may seriously degrade the charge carrier mobility [99]. On the other hand, a controlled type of defects in TMDCs can have positive structural effects. For example, it is shown theoretically that the sulfur vacancies in MoS$_2$ can ease the semiconductor to metal transition or *n*-type doing can be achieved by creating Se vacancies through H$_2$/He plasma treatment [100, 101]. Therefore, to tune the 2D TMDCs properties, it is highly required to control the density and type of defects in 2D TMDCs films. The typical defects in TMDCs are chalcogen vacancies, since they have the lowest formation energy. As an example, the *S* vacancies are the main electron donors on the surface of MoS$_2$ 2D films [102,

103]. These electron donor sites induce localized states in the bandgap of MoS_2 film and contribute a key role in the charge transport by nearest-neighbor hopping or by variable range hopping [104]. Thus, the successful design of high-performance 2D-based TMDCs devices are connected to the correct understanding of charge transfer mechanism and mobility. The metal vacancies are the other main group of the structural defects in TMDCs films, which have strong influence on macroscopic properties of materials. Because of the deep trap state of the metal vacancies, the electron mobility and photoluminescence properties in 2D TMDCs film are highly affected by the presence of metal vacancies while the impact of shallow-trap states on the electron mobility is much weaker than that of metal vacancies [105, 106]. Consequently, it is possible to control the level of chalcogen and metal vacancies during the CVD process by using different techniques.

1.4 THE WAFER-SCALE CVD GROWTH OF CONTINUOUS 2D MATERIALS

The wafer-scale deposition of continues 2D films compatible with silicon micro-fabrication techniques is highly demanded for the electronic and optoelectronic targets. By deposition of ultrathin 2D uniform films over a silicon wafer, the batches of different devices can be fabricated on a singular substrate. This achievement is highly important since it is a fundamental step toward the industrialization of 2D materials in electronic industry. This is critically challenging that the wafer-scale deposited 2D films be ideally uniform and the thickness of the film be homogenous all over the wafer area. Although the spray coating of 2D materials make it possible to deposit 2D films over a large surface area, the performance of prepared electronic unit is poor and the fabricated devices suffer from low charge carrier mobility and small on/off current ratio [107]. Thus, the development and introduction of wafer-scale high-quality 2D films into electronic and optoelectronic industry is highly necessary.

There are several fundamental challenges in front of the wafer-scale growth of 2D films. They are as follows:

(i) The first challenge is attributed to continuous growth of single grain or large grain film. The granular films suffer from the intensified scattering phenomenon, which unfavorably degrades the electronic and optoelectronic properties of 2D devices.

(ii) The second challenge is attributed to the growth mechanism in CVD method. The uniformity of wafer-scale developed CVD films are not satisfying currently, and the thickness of deposited films are not uniform all over the desired area of wafer substrate, thus the 2D layer distribution may be random. This specific problem hinders the scalable CVD fabrication of instruments.

(iii) Still the wafer-scale mass production of ultrathin 2D films remains a major challenge. The growth rate in CVD technique is slow and the growth condition is considerably strict to financially response the requirement of a reliable mass production technique.

The following subsections discuss on various topics on continuous growth of various type of 2D films.

1.4.1 CONTINUOUS CVD GROWTH OF 2D TMDC FILMS ON RIGID SUBSTRATES

To deposit ultra-uniform continuous 2D film over a large surface area, the modified CVD techniques including the metal-organic chemical vapor deposition (MOCVD), low pressure chemical vapor deposition (LPCVD), inductively coupled plasma chemical vapor deposition (ICP-CVD) and the other innovative approaches are employed. Some of these methods are discussed in the next subsections.

1.4.1.1 MOCVD

MOCVD is highly interesting technique for the wafer-scale deposition of 2D films. It is because of the facile control of reaction parameters in CVD chamber via precise alteration of the precursor content during the injection into CVD chamber. For example, the successful continues wafer-scale (4-in.) depositions of MoS_2 and WS_2 films were achieved by using $Mo(CO)_6$ and $W(CO)_6$ and $(C_2H_5)_2S$ precursors in MOCVD technique where the H_2/Ar are used as carrier gas (Figure 1.6 a, b) [33]. The 2D films follow a layer-by-layer growth style. To confirm the uniformity of properties of CVD-grown 2D materials, 8100 high-performance FET devices based on 2D MoS_2 films were fabricated by using photolithography technique with the same backgate substrate (Figure 1.6c). A vertical arrangement of device was also fabricated with stacked structure. In this instrument, the external SiO_2 dielectric layer was deposited over the first MoS_2 based units (Figure 1.6d). The measurement of I_{DS} vs. the V_{DS} confirmed that the conductance change of both units is dependent on the backgate voltage confirming the consistent behavior of MoS_2 based FET. Testing the FET devices based on 2D MoS_2 film confirms the reliable compatibility of CVD technique for development of 2D based devices for two-dimensional and even three-dimensional structures.

1.4.1.2 Conventional CVD Techniques

Continuous growth of 2D TMDCs single crystal with the defined domain size is successfully possible by the conventional CVD techniques. The scalable growth of polycrystalline monolayer MoS_2 film on Si/SiO_2 substrate was achieved by using MoO_3 and S precursors in LPCVD technique. The developed monolayer films had the lateral size from several nanometers to 1.0 μm [108]. The deposited monolayer was used to fabricate FET devices with fabricated backgate electrodes. The unit performance was consistent in different FET instruments confirming the successful CVD process for developing large-scale films with uniform properties [108]. The other useful case was attributed to the selective large-scale growth of MoS_2, $MoSe_2$ and WS_2 films by the conventional CVD technique [109]. Another group has reported the wafer-scale deposition of continuous $MoSe_2$ films with controlled number of layers by tuning the CVD temperature and adjusting the flow rate of carrier gas during the conventional CVD process. Furthermore, the large-scale growth of atomically thin WSe_2 film was achieved by using the WSe_2 powder as the precursor

CVD of 2D Semiconductors

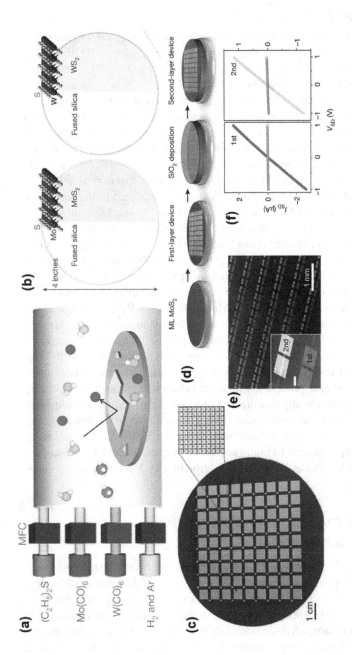

FIGURE 1.6 (a) and (b) The wafer-scale MOCVD growth of continues MoS$_2$ and WS$_2$ film. The graphical schematic of CVD device and the related precursors for MOCVD of 2D MoS$_2$ and WS$_2$ films. (c) The batch of 8100 FET devices based on 2D MoS$_2$ film over Si/SiO$_2$ wafer. The inset shows the enlarged image of one square containing 100 devices. (d) The sequential fabrication stage of stacked MoS$_2$ based device. (e) The image demonstrates the MoS$_2$ FET devices on first and second layers. (f) The I_{SD}-V_{SD} of two neighboring FET devices on two different layers. Reprinted with permission from Kang, K., Xie, S., Huang, L., Han, Y., Huang, P. Y., Mak, K. F., ... Park, J. (2015). High-mobility three-atom-thick semiconducting films with wafer-scale homogeneity. *Nature, 520*(7549), 656–660.

and the final structure was found to be polycrystalline with numerous numbers of individual domains of 2D films [110].

1.4.1.3 Other Thin-Film Deposition Method

The wafer-scale synthesis of 2D TMDC films are not merely restricted to the CVD technique, there are several innovative approaches for development of 2D films. One of the well-known strategies for the large-scale synthesis of 2D TMDC films is the decomposition or sulfurization of Mo or W foil, MoO_3 or WO_3 or their salts [111]. The predeposition methods can be done by the several different techniques including thermal deposition, atomic layer deposition, magnetron sputtering, electron beam deposition, and spin coating. In one of this methods, the post annealing process of thermally deposited MoO_3 powder at 500°C in a hydrogen atmosphere and the following sulfurization at 1000°C in sulfur vapor resulted in the growth of MoS_2 2D films over the sapphire substrate [111]. In another study, the conventional sulfurization or selenization of Mo film in a typical CVD system resulted in the formation of large-area vertically aligned MoS_2 or $MoSe_2$ 2D films. The thickness of final 2D film can be tuned by controlling the thickness of electron beam deposited Mo film resulting in facile control of the numbers of MoS_2 or $MoSe_2$ 2D films [112]. The same strategy was employed to synthesis the MoS_2 2D films by sulfurization of MoO_3 thin film at 650°C [113]. The MoO_3 was deposited by sputtering technique. Thus, the final thickness of 2D films can be tuned by adjusting the sputtering time. In another approach, the pulsed laser deposited (PLD) MoO_3 film was sulfurized and the final structure was a large-area MoS_2 film with a controlled number of layers ranging from 1 to 10 fundamental layers [114]. The number of MoS_2 2D films was controlled by controlling the sulfur particle size at the target composition and also by tuning the parameters of the PLD technique [114]. Apart from ultraprecise deposition techniques, there are other approaches, which are not considered as the technically precise methods. For example, dip coating and spin coating are simple techniques to prepare uniform thin films. However, the desired atomistic level of thickness control in these techniques is not actually possible to be achieved. Thus, the discussions of this chapter are merely restricted to high-precise techniques. Table 1.1 summarized some of the main achievements in wafer-scale growth and deposition of continues 2D TMDC films [115].

1.4.2 Continuous CVD Growth of 2D TMDC Films on Flexible Substrates

The advent of flexible devices opens up a novel field in electronic instruments with tremendous potential applications in wearable sensors, portable and flexible electronic and optoelectronics. The roll-up displays, flexible and transparent optoelectronic and biosensors and skins are all made based on flexible electronics. The 2D-based flexible and transparent devices can be fabricated by using transparent films, such as polyester, polyethylene naphthalene, polyether sulfone, and polyimide. To deposit ultrathin uniform 2D films over transparent substrates the main strategy is the direct CVD growth of 2D films on the surface.

CVD of 2D Semiconductors

TABLE 1.1

Continues Large-Area Growth of 2D TMDCs Films via CVD

Method	Materials/substrate	Growth parameters	Size of the film
MOCVD	Monolayer MoS_2/ WS_2/silicon	0.01 sccm $Mo(CO)_6$/W $(CO)_6$; 0.4 sccm $(C_2H_5)_2S$; 5 sccm H_2; 150 sccm Ar; 7.5 Torr, 550°C; 26 h	4-in. wafer scale; grain size ~1 μm
Vapor reaction	Monolayer MoS_2	35 mg of MoO_3, 600 mg of sulfur; 750–800°C; 130 sccm Ar; 0.67 Torr	wafer scale; grain size 20–1000 nm
Vapor reaction	$MoSe_2$	300 mg of Se; 20 mg of MoO_3; Si/SiO_2; 750°C; 20 min; H_2/Ar (15%)	2×2 cm^2; grain size 0.3–0.7 mm
Vapor transport	Monolayer/few-layer WSe_2/SiO_2/Si	WSe_2 powder, 1060°C for source T, 765°C for substrate T, 100 sccm, 40 min	1 cm^2
Thermal deposition	MoS_2/c-plane Sapphire	MoO_3 film deposition method; Ar/H_2 (4:1), 1 Torr, 500°C, 1 h; sulfur, 1000°C, 30 min, 70 sccm Ar, 600 Torr	2-in. wafer-scale; trilayer
Electron beam evaporation	1T′ $MoTe_2$/SiO_2/Si	1 nm Mo/MoO_3 film by electron beam evaporation; Te with molecular sieves, 3 sccm Ar+4 sccm H_2, 700°C, 1 h; atmospheric pressure	Wafer-scale film; 3.1 nm thickness
Magnetron sputtering	MoS_2/SiO_2/Si	MoO_3: RF magnetron sputtering at 300°C; sulfurization at 650°C for 1 h, 100 sccm Ar, 2×10^{-2} Torr	Wafer-scale film (1×9 cm^2) bilayer, trilayer, fewlayers
PLD	MoS_2 on various substrates	PLD targets: MoS_2/S ratio: 1:1, particle size: S(20 μm), MoS_2 (43 μm); deposition T: 700°C	Size up to substrate scale, bilayer
Thermal decomposition	MoS_2/polyimide Transparent Substrate	1.25 wt % $(NH_4)_2MoS_4$ in ethylene glycol; 280°C 30 min, 450°C 30 min; H_2/N_2 100 sccm; 1.8 Torr	Wafer-scale
Dip coating	MoS_2/SiO_2/Si	1.25 wt % $(NH_4)_2MoS_4$ in DMF, pull rate of 0.5 mm/s, sulfurization at 500 and 1000°C, 1 Torr, Ar/H_2=4:1	Wafer-scale, trilayer MoS_2 crystal domains larger Than 160 nm

Source: Reprinted and reproduced with permission from [115].

The employment of polymeric-based substrates for direct growth of 2D materials faces several following challenges. These substrates cannot stand high temperatures, their surface roughness is too high for uniform deposition of 2D films and their low thermal expansion made them non-preferable substrates for conventional high temperature CVD growth of 2D films. To tackle the above-mentioned difficulties, several strategies can be employed such as lowering the CVD growth temperature by the substrate modification. One approach is the employment of specific organic molecules on the polymeric substrate to effectively reduce the CVD reaction barrier and finally to decrease the CVD growth temperature of TMDC films. For example, the perylene-3,4,9,10-tetracarboxylic acid tetrapotassium salt (PTAS) as a promoter was used during the deposition of 2D MoS_2 film to decrease the CVD growth temperature to 650°C [116].

Thermal decomposition of molybdenum components and salts is another strategy to develop wafer-scale 2D films. In a modified two-step thermal decomposition technique, wafer-scale MoS_3 was initially deposited over polyimide plastic (PI) substrate at 280°C followed by decomposition of thin film at 450°C (Figure 1.7 a) [117]. The final structure was ultrathin continuous film of MoS_2 with stable photodetection

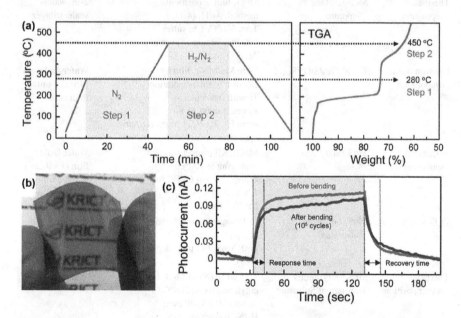

FIGURE 1.7 (a) The schematic of time dependence temperature profile of the two-step thermal decomposition process for synthesis of wafer-scale uniform MoS_2 layer accompanied by thermogravimetric analysis (TGA) of $(NH_4)_2MoS_4$. (b) The photograph of 2D MoS_2 film deposited on transparent flexible PI substrate. (c) The photocurrent of MoS_2 based device. Reprinted and reproduced with permission from Lee, S. S., An, K.-S., Choi, C.-J., Lim, J., Kim, S. J., Myung, S., ... Lim, J. (2016). Wafer-scale, homogeneous MoS_2 layers on plastic substrates for flexible visible-light photodetectors. *Advanced Materials*, 28(25), 5025–5030.

CVD of 2D Semiconductors

and photocurrent stability after several cases of bending (Figure 1.7 b, c). Although the thermal decomposition is a feasible approach to directly grow the large-scale 2D films, the decomposition temperature is still too high for many cases of flexible substrates.

1.5 THE CVD GROWTH OF HETEROSTRUCTURED 2D MATERIALS

Along with outstanding characteristics of 2D materials, the heterostructured 2D films have recently emerged as the frontier advancement for novel applications in materials science and engineering. From several different aspects, heterostructured 2D materials provide several fundamental new properties and characteristics originated from the specific charge transfer phenomena at the hetero-interfaces between two individual ultrathin films. These 2D based heterostructures exhibit novel physical properties related to electron-electron coupling and electron–phonon coupling, arisen from the layer-by-layer interactions [118, 119]. Consequently, such a developed heterostructures may deliver enhanced instrument performance compared with that of in separated 2D films. By selecting the well-matched 2D components the energy band diagram alignment and the charge carrier mobility can be modulated at hetero-interfaces, thus the requirements of different applications can be satisfied. The researcher and engineers well understand the importance of such a distinguished heterostructured design; thus, they start to build-up new 2D hetero-interfaces to open up countless opportunities for creation of novel structural materials aimed at development of high-performance devices [120].

There are generally two different types of 2D materials based heterostructures. The first one is vertical 2D structures, where different 2D materials are stacked layer by layer to each other in the vertical arrangement. The interactions between vertical heterostructures are not strong and mostly consider as the van der Waals interaction forces. Thus, there is no requirement for lattice matching between different components of layers. For the same reason, 2D heterostructures are always called van der Waals interaction heterostructures. The second type of 2D hetero-interfaced design is called lateral 2D heterostructures where different 2D materials are seamlessly joined to each other, usually by covalent bonds, in the same plane. Considering the typical physical interaction between 2D vertical films, several methods have been developed to fabricate 2D vertical heterostructures; including (i) mechanical exfoliation by scotch-tape-based method and multiple step aligned transfer of 2D films (ii) one-step or multistep CVD growth, (iii) multiple sequential deposition of solution containing 2D material and (iv) some innovative methods such as CVD of 2D materials over another 2D exfoliated films. On the other side, the type of hetero-interfaces in the lateral 2D heterostructures is usually sharp at atomistic level. Therefore, the regular methods are not capable of development of such a high-precise sharp interfaces. The atomistic level growth-based techniques are the most possible approaches for development of such a precise hetero-interfaces, among them CVD is one the most promising techniques to fabricate 2D lateral hetero-interfaces. The following sections will overview the recent progress in CVD based developed 2D heterostructures.

1.5.1 CVD Growth of Vertical 2D Heterostructures

2D materials cover a wide range of electronic properties from metallic to semiconducting and then to insulating. Thus, it is possible to design unlimited combination of 2D heterostructures. The thin-film deposition methods including CVD are also able to grow several different types of hetero-interfaces, including the (i) metal/semiconductor, (ii) semiconductor/semiconductor, (iii) semiconductor/insulator and finally (iv) metal/insulator vertical junctions.

1.5.1.1 CVD Growth of Metal/Semiconductor Vertical 2D Heterostructures

The 2D metal/semiconductor junctions always consist of graphene/TMDCs, graphene/phosphorous and TMDCs/TMDCs heterostructures. The 2D graphene/TMDCs based heterostructures are highly important and capable for fundamental applications since they take the full advantages of both heterostructured components. In this type of heterostructured 2D films, the high conductivity and work function tunability of transparent graphene are incorporated to the semiconducting characteristics of 2D TMDCs films with a wide range of bandgaps for application in electronic and optoelectronic instruments [121–123]. In one of the most common strategies, the graphene based 2D films are first deposited by CVD technique over the substrate and the process is followed by CVD growth of $(NH_4)_2MoS_4$ film [124]. The following thermal decomposition creates the graphene/MoS_2 heterostructures. Furthermore, the following MoS_2 also can be deposited by CVD over graphene substrate to develop high-quality hetero-interfaces with atomic scale resolution. Figure 1.8a depicts the SEM image of graphene/MoS_2 2D heterostructure and Figure 1.8 b and c respectively show the atomic scale image produced by scanning tunneling microscopy (STM) imaging [125]. The similar graphene/MoS_2 2D heterostructure was developed by two individual CVD methods where the MoS_2 2D films were developed as nanoribbons on the graphene surface. The fabricated photodetector based on graphene/MoS_2 2D heterostructure demonstrated high visible light photoresponsivity [126]. Although the CVD growth of 2D heterostructured graphene/MoS_2 was successful, the main drawback of this approach is attributed to the size of overlapped area between graphene and 2D MoS_2. Moreover, there is no serious evidence to confirm the successful wafer-scale CVD growth of vertical heterostructures.

1.5.1.2 CVD-Grown Semiconductor/Semiconductor Vertical 2D Heterostructures

The wide range bandgap 2D materials are the host of several various electronic characteristics. The hetero-interfaces between 2D materials can form three types of heterostructured energy band alignment (Figure 1.9 a) [115]. Type **I** heterostructure is briefly called straddling heterostructures because the valence band maximum and conduction band minimum of one of 2D films is located within the bandgap of the other 2D structure. In type **II** heterostructures the valence band maximum of heterostructure is attributed to one of 2D materials while the conduction band minimum is located in another 2D film. Type **II** hetero-interfaces are highly interesting structures for the charge transfer in optoelectronic applications, since it facilitates

CVD of 2D Semiconductors

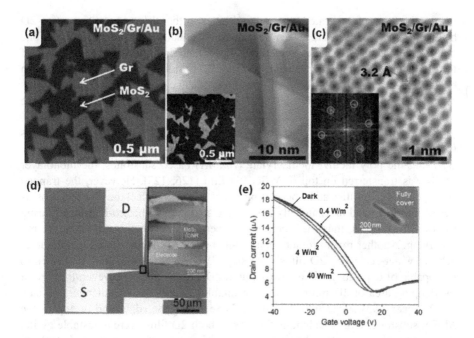

FIGURE 1.8 CVD growth of graphene/MoS$_2$ based metal/semiconductor vertical heterostructures. (a) The SEM image of as-grown MoS$_2$/graphene on Au foil. (b) The STM image of MoS$_2$/graphene heterostructure. (c) The STM image of the MoS$_2$/graphene with atomic-resolution quality and its corresponding FFT pattern (Inset). (d) The optical image of CVD-grown graphene/MoS$_2$ heterostructured photodetector device and (e) its performance under visible light illumination. (a–c) Reprinted and reproduced with permission from Shi, J. P., Liu, M. X., Wen, J. X., Ren, X., Zhou, X., Ji, Q., ... Zhang, Y. (2015). All chemical vapor deposition synthesis and intrinsic bandap observation of MoS$_2$/graphene heterostructures. *Advanced Materials, 27*(44), 7086–7092, (d, e) reprinted and reproduced with permission from Yunus, R. M., Endo, H., Tsuji, M., & Ag, H. (2015). Vertical heterostructures of MoS$_2$ and graphene nanoribbons by two step chemical vapor deposition for high-gain photodetectors. *Physical Chemistry Chemical Physics, 17*(38), 25210–25215.

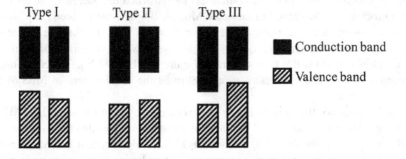

FIGURE 1.9 Different types of energy band alignment in 2D hetero-interfaces. Reprinted and reproduced with permission from Cai, Z., Liu, B., Zou, X., & Chen, H.-M. (2018). Chemical vapor deposition growth and applications of two dimensional materials and their heterostructures. *Chemical Reviews, 118*(13), 6091–6133.

the confinement of generated electron and holes in the different component of 2D heterostructures. In type **III** hetero-interfaces, the conduction band minimum of one of the components of 2D heterostructure is lower than the valence band maximum of the other one. Thus, the valence band of one of 2D films overlaps with the conduction band of another 2D film. This type of band alignment can also cause different levels of freedom for the charge transfer between the components of 2D heterostructure (Figure 1.9).

The most straightforward technique for CVD growth of heterostructured films is transfer of the as-grown individual TMDCs 2D film to another 2D film or substrate. By using the poly methyl methacrylate (*PMMA*) based substrates, the monolayer MoS_2 was transferred on top of WSe_2 2D film [126, 127]. However, the trapped contaminated impurities at the hetero-interfaces adversely affect the physical properties of interfaces and mechanically destroy the sharpness of hetero-interfaces during the transfer process, thus the electronic properties of 2D heterostructures are not satisfying. Another main problem is attributed to the difficulty of this technique to achieve wafer-scale size 2D films. One-step CVD approaches have been proven to be capable of deposition of heterostructured 2D TMDC films. However, the size of overlapping area of 2D films is not controllable. To solve this challenge, two-step CVD process was suggested to grow 2D WSe_2 along the edges and on top of the $MoSe_2$ substrate. In this technique the size of both 2D films were adjustable giving this opportunity to control the size of overlapping area of a triangle with 169 µm lateral faces (Figure 1.10) [128].

1.5.1.3 CVD Growth of Semiconductor/Insulator Vertical 2D Heterostructures

Hexagonal-boron nitride (h-BN) is an interesting material for quite numerous applications. The material is intrinsically chemically inert, with ultra-flat surface in the atomic range, free of dangling bonds and has the surface state free property. Moreover, the electron-phonon interactions in h-BN are weak originated from the small phonon energy in this nanostructure [115]. The combination of these characteristics put h-BN in outstanding position to be used as the ideal substrate or protective layer for 2D channel materials. CVD is considered as promising technique to make high-quality h-BN/TMDC vertical 2D heterostructures. Recently, an all-CVD based process was developed to fabricate MoS_2/h-BN heterostructure [129]. Ni-Ga/ Mo foil was employed as subtest and sulfur powders were used as the precursors (Figure 1.11a) to deposit high-quality single-crystal MoS_2 domains (200 µm²) on the CVD-grown 2D h-BN film substrate (Figure 1.11b, c) [129]. The SEM image accompanied by TEM characterization confirmed the development of MoS_2/h-BN heterostructures.

The h-BN is not the only insulator materials employed in insulator/2D TMDCs interfaces. Mica with hexagonal in-plane lattice structure has also been used extensively as substrate for the CVD growth of TMDC/mica vertical heterostructures. Because of atomic flatness, surface inertness and its lattice structure, mica is a suitable option for the epitaxial growth of TMDC materials since they have the same lattice symmetry [130–132]. The appropriate flexibility and transparency are the other

CVD of 2D Semiconductors

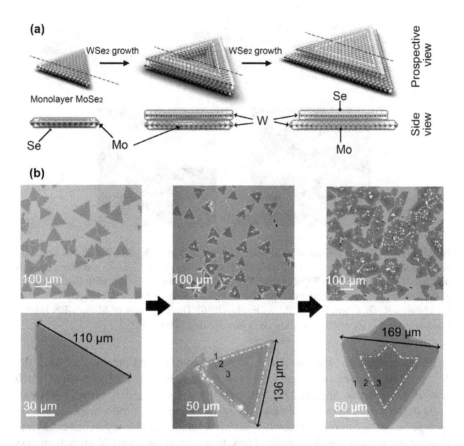

FIGURE 1.10 (a) The graphical scheme shows the CVD deposition of WSe$_2$ over monolayer MoSe$_2$. (b) The optical images of the sequential development stages for fabrication of WSe$_2$/MoSe$_2$ 2D heterostructures (from left to right). Reprinted and reproduced with permission from Gong, Y. J., Lei, S. D., Ye, G. L., Li, B., He, Y., Keyshar, K., ... Ajayan, P. M. (2015). Two-step growth of two dimensional WSe$_2$/MoSe$_2$ heterostructures. *Nano Letters*, *15*(9), 6135–6141.

main privileges of mica to be selected as the 2D heterostructure components in flexible and transparent optoelectronic devices.

1.5.1.4 CVD Growth of Metal/Insulator Vertical 2D Heterostructures

The h-BN/graphene heterostructure is the most common developed metal/insulator hetero-interfaces, since this typical 2D film is one of the main options for theoretical examination of high number of physical phenomena including Hofstadter's butterfly effect, as well as achieving high-performance graphene electronic instruments. CVD is still the most prevailing method since it provides the capability of development of large-area h-BN/graphene hetero-interfaces with acceptable thickness uniformity. The most common approach is the CVD growth of graphene over the h-BN layers. The CVD approach faces several difficulties arisen from the inert catalytic activity

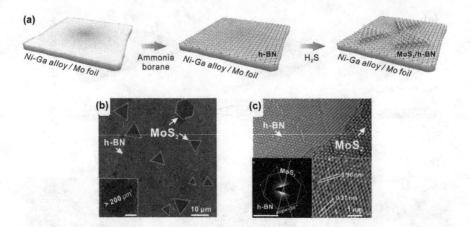

FIGURE 1.11 The graphical scheme showing the CVD growth of TMDC/h-BN semiconductor/insulator vertical heterostructures. (b) The SEM image of CVD-grown MoS$_2$ crystals over h-BN 2D substrate. (c) TEM image of MoS$_2$/h-BN heterostructures with SAED pattern of heterostructures. Reprinted and reproduced with permission from Fu, L., Sun, Y. Y., Wu, N., Mendes, R. G., Chen, L., Xu, Z., ... Fu, L. (2016). Direct growth of MoS$_2$/h-BN heterostructures via a sulfide-resistant alloy. *ACS Nano*, *10*(2), 2063–2070.

of h-BN substrate and the low decomposition rate of carbon precursors over h-BN 2D film. The proposed technique for solving the challenges of CVD growth of h-BN/graphene heterostructure consists of the deposition of graphene films over h-BN substrate through decomposition of nickelocene. In this process, the decomposed carbon rings serve as the carbon source, while the nickel atoms serve as a gaseous catalyst to increase the growth rate of graphene films. However, at one-step CV growth process the control of the quality of hetero-interfaces in the vertical h-BN/graphene heterostructure is a challenging task, since the deposition of graphene and h-BN is always accompanied by the formation of mixed domains with uncontrolled compositions.

1.5.2 CVD Growth of Lateral 2D Heterostructures

The lateral 2D heterostructures refer to a group of heterostructured materials where the components of heterostructure are attached together at the atomic level. The strict lattice matching is highly necessary for development of the 2D lateral hetero-interfaces. The following lines overview some of the CVD-grown lateral 2D heterostructures.

1.5.2.1 CVD Growth of Semiconductor/Semiconductor Lateral 2D Heterostructures

CVD is to be proved as one of the most reliable techniques to deposit the heterostructured TMDC films with the lateral structure. The one-step simple deposition of MoS$_2$/WS$_2$ lateral 2D heterostructures was reported when the sulfur, molybdenum trioxide and tungsten precursors are used during CVD process (Figure 1.12a, b). The developed hetero-interfaces were atomically sharp with atomic planes sharing the same

CVD of 2D Semiconductors

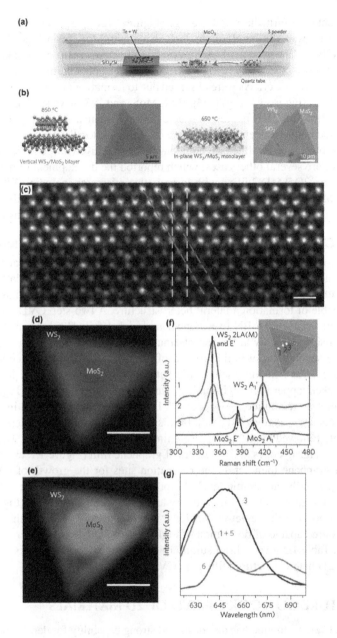

FIGURE 1.12 (a) and (b) The schematic of CVD growth of semiconductor/semiconductor lateral 2D heterostructures. (c) Atomic-resolution contrast STEM images of in-plane interface between WS_2 and MoS_2. (d) The Raman intensity mapping of 2D heterostructured 2D film at 351 cm^{-1} (yellow) and 381 cm^{-1} (purple). (e) The combined PL intensity mapping at 630 nm (orange) and 680 nm (green). (f) The Raman and (g) PL spectra of WS_2 / MoS_2 heterostructures. Reprinted and reproduced with permission from Gong, Y., Lin, J., Wang, X., Shi, G., Lei, S., Lin, Z., ... Ajayan, P. M. (2014). Vertical and in-plane heterostructures from WS_2/MoS_2 monolayers. *Nature Materials*, *13*(12), 1135–1142.

crystal orientation (armchair and zigzag) (Figure 1.12c) [133]. The final morphology of the film was a triangular core-shell shape with MoS_2 inside and WS_2 outside (Figure 1.12b, d). The growth of MoS_2/WS_2 lateral heterostructure was confirmed by Raman mapping (respectively in Figure 1.12d, e). Strong photoluminescence phenomenon and photovoltaic effects were observed due to formation of p-n junction between MoS_2 and WS_2 (Figure 1.12f, g) [133]. Both MoS_2 and WS_2 have similar lattice constant values thus the CVD process was able to fairly create a sharp hetero-interfaces. The same technique was employed to successfully develop the lateral 1H MoS_2/1T' $MoTe_2$ hetero-interfaces while two components had relatively large lattice mismatch [134]. There are several other cases, which reported the development of hetero-interfaces between 2D TMDC films with large lattice mismatch including $WSe_2/MoSe_2$ [135] and MoS_2/WS_2 [136] 2D heterostructures confirming the capability of CVD method for deposition of semiconductor/semiconductor lateral 2D heterostructures.

1.5.2.2 CVD Growth of Metal/Insulator Lateral 2D Heterostructures

The hetero-interfaces between graphene and h-BN are the most common example of the metal/insulator 2D heterostructures. Considering the similar lattice structure, the lattice mismatch between h-BN and graphene is only 1.7%. Thus, they are highly capable of formation a lateral heterostructure. A two-step CVD method was employed to first grow h-BN over copper and nickel foils [137]. Then the surface was etched selectively by argon ion plasma to remove the h-BN from the selected region. The second step was the growth of graphene films at the selective etched area (Figure 1.13a). In this method the hetero-interfaces are formed horizontally over the nickel/copper with high-quality sharp atomic-level boundary (Figure 1.13b) [137]. In another technique, a two-step CVD technique was employed in which the feeding sequence was reversed [138]. In this approach, graphene was deposited over Cu substrate followed by the selective etching of graphene with H_2 gas. This process creates fresh zigzag edges at the corners of graphene film (Figure 1.13c). The zigzag corners of graphene then act as the nucleation sites for the growth of CVD h-BN films [138]. Finally, an atomic-level abrupt interface was made at the graphene/h-BN hetero-interfaces to form high-quality lateral 2D heterostructures (Figure 1.13d, and e). Considering the reviewed sections, it was generally accepted that the CVD techniques are capable of deposition of high-quality 2D lateral and vertical heterostructures. Table 1.2 gives a brief summary of the typical techniques for continuous growth of 2D heterostructured films via CVD method [115].

1.6 FUTURE OUTLOOK OF CVD OF 2D MATERIALS

So far, the CVD technique has demonstrated strong capability for deposition of ultrathin semiconductor films. The present chapter especially focused on CVD deposition of TMDCs and their heterostructures. The key parameters of CVD growth were reviewed and the challenges were discussed. Furthermore, capability of CVD for the wafer-scale growth of ultrathin semiconductor films was investigated. The state-of-the-art methods for deposition of ultrathin 2D films over the rigid and transparent substrates were subsequently summarized to provide an overview on our current technical knowledge for fabrication of the 2D based devices. Finally, the current strategies

CVD of 2D Semiconductors

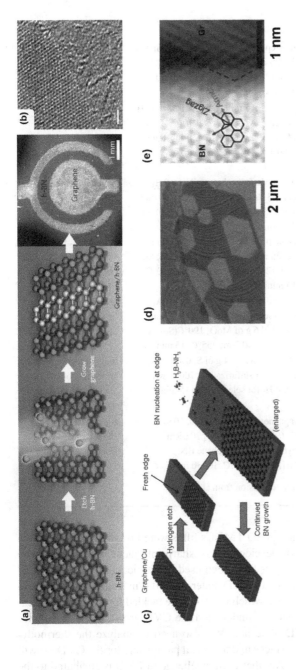

FIGURE 1.13 The two-step CVD growth of graphene/h-BN lateral 2D heterostructures. (a) The schematic shows the fabrication stage of in-plane graphene/h-BN heterostructure. (b) The STEM image of heterointerfaces between graphene and h-BN films. (c) The graphical demonstration of sequential stage of epitaxial growth of h-BN films on zigzag graphene edges. (d) The full coverage of etched holes covering the etched regions in the lack of graphene. (e) The STM atomic-resolution image of graphene/h-BN boundary. The honeycomb atomic structure and zigzag boundaries are clearly seen in the STM image of lateral hetero-interfaces. (a, b) Reprinted and reproduced with permission from Liu, Z., Ma, L. L., Shi, G., Zhou, W., Gong, Y., Lei, S., ... Ajayan, P. M. (2013). In-plane heterostructures of graphene and hexagonal boron nitride with controlled domain sizes. *Nature Nanotechnology, 8*(2), 119–124, (c, d and e) reprinted and reproduced with permission from Cai, Z., Liu, B., Zou, X., & Chen, H.-M. (2018). Chemical vapor deposition growth and applications of two dimensional materials and their heterostructures. *Chemical Reviews, 118*(13), 6091–6133.

TABLE 1.2
The Continuous Growth of 2D Heterostructured Films via CVD

Method	Materials/ substrate	Methods	
Vertical heterostructures	MoS_2/WS_2	one-step CVD: 10 mg of W + 100 mg of Te, 25 mg of MoO_3, 500 mg of S; Si/ SiO_2 (285 nm); 100 sccm Ar; 850°C, 15 min; atmospheric	triangular MoS_2 and WS_2; type II band alignment
Vertical heterostructures	$MoS_2/$ graphene	graphene: Au foil, AP, CH_4 1.5 sccm, H_2 30 sccm, Ar 200 sccm, 970°C MoS_2: MoO_3(530°C), S (102°C), substrate (680°C), Ar/H_2 (50/5 sccm)	monolayer graphene film at bottom; monolayer layer MoS_2 at top; weak n-doping level
Lateral heterostructures	$MoS_2/$ h-BN	Ni–Ga/Mo foil substrate, NH_3–BH_3 (110–130°C), Ar/H_2 75 sccm, 30 min, 1000°C; 4 sccm H_2S precursor, 680°C, 25 min, 10 sccm Ar	with whole sizes up to 200 μm^2
Lateral heterostructures	MoS_2/WS_2	one-step CVD: 10 mg of W + 100 mg of Te, 25 mg of MoO_3, 500 mg of S; Si/ SiO_2 (285 nm); 100 sccm Ar; 650°C, 15 min; atmospheric	WS_2–MoS_2 interface roughness is four unit cells with a width of 15 nm
Lateral heterostructures	$MoS_2/$ WSe_2	step 1: 0.6 g of WO_3 260°C (Se), Ar/H_2 (90/6 sccm) 20 Torr, 925°C, 15 min; step 2: 0.6 g of MoO_3 190°C(Se), Ar (70 sccm) 40 Torr, 755°C, 15 min	junction depletion width is ~320 nm, type II band alignment
Lateral heterostructures	$MoS_2/$ $MoSe_2$	0.7 g of MoO_3, 0.4 g of S, 0.6 g of Se, 750°C for sulfurization, 700°C with 5 sccm H_2 for selenization, 15 min	triangular geometry thickness of 0.8 nm interface transition in scale of ~40 nm
Lateral heterostructures	graphene/ BN	graphene: Cu foil, APCVD, 1050°C, $Ar/$ H_2 (930/60 sccm) CH_4 20 sccm hydrogen etch graphene BN: NH_3–BH_3, 120°C, 10–30 min	zigzag oriented boundaries; sharp interface boundary with width of 0.5 nm

Source: Reprinted and reproduced with permission from [115].

for the growth of vertical and lateral 2D TMDCs films were investigated to show the capability of CVD technique for development of sharp interfaces with atomic-level precision. While the CVD techniques have been used for fabrication of 2D films and their heterostructures, the CVD growth of 2D materials still is in its infant stage. There are still several remained challenges, which needed to be tackled for further modification of the technique. We still need to understand the CVD growth mechanisms for development of the different 2D materials. We also need to analyze the thermodynamic of formation of hetero-interfaces and technical parameters for the CVD growth of each individual 2D materials. Another main challenge of CVD is attributed to the evaluation of the capability of CVD technique for wafer-scale large-area deposition of

CVD of 2D Semiconductors

ultrathin 2D films over rigid and flexible substrates. Technically, it is quiet challenging procedure to deposit continues 2D single-crystal films with precise thickness controlability over a large-area of the wafer. Particular attentions are required to be allocated for further development of CVD growth of 2D films. There are several main strategies as road map for development of CVD techniques of 2D materials, including:

i. *The Development of the Current CVD Growth Technology*
There are several cases, which should be investigated. Our knowledge about the CVD precursors needs to be improved, especially when the wafer-scale growth of high-quality continues films are targeted. Currently, chalcogenide vapor phases are considered as the reliable options to be employed as the CVD precursors. It returns to the technological capability of current CVD instruments for controlled feeding of gaseous precursors into CVD chambers. Furthermore, the other type of CVD technologies, including PECVD and ICP-CVD can be employed and modified. The utilization of other thermal energy sources, such as lasers, electric or magnetic field assisted CVD techniques can be further investigated to evaluate the new opportunities for precise control of parameters of CVD technique. Furthermore, the novel in-situ characterization techniques should be developed to disclose the key information on growth mechanisms of 2D materials.

ii. *Deposition of High-Quality 2D Films*
High-quality 2D films and their heterostructures are critically important for high-performance applications of 2D materials in electronics and optoelectronics. The defect characterization techniques and related methods, including STEM imaging and spectroscopic methods are required to be improved toward the high-precise and fast characterization techniques.

iii. *Wafer-Scale Deposition of Continuous 2D Films*
Numerous efforts have been devoted during last few years to grow single-crystal 2D films over large-area substrates. The wafer-scale growth of 2D films is now the bottle neck challenge and the main obstacle in front of the industrialization and the application of 2D semiconductors. There are several proposed strategies with their own individual pros and cons. One example of this strategy is the single-crystal nucleation and continuous growth of a single-crystal 2D film. This approach requires precise adjustment of the local-feeding regime of the metal-organic precursors or liquid precursors.

The substrate is the other main challenge in the wafer-scale deposition of 2D semiconductor films. The CVD growth of 2D films over wafer substrates are limited to some specific substrates such as c-plane sapphire. It is highly required to develop novel substrates for the wafer-scale growth of 2D films. To this aim the characteristics of van der Waals epitaxial growth of the different substrates should be investigated considering the reaction chemistry of chalcogenides on the selected substrates. Nowadays, it is proved that the liquid substrates have a great potential to be promoted for the scalable growth of ultrathin 2D TMDC films. Further explorations are highly required to find the appropriate substrates or alternatively to improve the present options.

iv. *Exploration of Novel 2D Materials*

The 2D graphene aside, the exploration of other 2D semiconductor materials, including TMDCs, 2D ultrathin oxide films, and 2D insulator films are experiencing their early age of developments. Considering the numerous numbers of materials with layered structures, there is an unlimited potential for exploration and expansion of the library of 2D materials. For example, recently 2D materials such as PtS_2, ReS_2, Cu_2S, Bi_2Se_3, etc. are synthesized and investigated, which have demonstrated unique electronic, spintronic and valleytronics properties. The computational materials science has predicted numerous 2D materials with exotic properties. The manipulation of present 2D materials is one of the main strategies helping to create new 2D films with novel properties. For instance, by selectively stacking different atoms in a monolayer crystal of MoS_2 and $MoSe_2$ 2D films, novel MoSSe 2D crystalline material is synthesized with promising spintronic properties [76]. In another example, by the replacement of an oxygen atom with a Se atom in Bi_2O_3 film, new layered Bi_2O_2Se films were prepared with a high charge carrier mobility and stable performance in air atmosphere [139]. It should be stressed that the air stability is one of the challenging subjects in the development of TMDC 2D films. Therefore, the alteration and modification of present 2D materials are some of the main directions and subjects in the further exploration of the 2D materials science.

v. *Low-Temperature CVD of 2D Films*

The advent of transparent and flexible lightweight electronic devices in recent years has opened up new opportunities toward the employment of 2D films in these novel instruments. CVD can be one of the most promising and capable methods for fabrication of this type of devices. The low stability of polymeric flexible films in high temperature CVD chambers is a serious challenge for further development of this technique. To this aim two main topics should be well addressed. (1) Novel or improved pseudo-catalysts should be introduced for the low-temperature CVD growth of 2D films. Furthermore, (2) this is also necessary to find new substrates that can bear high temperature of CVD chamber.

vi. *Development of Heterostructured 2D Films*

While considerable advancement has been achieved during last years for development of 2D heterostructured films, there is still more unknown area that should be explored to facilitate the development of 2D heterostructured films. In case of the vertical heterostructured 2D films, it is highly required to develop all-CVD approach to grow high-quality vertical heterostructures. Alternatively, in case of the lateral 2D heterostructures, because of the matching lattice requirement, the number of possible options for 2D films is restricted. To be able to grow high-quality 2D structures with atomically sharp junctions at the interfaces of 2D films, both technological parameters of CVD techniques and the library of 2D materials should be improved simultaneously. The proposed road map for further development of CVD techniques and related approaches for successful deposition of 2D materials are demonstrated in Figure 1.14.

CVD of 2D Semiconductors 33

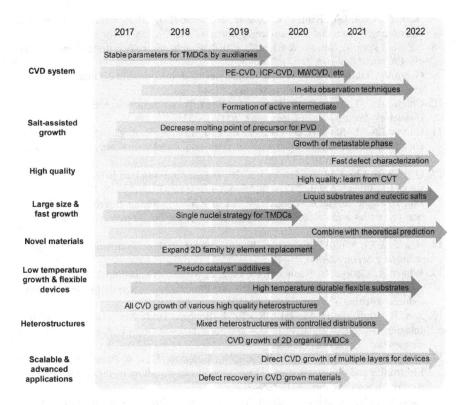

FIGURE 1.14 The proposed road map for development of scientific and technological aspects of CVD techniques aiming to the large-scale deposition of novel 2D films. Reprinted with permission from Cai, Z., Liu, B., Zou, X., & Chen, H.-M. (2018). Chemical vapor deposition growth and applications of two dimensional materials and their heterostructures. *Chemical Reviews*, *118*(13), 6091–6133.

REFERENCES

1. Velick, M. & Toth, P. S. (2017). From two-dimensional materials to their heterostructures: An electrochemist's perspective. *Applied Materials Today*, *8*, 68–103.
2. Frantzeskakis, E., Rödel, T. C., Fortuna, F., & Santander-Syro, A. F. (2017). 2D surprises at the surface of 3D materials: Confined electron systems in transition metal oxides. *Journal of Electron Spectroscopy*, *219*, 16–28.
3. Akinwandeh, D., Huyghebaert, C., Wang, C. H., Serna, M. I., Goossens, S., Li, L. J., ... Koppens, F. H. L. (2019). Graphene and two-dimensional materials for silicon technology. *Nature*, *573*(7775), 507–518.
4. Castellanos-Gomez, A. (2016). Why all the fuss about 2D semiconductors? *Nature Photonics*, *10*, 202–204.
5. Gupta, A., Sakthivel, T., & Seal, S. (2015). Recent development in 2D materials beyond graphene. *Progress in Materials Science*, *73*, 44–126.
6. Allen, M. J., Tung, V. C., & Kaner, R. B. (2010). Honeycomb carbon: A review of graphene. *Chemical Reviews*, *110*(1), 132–145.

7. Kim, K., Choi, J.-Y., Kim, T., Cho, S.-H., & Chung, H.-J. (2011). A role for graphene in silicon-based semiconductor devices. *Nature, 479*(7373), 338–344.

8. Xu, M., Liang, T., Shi, M., & Chen, H. (2013). Graphene-like two-dimensional materials. *Chemical Reviews, 113*(5), 3766–3798.

9. Gupta, A., Sakthivel, T., & Seal, S. (2015). Recent development in 2D materials beyond graphene. *Progress in Materials Science, 73*, 44–126.

10. Chhowalla, M., Jena, D., & Zhang, H. (2016). Two-dimensional semiconductors for transistors. *Nature Reviews Materials, 1*(11), 16052.

11. Chhowalla, M., Shin, H. S., Eda, G., Li, L.-J., Loh, K. P., & Zhang, H. (2013). The chemistry of two-dimensional layered transition metal dichalcogenide nanosheets. *Nature Chemistry, 5*(4), 263–275.

12. Qian, X. F., Liu, J. W., Fu, L., & Li, J. (2014). Quantum spin Hall Effect in two-dimensional transition metal dichalcogenides. *Science, 346*(6215), 1344–1347.

13. Liu, Y., Zhang, S., He, J., Wang, Z. M., & Liu, Z. (2019). Recent progress in the fabrication properties and devices of heterostructures based on 2D materials. *Nano-Micro-Letters, 11*(1), 13.

14. Niu, T., & Li, A. (2015). From two-dimensional materials to heterostructures. *Progress in Surface Science, 90*(1), 21–45.

15. Winter, A., George, A., Neumann, C., Tang, Z., Mohn, M. J., Biskupek, J. & Turchanin, A.,... (2018). Lateral heterostructures of two-dimensional materials by electron-beam induced stitching. *Carbon, 128*, 106–116.

16. Zan, R., Ramasse, Q. M., Jalil, R., Tu, J., Bangert, U., & Novoselov, K. S. (2017). Imaging two-dimensional materials and their heterostructures. *Journal of Physics: Conference Series, 902*, 012028.

17. Wang, Q. H., Kalantar-Zadeh, K., Kis, A., Coleman, J. N. & Strano, M. S. (2012). Electronics and optoelectronics of two-dimensional transition metal dichalcogenides. *Nature Nanotechnology, 7*(11), 699–712.

18. Mak, K. F., Lee, C., Hone, J., Shan, J. & Heinz, T. F. (2010). Atomically thin MoS_2: A new direct-gap semiconductor. *Physical Review Letters, 105*(13), 136805.

19. Tan, C., Cao, X., Wu, X. J., He, Q., Yang, J., Zhang, X., … Zhang, H. (2017). Recent advances in ultrathin two-dimensional nanomaterials. *Chemical Reviews, 9*(9), 6225–6231.

20. Shi, Y. M., Li, H. N., & Li, L. J. (2015). Recent advances in controlled synthesis of two-dimensional transition metal dichalcogenides via vapor deposition techniques. *Chemical Society Reviews, 44*(9), 2744–2756.

21. Zhang, L., Gong, T., Wang, H., Guo, Z. & Zhang, H. (2019). Memristive devices based on emerging two-dimensional materials beyond graphene. *Nanoscale, 11*(26), 12413–12435.

22. Liu, T., Liu, S., Tu, K. H., Schmidt, H., Chu, L., Xiang, D., … Garaj, S. (2019). Crested two-dimensional transistors. *Nature Nanotechnology, 14*(3), 223–226.

23. Tong, X., Liu, K., Zeng, M., & Fu, L. (2019). Vapor-phase growth of high-quality wafer-scale two-dimensional materials. *InfoMat, 1*(4), 460–478.

24. Liu, H. F., Wong, S. L., & Chi, D. Z. (2015). CVD growth of MoS_2-based two-dimensional materials. *Chemical Vapor Deposition, 21*(10–11), 241–259.

25. Zhang, Y., Zhang, L., & Zhou, C. (2013). Review of chemical vapor deposition of graphene and related applications. *Accounts of Chemical Research, 46*(10), 2329–2339.

26. Ji, Q. Q., Zhang, Y., Zhang, Y. F., & Liu, Z. F. (2015). Chemical vapor deposition of group-VIB metal dichalcogenide monolayers: Engineered substrates from amorphous to single crystalline. *Chemical Society Reviews, 44*(9), 2587–2502.

27. Wu, T. R., Zhang, X. F., Yuan, Q. H., Xue, J., Lu, G., Liu, Z., Jiang, M. (2016). Fast growth of inch-sized single-crystalline graphene from a controlled single nucleus on Cu-Ni alloys. *Nature Materials, 15*(1), 43–48.

28. Shi, Y. M., Li, H. N., & Li, L. J. (2015). Recent advances in controlled synthesis of two-dimensional transition metal dichalcogenides via vapor deposition techniques. *Chemical Society Reviews, 44*(9), 2744–2756.

29. Shinde, N. B., Francis, B., Rao, R., Ryu, B. D., Chandramohan, S., & Eswaran, S. K. (2019). Rapid wafer-scale fabrication with layer-by-layer thickness control of atomically thin MoS_2 films using gas-phase chemical vapor deposition. *APL Materials, 7*(8), 081113.

30. Yu, H., Liao, M., Zhao, W., Liu, G., Zhou, X. J., Wei, Z., ... Zhang, G. (2017). Wafer-scale growth and transfer of highly-oriented monolayer MoS_2 continuous films. *ACS Nano, 11*(12), 12001–12007.

31. Wei, D. C., Liu, Y. Q., Wang, Zhang, H., Huang, L., & Yu, G. (2009). Synthesis of N-doped graphene by chemical vapor deposition and its electrical properties. *Nano Letters*, 1752–1758.

32. Lv, R., Li, Q., Botello-Méndez, A. R., Hayashi, Wang, B., Berkdemir, A., ... Terrones, M. (2012). Nitrogen-doped graphene: Beyond single substitution and enhanced molecular sensing. *Scientific Reports, 2*, 586.

33. Kang, K., Xie, S., Huang, L., Han, Y., Huang, P. Y., Mak, K. F., ... Park, J. (2015). High-mobility three-atom-thick semiconducting films with wafer-scale homogeneity. *Nature, 520*(7549), 656–660.

34. He, Y. M., Sobhani, A., Lei, S., Zhang, Z., Gong, Y., Jin, Z., ... Ajayan, P. (2016). Layer engineering of 2D semiconductor junctions. *Advanced Materials, 28*(25), 5126–5132.

35. Yu, Y., Li, C., Liu, Y., Su, L., Zhang, Y., & Cao, L. (2013). Controlled scalable synthesis of uniform, high-quality monolayer and few-layer MoS_2 films. *Scientific Reports, 3*, 1866.

36. Elías, A. L., Perea-López, N., Castro-Beltrán, A., Berkdemir, A., Lv, R., Feng, S., ... Terrones, M. (2013). Controlled synthesis and transfer of large-area WS_2 sheets: From single layer to few layers. *ACS Nano, 7*(6), 5235–5242.

37. Liu, B., Fathi, M., Chen, L., Abbas, A., Ma, Y., & Zhou, C. (2015). Chemical vapor deposition growth of monolayer WSe_2 with tunable device characteristics and growth mechanism study. *ACS Nano, 9*(6), 6119–6127.

38. Ji, Q., Zhang, Y., Gao, T., Zhang, Y., Ma, D., Liu, M., ... Liu, Z. (2013). Epitaxial monolayer MoS_2 on mica with novel photoluminescence. *Nano Letters, 13*(8), 3870–3877.

39. Gong, Y. J., Li, B., Ye, G. L., Yang, S., Zou, X., Lei, S., ... Ajayan, P. M. (2017). Direct growth of MoS_2 single crystals on polyimide substrates. *2D Materials, 4*(2), 021028.

40. Zhang, Z. P., Ji, X. J., Shi, J. P., Zhou, X., Zhang, S., Hou, Y., ... Zhang, Y. (2017). Direct chemical vapor deposition growth and band-gap characterization of MoS_2/h-BN van der Waals heterostructures on Au foils. *ACS Nano, 11*(4), 4328–4336.

41. Zhou, X. B., Shi, J. P., Qi, Y., Liu, M., Ma, D., Zhang, Y., ... Zhang, Y. (2016). Periodic modulation of the doping level in striped MoS_2 superstructures. *ACS Nano, 10*(3), 3461–3468.

42. Sanne, A., Ghosh, R., Rai, A., Movva, H. C. P., Sharma, A., Rao, R., ... Banerjee, S. K. (2015). Top-gated chemical vapor deposited MoS_2 field-effect transistors on Si_3N_4 substrates. *Applied Physics Letters, 106*(6), 062101.

43. Lee, Y.-H., Yu, L., Wang, H., Fang, W., Ling, X., Shi, Y., ... Kong, J. (2013). Synthesis and transfer of single-layer transition metal disulfides on diverse surfaces. *Nano Letters, 13*(4), 1852–1857.

44. Zhang, Y., Zhang, Y. F., Ji, Q. Q., Ju, J., Yuan, H., Shi, J., ... Liu, Z. (2013). Controlled growth of high-quality monolayer WS_2 layers on sapphire and imaging its grain boundary. *ACS Nano, 7*(10), 8963–8971.

45. Chen, L., Liu, B. L., Ge, M. Y., Ma, Y., Abbas, A. N., & Zhou, C. (2015). Step-edge-guided nucleation and growth of aligned WSe_2 on sapphire via a layer-over-layer growth mode. *ACS Nano, 9*(8), 8368–8375.

46. Fu, Q., & Wagner, T. (2007). Interaction of nanostructured metal over layers with oxide surfaces. *Surface Science Reports, 62*(11), 431–498.
47. Netzer, F. P., Allegretti, F., & Surnev, S. (2010). Low-dimensional oxide nanostructures on metals: Hybrid systems with novel properties. *Journal of Vacuum Science and Technology. Part B, 28*(1), 1–16.
48. Nagel, M., Biswas, I., Peisert, H., & Chassé, T. (2007). Interface properties and electronic structure of ultrathin manganese oxide films on Ag (001). *Surface Science 601*: 4484–87.
49. Netzer, F. P. (2010). Small and beautiful-the novel structures and phases of nano-oxides. *Surface Science, 604*(5–6), 485–489.
50. Dahal, A., & Batzill, M. (2015). Growth from behind: Intercalation-growth of two dimensional FeO moiré structure underneath of metal-supported graphene. *Scientific Reports, 5*, 11378.
51. Chang, C., Sankaranarayanan, S., Ruzmetov, D., Engelhard, M. H., Kaxiras, E., & Ramanathan, S. (2010). Compositional tuning of ultrathin surface oxides on metal and alloy substrates using photons: Dynamic simulations and experiments. *Physical Review. Part B, 81*(8), 085406.
52. Shaikhutdinov, S., & Freund, H. J. (2012). Ultrathin oxide films on metal supports: Structure-reactivity relations. *Annual Review of Physical Chemistry, 63*, 619–633.
53. Nilius, N. Properties of oxide thin films and their adsorption behavior studied by scanning tunneling microscopy and conductance spectroscopy. (2009). *Surface Science Reports, 64*(12), 595–559.
54. Ma, S., Rodriguez, J., & Hrbek, J. (2008). STM study of the growth of cerium oxide nanoparticles on Au(111). *Surface Science, 602*(21), 3272–3278.
55. Sindhu, S., Heiler, M., Schindler, K. M., & Neddermeyer, H. (2003). A photoemission study of CoO-films on Au(111). *Surface Science, 541*(1–3), 197–206.
56. Barcaro, G., Cavaliere, E., Artiglia, L., Sementa, L., Gavioli, L., Granozzi, G., & Fortunelli, A. (2012). Building principles and structural motifs in TiO_x ultrathin films on a (111) substrate. *Journal of Physical Chemistry C, 116*(24), 13302–13306.
57. Tauster, S. J., Fung, S. C., Baker, R. T. K., & Horsley, J. A. (1981). Strong-interactions in supported-metal catalysts. *Science, 211*(4487), 1121–1125.
58. Ritter, M., Ranke, W., & Weiss, W. (1998). Growth and structure of ultrathin FeO films on Pt (111) studied by STM and LEED. *Physical Review. Part B, 57*(12), 7240–7251.
59. Nahm, T. U. (2016). Study of high-temperature oxidation of ultrathin Fe films on Pt(100) by using X-ray photoelectron spectroscopy. *Journal of the Korean Physical Society, 68*(10), 1215–1220.
60. Ragazzon, D., Schaefer, A., Farstad, M. H., Walle, L. E., Palmgren, P., Borg, A., … Sandell, A. (2013). Chemical vapor deposition of ordered TiO_x nanostructures on Au(111). *Surface Science, 617*, 211–217.
61. Seifert, J., Meyer, E., Winter, H., & Kuhlenbeck, H. (2012). Surface termination of an ultrathin V_2O_3-film on Au (111) studied via ion beam triangulation. *Surface Science, 606*(9–10), L41–L44.
62. Li, M., & Altman, E. I. (2014). Cluster-size dependent phase transition of Co oxides on Au(111). *Surface Science, 619*, L6–L10.
63. Quek, S. Y., Biener, M. M., Biener, J., Friend, C. M., & Kaxiras, E. (2005). Tuning electronic properties of novel metal oxide nanocrystals using interface interactions: MoO_3 monolayers on Au(111). *Surface Science, 577*(2–3), L71–L77.
64. Pan, Y., Benedetti, S., Nilius, N., & Freund, H. (2011). Change of the surface electronic structure of Au(111) by a monolayer MgO(001) film. *Physical Review. Part B, 84*(7), 075456.
65. Stavale, F., Pascua, L., Nilius, N., & Freund, H. J. (2013). Morphology and luminescence of ZnO films grown on a Au(111) support. *Journal of Physical Chemistry C, 117*(20), 10552–10557.

66. Zhuiykov, S., Hyde, L., Hai, Z., Akbari, M. K., Kats, E., Detavernier, C., ... Xu, H. (2017). Atomic layer deposition-enabled single layer of tungsten trioxide across a large area. *Applied Materials Today*, *6*, 44–53.
67. Wu, C., & Castel, M. R. (2016). Ultra-thin oxide film on Au 111 substrate. In F. P. Netzer & A. Fortunelli (Eds.), *Oxide material at the two-dimensional limit*, Springer Series in Materials Science, Switzerland, 234, 149–168. doi: 10.1007/978-3-319-28332-6-5.
68. Tauster, S. J., Fung, S. C., & Garten, R. L. (1978). Strong metal-support interactions. Group 8 noble metals supported on titanium dioxide. *Journal of the American Chemical Society*, *100*(1), 170–175.
69. Wu, C., Marshall, M. S. J., & Castell, M. R. (2011). Surface structures of ultrathin TiO_x films on Au(111). *Journal of Physical Chemistry C*, *115*(17), 8643–8652.
70. Deng, X., & Matranga, C. (2009). Selective growth of Fe_2O_3 nanoparticles and islands on Au (111). *Journal of Physical Chemistry C*, *113*(25), 11104–11109.
71. Li, M., & Altman, E. I. (2014). Shape, morphology, and phase transitions during Co oxide growth on Au (111). *Journal of Physical Chemistry C*, *118*(24), 12706–12716.
72. Tao, J., Luttrell, T., & Batzill, M. (2011). A two-dimensional phase of TiO_2 with a reduced bandgap. *Nature Chemistry*, *3*(4), 296–300.
73. Wei, D., Peng, L., Li, M., Mao, H., Niu, T., Han, C., ... Wee, A. T. (2015). Low temperature critical growth of high quality nitrogen doped graphene on dielectrics by plasma-enhanced chemical vapor deposition. *ACS Nano*, *9*(1), 164–171.
74. Chen, Z., Ren, W., Gao, L., Liu, B., Pei, S., & Cheng, H. M. (2011). Three-dimensional flexible and conductive interconnected graphene networks grown by chemical vapour deposition. *Nature Materials*, *10*(6), 424–428.
75. Gao, L., Ren, W., Zhao, J., Ma, L. P., Chen, Z., & Cheng, H. M. (2010). Efficient growth of high-quality graphene films on Cu foils by ambient pressure chemical vapor deposition. *Applied Physics Letters*, *97*, 183109.
76. Lu, A.-Y., Zhu, H., Xiao, J., Chuu, C.-P., Han, Y., Chiu, M. H., ... Li, L. J. (2017). Janus monolayers of transition metal dichalcogenides. *Nature Nanotechnology*, *12*(8), 744–749.
77. Zhang, Y., Ji, Q. Q., Wen, J. X., Li, J., Li, C., Shi, J., ... Zhang, Y. (2016). Monolayer MoS_2 dendrites on a symmetry-disparate $SrTiO_3$ (001) substrate: Formation mechanism and interface interaction. *Advanced Functional Materials*, *26*(19), 3299–3205.
78. Xia, J., Zhu, D. D., Wang, L., Huang, B., Huang, X., & Meng, X. (2015). Large-scale growth of two-dimensional SnS_2 crystals driven by screw dislocations and application to photodetectors. *Advanced Functional Materials*, *25*(27), 4255–4261.
79. Jin, S., Bierman, M. J., & Morin, S. A. (2010). A new twist on nanowire formation: Screw-dislocation-driven growth of nanowires and nanotubes. *Journal of Physical Chemistry Letters*, *1*(9), 1472–1480.
80. Meng, F., Morin, S. A., Forticaux, A., & Jin, S. (2013). Screw dislocation driven growth of ganomaterials. *Accounts of Chemical Research*, *46*(7), 1616–1626.
81. Chen, L., Liu, B., Abbas, A. N., Ma, Y., Fang, X., Liu, Y., & Zhou, C. (2014). Screw-dislocation-driven growth of two-dimensional few-layer and pyramid-like WSe_2 by sulfur-assisted chemical vapor deposition. *ACS Nano*, *8*(11), 11543–11551.
82. Wu, J. J., Hu, Z. L., Jin, Z. H., Lei, S., Guo, H., Chatterjee, K., ... Ajayan, P. M. (2016). Spiral growth of $SnSe_2$ crystals by chemical vapor deposition. *Advanced Materials Interfaces*, *3*(16), 1600383.
83. Gao, Y., Liu, Z. B., Sun, D. M., Huang, L., Ma, L. P., Yin, L. C., ... Ren, W. (2015). Large-area synthesis of high-quality and uniform monolayer WS_2 on reusable Au foils. *Nature Communications*, *6*, 8569.
84. Xu, Z. Q., Zhang, Y. P., Lin, S. H., Zheng, C., Zhong, Y. L., Xia, X., ... Bao, Q. (2015). Synthesis and transfer of large-area monolayer WS_2 crystals: Moving toward the recyclable use of sapphire substrates. *ACS Nano*, *9*(6), 6178–6187.

85. Lan, Y. W., Torres, C. M., Tsai, S. H., Zhu, X., Shi, Y., Li, M. Y., ... Wang, K. L. (2016). Atomic-monolayer MoS_2 band-to-band tunneling field-effect transistor. *Small, 12*(41), 5676–5683.

86. Yu, Y., Huang, S.-Y., Li, Y., Steinmann, S. N., Yang, W., & Cao, L. (2014). Layer-dependent electrocatalysis of MoS_2 for hydrogen evolution. *Nano Letters, 14*(2), 553–558.

87. Zhou, X., Gan, L., Tian, W., Zhang, Q., Jin, S., Li, H., ... Zhai, T. (2015). Ultrathin $SnSe_2$ flakes grown by chemical vapor deposition for high-performance photodetectors. *Advanced Materials, 27*(48), 8035–8041.

88. Babu, G., Masurkar, N., Al Salem, H., & Arava, L. M. R. (2017). Transition metal dichalcogenide atomic layers for lithium polysulfides electrocatalysis. *Journal of the American Chemical Society, 139*(1), 171–178.

89. Xia, M., Yin, K., Capellini, G., Capellini, G., Niu, G., Gong, Y., ... Xie, Y. H. (2015). Spectroscopic signatures of AA′ and AB stacking of chemical vapor deposited bilayer MoS_2. *ACS Nano, 9*(12), 12246–12254.

90. Xu, X., Zhang, Z., Dong, J., Yi, D., Niu, J., Wu, M., ... Liu, K. (2017). Ultrafast epitaxial growth of meter-sized single-crystal graphene on industrial Cu foil. *Scientific Bulletin, 62*(15), 1074–1080.

91. Einax, M., Dieterich, W., & Maass, P. (2013). Colloquium: Cluster growth on surfaces: Densities, size distributions, and morphologies. *Reviews of Modern Physics, 85*(3), 921–939.

92. Cao, D., Shen, T., Liang, P., Chen, X. S., & Shu, H. B. (2015). Role of chemical potential in flake shape and edge properties of monolayer MoS_2. *Journal of Physical Chemistry C, 119*(8), 4294–4301.

93. Wang, S. S., Rong, Y. M., Fan, Y., Pacios, M., Bhaskaran, H., He, K., & Warner, J. H. (2014). Evolution of monolayer MoS_2 crystals grown by chemical vapor deposition. *Chemistry of Materials, 26*(22), 6371–6379.

94. Wang, Z. Q., Ning, S. C., Fujita, T., Hirata, A., & Chen, M. W. (2016). Unveiling three-dimensional stacking sequences of 1T phase MoS_2 monolayers by electron diffraction. *ACS Nano, 10*(11), 10308–10316.

95. Lin, Y. C., Dumcenco, D. O., Huang, Y. S., & Suenaga, K. (2014). Atomic mechanism of the semiconducting-to-metallic phase transition in single-layered MoS_2. *Nature Nanotechnology, 9*(5), 391–396.

96. Umrao, S., Jeon, J., Jeon, S. M., Choi, Y. J., & Lee, S. (2017). A homogeneous atomic layer MoS_2(1-x)Se_2x alloy prepared by low-pressure chemical vapor deposition, and its properties. *Nanoscale, 9*(2), 594–603.

97. Yu, P., Lin, J. H., Sun, L. F., Le, Q. L., Yu, X., GVao, G., ... & Liu, F. (2017). Metal-semiconductor phase-transition in $WSe_{2(1-x)}Te_{2x}$ monolayer. *Advanced Materials, 29*, 1603991.

98. Ma, Q., Isarraraz, M., Wang, C. S., Preciado, E., Klee, V., Bobek, S., ... Bartels, L. (2014). Post growth tuning of the bandgap of single-layer molybdenum disulfide films by sulfur/selenium exchange. *ACS Nano, 8*(5), 4672–4677.

99. Najmaei, S., Yuan, J. T., Zhang, J., Ajayan, P., & Lou, J. (2015). Synthesis and defect investigation of two-dimensional molybdenum disulfide atomic layers. *Accounts of Chemical Research, 48*(1), 31–40.

100. Wang, S. S., Lee, G. D., Lee, S., Yoon, E., & Warner, J. H. (2016). Detailed atomic reconstruction of extended line defects in monolayer MoS_2. *ACS Nano, 10*(5), 5419–5430.

101. Tosun, M., Chan, L., Amani, M., Roy, T., Ahn, G. H., Taheri, P., ... Javey, A. (2016). Air-stable n-doping of WSe_2 by anion vacancy formation with mild plasma treatment. *ACS Nano, 10*(7), 6853–6860.

CVD of 2D Semiconductors

102. Kim, I. S., Sangwan, V. K., Jariwala, D., Wood, J. D., Park, S., Chen, K. S., ... Lauhon, L. J. (2014). Influence of stoichiometry on the optical and electrical properties of chemical vapor deposition derived MoS_2. *ACS Nano, 8*(10), 10551–10558.

103. Amani, M., Lien, D.-H., Kiriya, D., Xiao, J., Azcatl, A., Noh, J., ... Javey, A. (2015). Near-unity photoluminescence quantum yield in MoS_2. *Science, 350*(6264), 1065–1068.

104. Qiu, H., Xu, T., Wang, Z. L., Ren, W., Nan, H., Ni, Z., ... Wang, X. (2013). Hopping transport through defect-induced localized states in molybdenum disulphide. *Nature Communications, 4*, 2642.

105. Yu, Z., Pan, Y., Shen, Y., Wang, Z., Ong, Z. Y., Xu, T., ... Wang, X. (2014). Towards intrinsic charge transport in monolayer molybdenum disulfide by defect and interface engineering. *Nature Communications, 5*, 5290.

106. Han, H. V., Lu, A. Y., Lu, L. S., Huang, J. K., Li, H., Hsu, C. L., ... Shi, Y. (2016). Photoluminescence enhancement and structure repairing of monolayer $MoSe_2$ by hydrohalic acid treatment. *ACS Nano, 10*(1), 1454–1461.

107. Kelly, A. G., Hallam, T., Backes, C., Harvey, A., Esmaeily, A. S., Godwin, I., ... Coleman, J. N. (2017). All printed thin-film transistors from networks of liquid-exfoliated nanosheets. *Science, 356*(6333), 69–73.

108. Zhang, J., Yu, H., Chen, W., Tian, X., Liu, D., Cheng, M., ... Zhang, G. (2014). Scalable growth of high-quality polycrystalline MoS_2 monolayers on SiO_2 with tunable grain sizes. *ACS Nano, 8*(6), 6024–6030.

109. Gong, Y. J., Ye, G. L., Lei, S. D., Shi, G., He, Y., Lin, J., ... Ajayan, P. M. (2016). Synthesis of millimeter-scale transition metal. dichalcogenides single crystals. *Advanced Functional Materials, 26*(12), 2009–2015.

110. Zhou, H. L., Wang, C., Shaw, J. C., Cheng, R., Chen, Y., Huang, X., ... Duan, X. 2015. Large area growth and electrical properties of P-type WSe_2 atomic layers. *Nano Letters, 15*(1), 709–713.

111. Lin, Y. C., Zhang, W. J., Huang, J. K., Liu, K. K., Lee, Y. H., Liang, C. T., ... Li, L. J. (2012). Wafer-scale MoS_2 thin layers prepared by MoO_3 sulfurization. *Nanoscale, 4*(20), 6637–6641.

112. Kong, D. S., Wang, H. T., Cha, J. J., Pasta, M., Koski, K. J., Yao, J., & Cui, Y. (2013). Synthesis of MoS_2 and $MoSe_2$ films with vertically aligned layers. *Nano Letters, 13*(3), 1341–1347.

113. Yu, F. F., Liu, Q. W., Gan, X., Hu, M., Zhang, T., Li, C., ... Lv, R. (2017). Ultrasensitive pressure detection of few-layer MoS_2. *Advanced Materials, 29*(4), 1603266.

114. Serna, M. I., Yoo, S. H., Moreno, S., Xi, Y., Oviedo, J. P., Choi, H., ... Quevedo-Lopez, M. A. (2016). Large-area deposition of MoS_2 by pulsed laser deposition with in situ thickness control. *ACS Nano, 10*(6), 6054–6061.

115. Cai, Z., Liu, B., Zou, X., & Chen, H.-M. (2018). Chemical vapor deposition growth and applications of two dimensional materials and their heterostructures. *Chemical Reviews, 118*(13), 6091–6133.

116. Ling, X., Lee, Y.-H., Lin, Y., Fang, W., Yu, L., Dresselhaus, M. S., & Kong, J. (2014). Role of the seeding promoter in MoS_2 growth by chemical vapor deposition. *Nano Letters, 14*(2), 464–472.

117. Lee, S. S., An, K.-S., Choi, C.-J., Lim, J., Kim, S. J., Myung, S., ... Lim, J. (2016). Wafer-scale, homogeneous MoS_2 layers on plastic substrates for flexible visible-light photodetectors. *Advanced Materials, 28*(25), 5025–5030.

118. Novoselov, K. S., Mishchenko, A., Carvalho, A., & Castro Neto, A. H. (2016). 2D materials and van der Waals heterostructures. *Science, 353*(6298), aac9439.

119. Jin, C. H., Kim, J., Suh, J., Shi, Z., Chen, B., Fan, X., ... Wang, F. (2017). Interlayer electron phonon coupling in WSe_2/h-BN heterostructures. *Nature Physics, 13*(2), 127–131.

120. Liu, Y., Weiss, N. O., Duan, X. D., Cheng, H., Huang, Y., & Duan, X. (2016). van der Waals heterostructures and devices. *Nature Reviews Materials, 1*(9), 16042.

121. Li, Y., Qin, J. K., Xu, C. Y., Cao, J., Sun, Z., Ma, L., ... Zhen, L. (2016). Electric field tunable interlayer relaxation process and interlayer coupling in WSe_2/graphene heterostructures. *Advanced Functional Materials, 26*(24), 4319–4328.

122. Yu, W. J., Liu, Y., Zhou, H., Yin, A., Li, Z., Huang, Y., & Duan, X. (2013). Highly efficient gate-tunable photocurrent generation in vertical heterostructures of layered materials. *Nature Nanotechnology, 8*(12), 952–958.

123. Georgiou, T., Jalil, R., Belle, B. D., Britnell, L., Gorbachev, R. V., Morozov, S. V., ... Mishchenko, A. (2013). Vertical field-effect transistor based on graphene-WS_2 heterostructures for flexible and transparent electronics. *Nature Nanotechnology, 8*(2), 100–103.

124. Shi, J. P., Liu, M. X., Wen, J. X., Ren, X., Zhou, X., Ji, Q., ... Zhang, Y. (2015). All chemical vapor deposition synthesis and intrinsic bandap observation of MoS_2/graphene heterostructures. *Advanced Materials, 27*(44), 7086–7092.

125. Yunus, R. M., Endo, H., Tsuji, M., & Ag, H. (2015). Vertical heterostructures of MoS_2 and graphene nanoribbons by two step chemical vapor deposition for high-gain photodetectors. *Physical Chemistry Chemical Physics, 17*(38), 25210–25215.

126. Chiu, M. H., Zhang, C. D., Shiu, H. W., Chuu, C. P., Chen, C. H., Chang, C. Y., ... Li, L. J. (2015). Determination of band alignment in the single-layer MoS_2/WSe_2 heterojunction. *Nature Communications, 6*, 7666.

127. Furchi, M. M., Pospischil, A., Libisch, F., Burgdörfer, J., & Mueller, T. (2014). Photovoltaic effect in an electrically tunable van der Waals heterojunction. *Nano Letters, 14*(8), 4785–4791.

128. Gong, Y. J., Lei, S. D., Ye, G. L., Li, B., He, Y., Keyshar, K., ... Ajayan, P. M. (2015). Two-step growth of two dimensional WSe_2/$MoSe_2$ heterostructures. *Nano Letters, 15*(9), 6135–6141.

129. Fu, L., Sun, Y. Y., Wu, N., Mendes, R. G., Chen, L., Xu, Z., ... Fu, L. (2016). Direct growth of MoS_2/h-BN heterostructures via a sulfide-resistant alloy. *ACS Nano, 10*(2), 2063–2070.

130. Cui, F., Li, X., Feng, Q., Yin, J., Zhou, L., Liu, D., ... Xu, H. (2017). Epitaxial growth of large-area and highly crystalline anisotropic $ReSe_2$ atomic layer. *Nano Research, 10*(8), 2732–2742.

131. Zhou, Y., Nie, Y., Liu, Y., Yan, K., Hong, J., Jin, C., ... Peng, H. (2014). Epitaxy and photoresponse of two-dimensional GaSe crystals on flexible transparent mica sheets. *ACS Nano, 8*(2), 1485–1490.

132. Liu, Y., Tang, M., Meng, M., Wang, M., Wu, J., Yin, J., ... Peng, H. (2017). Epitaxial growth of ternary topological insulator Bi_2Te_2Se 2D crystals on mica. *Small, 13*(18), 1603572.

133. Gong, Y., Lin, J., Wang, X., Shi, G., Lei, S., Lin, Z., ... Ajayan, P. M. (2014). Vertical and in-plane heterostructures from WS_2/MoS_2 monolayers. *Nature Materials, 13*(12), 1135–1142.

134. Naylor, C. H., Parkin, W. M., Gao, Z., Berry, J., Zhou, S., Zhang, Q., ... Johnson, A. T. C. (2017). Synthesis and physical properties of phase-engineered transition metal dichalcogenide monolayer heterostructures. *ACS Nano, 11*(9), 8619–8627.

135. Gong, Y. J., Lei, S. D., Ye, G. L., Li, B., He, Y., Keyshar, K., ... Ajayan, P. M. (2015). Two-step growth of two dimensional WSe_2/$MoSe_2$ heterostructures. *Nano Letters, 15*(9), 6135–6141.

136. Yoo, Y. D., Degregorio, Z. P., & Johns, J. E. (2015). Seed crystal homogeneity controls lateral and vertical heteroepitaxy of monolayer MoS_2 and WS_2. *Journal of the American Chemical Society, 137*(45), 14281–14287.

137. Liu, Z., Ma, L. L., Shi, G., Zhou, W., Gong, Y., Lei, S., ... Ajayan, P. M. (2013). In-plane heterostructures of graphene and hexagonal boron nitride with controlled domain sizes. *Nature Nanotechnology, 8*(2), 119–124.
138. Liu, L., Park, J., Siegel, D. A., McCarty, K. F., Clark, K. W., Deng, W., ... Gu, G. (2014). Heteroepitaxial growth of two-dimensional hexagonal boron nitride templated by graphene edges. *Science, 343*(6167), 163–167.
139. Wu, J., Yuan, H., Meng, M., Chen, C., Sun, Y., Chen, Z., ... Peng, H. (2017). High electron mobility and quantum oscillations in non-encapsulated ultrathin semiconducting Bi_2O_2Se. *Nature Nanotechnology, 12*(6), 530–535.

2 Atomic Layer Deposition of Two-Dimensional Semiconductors

2.1 OVERVIEW OF ALD TECHNIQUE OF 2D MATERIALS

To take advantage of the properties of two-dimensional (2D) nanostructures, it is required to incorporate ultrathin 2D materials into practical devices. To fulfill this ambition, novel fabrication techniques should be developed to facilitate the fabrication of the instruments based on atomically thin 2D semiconductor films. For instance, the field effect transistors (FET) based on 2D materials, biosensors, photodetectors, memristors, and artificial synapses are some typical examples of tailoring of 2D nanostructures to the requirements of the practical instruments which can be employed for fundamental applications in nanoelectronic and optics [1]. Thus, the availability of reliable advanced fabrication techniques is an essential prerequisite for development of 2D semiconductor-based devices. Thin-film depositions are the most reliable methods for the large-scale development of electronic systems based on 2D semiconductor materials. However, the existing technologies face serious challenges, especially in the case of conformal deposition of ultrathin 2D films at nanoscale [2]. The most widely used conventional deposition techniques for fabrication of the thin films include: sputtering, gas evaporation, and chemical vapor deposition (CVD) and atomic layer deposition (ALD). Sputtering techniques are mostly suitable for the deposition of micron-sized to several-hundred-nanometer-thick films, and so far they have not been suitable for tailoring of 2D materials to fabricated devices [2, 3]. The sputtering technique uses a quite harsh environment; therefore, it is impossible to achieve the flawless 2D structures, even if the thickness of developed sputtered film reaches to nanoscale range.

Evaporation- and CVD-based techniques are the other main deposition methods to be employed for fabrication of ultrathin 2D films. Specifically, CVD techniques and related methods have demonstrated outstanding capabilities for deposition of a wide range of 2D materials with metallic, insulating and semiconducting characteristics, as outlined in Chapter 1. 2D transition metal dichalcogenides (TMDCs) semiconductor films were extensively deposited by CVD methods. By controlling the CVD technique parameters, the deposition of high-quality single crystal films was obtained. However, there are several challenges confronting CVD techniques in the large-scale wafer deposition of ultrathin 2D films. From a technical point of view, it is quite difficult to achieve high-quality deposition of 2D films with precise control of their thickness over the substrate area with high level of uniformity. At best, the substrate consists of several island-like domains with thickness ranging from few nanometers to several nanometers. These island-like structures are considered as uniformly deposited 2D nanostructures [4]. Furthermore, full coverage of the substrates with a high-aspect ratio is not attainable with these methods. The conformal deposition of 2D films over the substrates with a high level of roughness and geometrical complexity is almost impossible with CVD methods. The other drawback of CVD can be attributed to the high temperature deposition factors. The high temperature in a CVD chamber does not allow the deposition of semiconductor films over the flexible substrates, since most of the flexible films lose their physical stability and mechanical flexibility. Moreover, the number of 2D materials and especially 2D semiconductors synthesized by the CVD technique should be developed. Nevertheless, currently most developed CVD semiconductor films are from TMDC families, which are also predominantly sensitive to the atmospheric conditions. In addition, the CVD deposition of heterostructured 2D films is also highly challenging. Since the growth mechanism in CVD is based on the nucleation and growth of the solid phases, it is difficult to control the growth of the top 2D layer over the first CVD film. In this case, the second film usually starts to nucleate at several points with the lower nucleation barrier energy. Thus, the final structure is not as conformal and uniform as it is required for the high-tech precise electronic applications. There are also other CVD challenges, which were discussed in detail in Chapter 1. Thus, further scientific and technological developments are required to enable the reliable deposition technique for 2D films development [1–3].

On the contrary, ALD, the ultra-precise deposition technique, is based on controlled and self-limited chemical reactions on the surface of substrates. In practical implementation, it is similar to CVD, except that the precursors are not introduced into the reactor simultaneously but are typically separated in time by inert gas purges. As distinct from the other CVD techniques, in ALD the source vapors are pulsed into the reactor alternately, one at a time, separated by purging or evacuation periods. Each precursor exposure step saturates the surface with a (sub) monolayer of that precursor [4, 5]. Therefore, the ALD has a unique self-limiting growth mechanism with a number of distinguished features. In fact, the self-saturation mechanism of chemical reactions is the main ALD privilege, which assists the atomic scale thickness control of the ultrathin films. As a result, the reaction is limited to the monolayer of reactants adsorbed to the substrate [6, 7]. Excellent conformity and uniformity over geometrically complicated substrates can be achieved through this modification on a large spatial resolution. In general, one ALD cycle consists of four individual consecutive steps, including the precursor's injection into the ALD chamber, purging of chemically reacted products from the chamber, counterreactant exposure, and finally, the purging stage of remained residues from the chamber, as it is graphically demonstrated in Figure 2.1 [1]. With regard to the ALD mechanism and its sequential exposure of precursors and counterreactants, the chemical reactions are restricted to the substrate surface and strongly depended on the surface characteristics. To be clear, even on the substrate surface, when the surface properties of an individual area are different from the other part of the surfaces, the ALD growth behavior will also be different on those regions with various surface properties. ALD surface sensitivity is a strong tool to control the deposition characteristics including the growth rate and growth hindrance phenomena, so

ALD of 2D Semiconductors

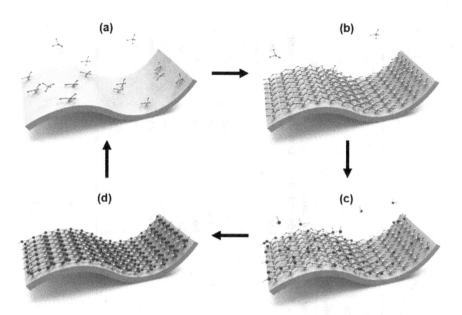

FIGURE 2.1 The schematic of consecutive ALD cycles: (a) exposure of surface to a precursor, (b) removal of unreacted product molecules by purging of inert gas, (c) exposure of a counterreactants, and (d) purging inert gas into chamber to remove unreacted counterreactant molecules. Reproduced with permission from Lee, Y. H., Zhang, X. Q., Zhang, W., Chang, M. T., Lin, C. T., Chang, K. D., ... Lin, T. W. (2012). Synthesis of large area MoS$_2$ atomic layers with chemical vapor deposition. *Advanced Materials*, 24(17), 2320–2325.

that it is possible to selectively deposit ultrathin films on the individual substrate. This ALD feature is known as area-selective ALD [8]. The self-limiting nature of ALD also allows the excellent composition control and fabrication of multilayer 2D nanostructures. Compared with parallel methods of the typical thin-film deposition techniques, ALD reactions are operated under much milder conditions, which are necessary and highly preferable for uniform deposition of 2D structures. Generally, there is no rival method competing with ALD technique in development of ultra-fine hetero-interfaces with atomistic levels of precision. The other outstanding characteristic of ALD is its slow growth rate, providing this technique with an opportunity to control semiconductor film deposition even in an angstrom level. Furthermore, it is possible to control the stoichiometry of ALD film by adjusting the deposition parameters. The same capability of ALD is employed to control the level of oxygen vacancy in the thin-film structure or even dope the ultrathin films with other type of dopants to control the conductivity and mobility of ALD films.

All of these ALD advantages turned this deposition method into one of the most promising techniques for the wafer-scale deposition of ultrathin 2D materials for practical applications and placed this technology in a unique position to address many challenges in development of 2D semiconductors [2]. Table 2.1 provides comparative information on both advantages and disadvantages of the existing thin-film fabrication methods [2]. Considering comparative information presented in Table 2.1,

TABLE 2.1
A Comparison between Various Techniques for Deposition of Ultrathin Films

Property	Chemical Vapor Deposition (CVD)	Molecular Beam Epitaxy (MBE)	Atomic Layer Deposition (ALD)	Pulsed Layer Deposition (PLD)	Evaporation	Sputtering
			Deposition Technique			
Deposition rate	Good	Fair	Poor	Good	Good	Good
Film density	Good	Good	Good	Good	Fair	Good
Lack of pinholes	Good	Good	Good	Fair	Fair	Fair
Thickness uniformity	Good	Fair	Good	Fair	Fair	Good
Sharp dopant profiles	Fair	Good	Good	Varies	Good	Poor
Step coverage	Varies	Poor	Good	Poor	Poor	Poor
Sharp interfaces	Fair	Good	Good	Varies	Good	Poor
Low substrate temp	Varies	Good	Good	Good	Good	Good
Smooth interfaces	Varies	Good	Good	Varies	Good	Varies
No plasma damage	Varies	Good	Good	Fair	Good	Poor

Source: Reprinted with permission from [2].

ALD of 2D Semiconductors

it is estimated that the ALD is highly capable technique for deposition of 2D materials with numerous significant advantages over the other deposition methods in terms of deposition process itself and control over various parameters of fabrication. In addition, the number of reported cases for ALD of 2D films are not as much as the CVD cases, there are examples which reported ALD of graphene, 2D boron nitride (BN), 2D boron halide (BH), 2D TMDCs and metal oxides. In this perspective, the next paragraphs concentrate on ALD of 2D semiconductor films.

2.2 ALD PARAMETERS

2.2.1 ALD WINDOW

As mentioned earlier, ALD is a surface sensitive method and its mechanisms are the function of precursors, reactant agents and surface properties of substrate, and furthermore depended on ALD fabrication parameters including process design, chamber pressure, and temperature and the plasma control ALD chamber [9, 10]. The ALD recipe is the detailed technical information of the sequential ALD procedures in which every single detail of ALD process is explained. Therefore, the developed ALD recipe of 2D materials is technically regarded as very valuable because it provides the sequence of events for repetitive fabrication of 2D structures with the same high-quality standards.

The uniformity and thickness controllability of ALD process is determined by the self-saturation mechanisms and phenomena on the substrate surface during ALD process [11–13]. Growth rate in ALD process is controlled by the range of processing temperatures in which the ALD reactions completed. This temperature range is known as an "ALD window," as exhibited in Figure 2.2a. To initiate and continue chemical reactions during the ALD process, an elevated ALD temperature is required. The appropriately selected ALD window reduces the number of precursors that do not participate in ALD reactions.

The inefficient and low temperature ALD process may result in the occurrence of physical adsorption of precursors on the surface substrate, which would consequently lead to precursor condensation on the substrate. However, the ALD window is restricted by excess condensation or thermal activation at the low temperature range and is generally bounded by the precursor degradation or desorption at high temperatures [14, 15]. The typical ALD window for successful ALD process is in the range of 200 ~ 400°C as presented in Figure 2.2 (b) [2]. Nevertheless, the ALD window is attributed to several different factors, including the carrier gas, the size of samples and other technical parameters, such as the characteristics of chemical components in ALD chamber and the way they react to the surface. The self-controlled growth in ALD window is the consequence of chemisorption. In some cases, due to lack of thermal energy, the ALD process is assisted by using plasma. In fact, the plasma-enhanced atomic layer deposition (PE-ALD) has expanded the number of employed precursors in ALD fabrication. In deposition of oxide films, PE-ALD is typically employed to circumvent the usage of H_2O as oxygen precursor and to increase the ALD growth rate [16, 17]. Furthermore, a dense thin film is expected to grow by using plasma during ALD process. However, it is a technologically challenging target to conformaly deposit continuous 2D films. The difficulty

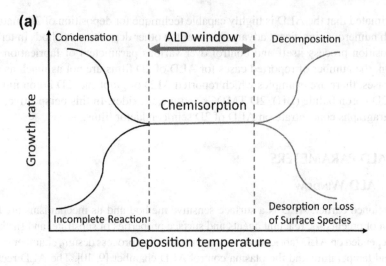

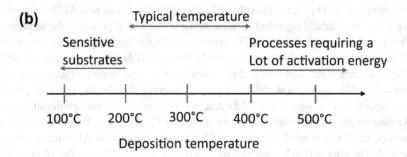

FIGURE 2.2 (a) The dependence of the ALD temperature on growth rate or ALD deposition and (b) typical temperature range for the most common ALD precursors. Reproduced with permission from Zhuiykov, S., Kawaguchi, T., Hai, Z., Karbalaei Akbari, M., & Heynderickx, P. M. (2017). Interfacial engineering of two-dimensional nano-structured materials by atomic layer deposition. *Applied Surface Science*, *392*, 231–243.

of deposition of high-ratio 2D films originates from the nonuniform generation of ozone during PE-ALD process which can be exacerbated due to inefficient mean free path of plasma ions [16–18].

2.2.2 ALD Precursors

The precursors in ALD process can be classified into three major types including, *inorganic, metal-organic* and *organometallic* precursors. The inorganic precursors do not contain any carbon components, while the metal-organic precursors possess organic ligands without any metal to carbon bonds. The other group is organometallic precursors, which contain organic ligands wherein metal to carbon bands exist [2, 9]. The successful design of an ALD process is highly related to the correct selection of ALD precursors [9].

ALD of 2D Semiconductors

As a chemical product, processors are required to be synthesized by easy routes and should be available conveniently for practical ALD targets. Self-decomposition, volatility, dissolution and etching effects on substrates, sufficient purity and byproducts are the main parameters, all of which should be considered during synthesis and the selection of ALD precursors. Despite the significant progress in synthesis of novel chemical compounds, the fulfillment of requirements of the suitable ALD precursor for conformal deposition of 2D nanostructures requires a great deal of works. The growth rate saturation as the function of precursor pulse duration is directly attributed to the temperature of ALD window. To deposit high-quality 2D nanostructures, self-saturation should be controlled during the ALD process. Each ALD precursor shows a specific sticking coefficient, which can be changed on various substrates. By altering the ALD parameters, the sticking coefficient directly affects the precursor purge time. For instance, the growth per cycle (GPC) values for an insufficient purge time can be twice as high as the obtained values for optimized ALD recipe [19, 20]. The similar difficulties should be taken into account when the second reactant precursor comes into the action in CVD chamber. For example, in the case of ALD of metal oxide films, sufficient pulse time for H_2O injection into CVD chamber should be chosen to guarantee the deposition of high-quality stoichiometric 2D oxide nanofilms. The lack of sufficient oxidant agents changes the growth regime and affects the composition of ALD films [19]. Another important parameter is the reactivity of precursors. This is more important when one can consider the fact that most of the strategies on design of ALD recipe are based on the thermally driven chemical reactions [21–23]. In this case, the reactivity becomes more critical issue when the thermally stable precursors are used or when the low temperature deposition is required. Considering the ALD variables, the correct selection of ALD precursors is considered as one of the main components of ALD recipe for deposition of 2D materials. The next sections will provide an overview of the ALD of ultrathin 2D films of metal chalcogenides and metal oxides.

2.3 ALD OF 2D METAL CHALCOGENIDE FILMS

Layered metal dichalcogenides (MX_2, X = S, Se, Te) have attracted tremendous attention due to their promising optical and electrical characteristics. This family of 2D materials is composed of the metal layer covalently bonded with two separate surrounding chalcogen atoms. The final film is a sandwiched structure, which is considered as one single layer materials. The layer-by-layer deposition mechanism of ALD makes this technology as a capable method for deposition of layered 2D films. However, most of the ALD films do not show the layered structures. At present, the number of ALD-deposited binary, ternary and doped chalcogenides is more than 25 different layered materials. Among them, only a few specific cases are layered materials with rhombohedral, orthorhombic or hexagonal structures, including MoS_2, ZnS, WS_2, WSe_2, SnS_x, GaS, γ-InSe, Bi_2Te_3, Bi_2Se_3, Bi_2S_3, Sb_2Te_3, Sb_2Se_3, β-NiS, and FeSx. The layered ZnS was the first member of the metal chalcogenide family deposited by ALD [24]. Today, the ALD chemistry of metal sulfides is well documented [25]. It was in 2015 when for the first time the number of MoS_2 layers was controlled by tuning the parameters of CVD technique [26]. A bit later, the ALD synthesis of TMDCs monolayer was also reported [27–31]. The self-limiting layer synthesis (LSL) of MoS_2 monolayer film was facilitated based on the typical ALD half-reactions [27, 28, 32]. Here, the practical cases of ALD of stable 2D films of metal dichalcogenides are presented.

2.3.1 ALD OF 2D MoS_2

As one of the most investigated and stable 2D materials, MoS_2 was one of the first TMDCs deposited by ALD technique. This material displays hexagonal structure with covalent S–Mo–S atom arrangement along the z-axis and van der Waals inter-action between layers. The synthesis of 2D MoS_2 film was first realized after the sulfidation of ALD MoO_3 film at 600°C for 10 min under S vapors [33]. The other studies have investigated the importance of MoO_x morphology, composition, stoi-chiometry as well as sulfidation parameters on the final characteristics of MoS_2 film [34, 35]. It was confirmed that the quality of MoS_2 film coverage, uniformity, thick-ness, crystallinity and nucleation sites of MoO_x considerably affected the uniformity of ultrathin molybdenum sulfide film. It was well understood that the tuning of ALD parameters and the pretreatment of substrate can directly influence the coverage of resulting molybdenum sulfide 2D films. It was found that the full coverage is created when the ultra-uniform thin MoO_x film was deposited rather than from particulate MoO_x films. Furthermore, multistep annealing facilitates controlled formation of MoS_2 films with sequential reduction of Mo, leading to incorporation of S atoms. Multistep reactions also allow coverage control and phase transformation from crys-talline MoS_2 to the 2H-phase at higher post-treatment temperatures [35].

The first report of direct ALD deposition of MoS_2 was attributed to 2014, when $MoCl_5$ and H_2S were employed as the precursors in ALD system to coat a 2-inch sapphire wafer at 300°C. Continues film was deposited after using 10 ALD cycles, since the island growth did not let to achieve a uniform film when the number of ALD cycles was less than 10. However, structural characterization (TEM, XRD and Photoluminescence spectra) of MoS_2 films suggested the as-deposited film suffers from insufficient quality. Postannealing treatment at 800°C improved the crystallin-ity of ALD MoS_2 film in the expense of losing thickness uniformity, thus 20 ALD cycles were employed to guarantee the wafer-scale full coverage of MoS_2 film after annealing at higher temperatures. Considering the ALD window, it was found that the ALD process in 350 to 450°C permits obtaining higher-quality sulfide layers with growth rate of 0.87 Å/cycles [36].

It was experimentally shown that the ALD in the range of 430–470°C leads to deposition of hexagonal MoS_2 plane films with preferential parallel orientation toward the substrate with sulfur sub-stoichiometry (S/Mo = 1.4) [37]. The stoichiom-etry of as-deposited film can be improved by further annealing under H_2S or sulfur gas phase to reach the S/Mo ratio of 2. The results were confirmed by XPS analysis and Raman spectroscopy. The sharpening of Raman bands confirmed the improve-ment of stoichiometry of MoS_x film.

The controlled ALD fabrication of mono-, bi-, and trilayer MoS_2 films was reported where the $MoCl_5$ and H_2S were used as precursors in the self-limiting layer synthesis (SLS) approach [38]. In this novel approach, the inertness of the basal 2D TMDC layer was used to restrain the additional layer deposition. Once the first crystalline layer of TMDCs film was formed, the further growth of MoS₂ film was suppressed. After the deposition of the first layer the remained number of anchors is limited, thus the reactivity of 2D film was declined and further growth of 2D film is possible only through the physisorption of MoCl₅ precursors. The precursor desorption was facilitated by the increase of ALD temperature. The AFM studies confirmed the impact of ALD temperature on the thickness of MoS₂ films (Figure 2.3a,

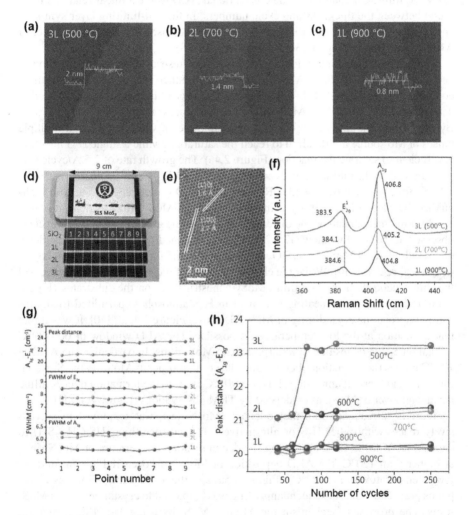

FIGURE 2.3 (a) The AFM image of (a) tri-, (b) bi- and (c) mono-layers of MoS₂ films over SiO₂ substrate. The scale bar is = 0.5 μm. (d) The large-scale deposition of 2D MoS₂ films over SiO₂ substrate. (e) The crystalline structure of 2D MoS₂ film depicted by HRTEM image (scale bar is 2 nm). (f) Raman spectra of MoS₂ films and the effect of layer numbers on (g) FWHM and E^1_{2g} and A_{1g} modes for several cases of measurements on mono-, bi-, and trilayer MoS₂. (h) Raman peak for MoS₂ with various SLS cycles and growth temperatures. Reproduced with permission from Kim, Y., Song, J. G., Park, Y. J., Ryu, G. H., Lee, S. J., Kim, J. S., ... Kim, H. (2016). Self-limiting layer synthesis of transition metal dichalcogenides. *Scientific Reports*, 6, 18754.

b, and c). Indeed, the different growth mechanism is involved in SLS, which is the temperature determined mechanism. As evidenced by AFM observations, 2D films with three, two and monolayer thickness were fabricated over the large-area substrates (Figure 2.3d) with well-crystalline structure (Figure 2.3e) at deposition temperatures of, 500°C, 700°C, and 900°C, respectively. Raman studies (Figure 2.3f, g) were employed to evaluate the effect of layer thickness on the characteristics of ALD 2D MoS_2 films. On contrary to the conventional ALD where a linear relationship is found between the thickness and cycle numbers, in the self-limiting layer synthesis approach the number of deposited layers does not increase after a certain number of ALD cycles (Figure 2.3g, h) [38].

The library of precursors for ALD MoS_2 is not restricted merely to the molybdenum halide. In particular, $Mo(CO)_6$ reacts with either H_2S or dimethyl disulfide compounds (DMDS, CH_3-S-S-CH_3) are the other type of the precursors employed to deposit 2D MoS_2 films. The ALD window of the process ranges from 155 to 170°C. By monitoring the mass gain during the ALD cycles it was revealed that multiple pulses of Mo source are required to reach the saturation, while a single H_2S injection is sufficient to saturate the reaction (Figure 2.4a). The growth rate of 2.5 Å/cycle was measured when the multiple pulses are employed in ALD process compared with 0.75 Å/cycle for a single pulse case [39]. Recently, PE-ALD was employed to deposit high-quality 2D MoS_2 films [40]. By using H_2S plasma, h-MoS_2 film was deposited in a self-limiting mechanism on the different substrates. The growth rate of 0.5 Å/cycle is observed at the ALD window (175–200°C), accompanied by a short nucleation delay on growth of ALD MoS_2 on Si/SiO_2 film, as well as the formation of MoS_2 nanoparticles (Figure 2.4b) [40]. While the characterization of the as-deposited film revealed the direct deposition of crystalline MoS_2 nanostructures on the substrates (Figure 2.4b), the further postannealing treatment in H_2S atmosphere permitted to significantly improve the crystallinity of MoS_2 2D films (Figure 2.4c,d) [40]. It was found that deposition at the low temperature is possible. The ALD window of 60–120°C substantiates the deposition of amorphous 2D MoS_2 films. In particular, the ALD at 100°C allows the formation of continuous layer of amorphous MoS_2 on the Si wafer. Further rapid thermal annealing (RTA) at 900°C leads to the formation of crystalline hexagonal layer of MoS_2, as evidenced by TEM observations (Figure 2.4e) [41].

Tetrakis (dimethylamido) molybdenum ($Mo(NMe_2)_4$ was employed to successively reacts with either H_2S or alternatively 1, 2 ethanedithiol ($HS(CH_2)_2SH$) to deposit 2D MoS_2 films. Considering the decomposition temperature of $Mo(NMe_2)_4$ at higher than 120°C, the ALD deposition of MoS_2 film occurred at considerably lower temperatures (T < 120°C) (Figure 2.5a) and the as-deposited film was amorphous [42]. Thus, the subsequent annealing was employed to crystalize the 2D MoS_2 films. The proposed mechanism for ALD of MoS_2 by using $Mo(NMe_2)_4$ and 1,2 ethanedithiol is investigated with mass spectroscopy analysis. It was revealed the mass gain during the injection of Mo precursors into chamber was followed by the mass lost during the injection of the sulfur source into ALD chamber. The mass reduction was due to the removal of the remaining NMe_2 ligands on the surface. The main advantage of the low temperature ALD technique is the capability of coating polymer substrates. This also makes it possible to pattern the films by using a lift-off method (Figure 2.5b). The postannealing treatment at 1000°C facilitated the

ALD of 2D Semiconductors 53

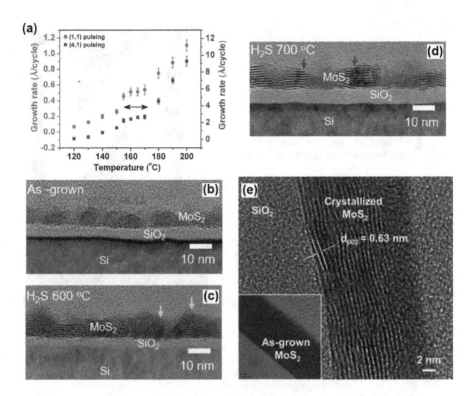

FIGURE 2.4 (a) GPC of MoS$_2$ film using Mo(CO)$_6$ and H$_2$S as a function of the deposition temperature and pulse lengths of Mo and source. (b) The cross-sectional image of as-deposited MoS$_2$ film, (c) annealed at 600°C and (d) 700°C in H$_2$S atmosphere. (e) The cross-sectional TEM image of MoS$_2$ film deposited by using Mo(CO)$_6$ and DMDS precursors and annealed at 900°C, the inset demonstrates the as-deposited film. (a) Reproduced with permission from Nandi, D. K., Sen, U. K., Choudhury, D., Mitra, S., & Sarkar, S. K. (2014). Atomic layer deposited MoS$_2$ as a carbon and binder free anode in Li-ion battery. *Electrochimica Acta, 146*, 706–713, (b, c, and d) are reproduced with permission from Oh, S., Kim, J. B., Song, J. T., Oh, J., & Kim, S.-H. (2017). Atomic layer deposited molybdenum disulfide on Si photocathodes for highly efficient photoelectrochemical water reduction reaction. *Journal of Materials Chemistry A, 5*(7), 3304–3310, and (e) is Reproduced with permission from Jin, Z., Shin, S., Kwon, D., H., Han, S.-J., & Min, Y.-S. (2014). Novel chemical route for atomic layer deposition of MoS$_2$ thin film on SiO$_2$/Si substrate. *Nanoscale, 6*, 14453–14458.

deposition of polycrystalline 2D MoS$_2$ film with strong photoluminescence effect and sharp Raman peaks [42]. Furthermore, no remaining carbon-based ligand or surface oxidation is noted. On the other hand, the deposition of MoS$_2$ by 1,2 ethanedithiol consists of a two-step process whereby the Mo(IV) thiolate layer is formed at 50°C. The mechanism of ALD reactions is depicted in Figure 2.5c, where the 1,2 ethanedithiol breaks to Mo–N bonds and provides S-H species. The ALD mechanism is evidenced by IR spectroscopy and elemental analysis during ALD process. Postannealing led to thermal conversion of amorphous film to the defect free crystalline hexagonal MoS$_2$ (Figure 2.5d) [42].

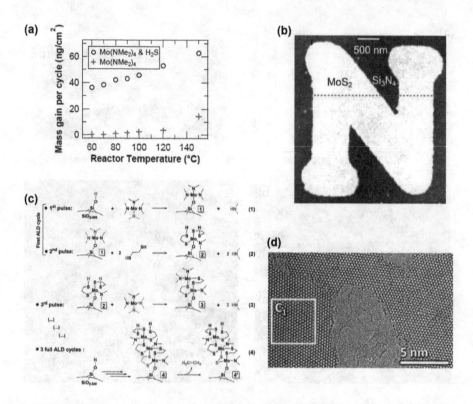

FIGURE 2.5 (a) The mass gain per cycle versus reactor temperature with different type of precursors. The ALD process was based on employment of Mo(NMe$_2$)$_4$ reacting with H$_2$S (circles) and Mo(NMe$_2$)$_4$ CVD (markers). (b) The AFM mage of patterned MoS$_2$ film on Si/Si$_3$N$_4$ substrate using ALD, after electron-beam lithography and lift-off. (c) The reaction ALD mechanism when Mo(NMe$_2$)$_4$ and 1,2 ethanedithiol are used as precursors. (d) TEM image of ALD film annealing at 800°C for 30 min under argon. (a and b) are reprinted and reproduced with permission from Jurca, T., Moody, M. J., Henning, A. Emery, J. D., Wang, B., Tan, J. M., ... Marks, T. J. (2017). Low-temperature atomic layer deposition of MoS$_2$ films. *Angewandte Chemie, 129*, 5073–5077, and (c, d) are reproduced with permission from Cadot, S., Renault, O., Frégnaux, M., Rouchon, D., Nolot, E., Szeto, K., ... Quadrelli, E. A. (2017). A novel 2-step ALD route to ultra-thin MoS$_2$ films on SiO$_2$ through a surface organometallic intermediate. *Nanoscale, 9*(2), 538–546.

2.3.2 ALD OF 2D WS$_2$

The first report on ALD of WS$_2$ layered films was in 2004, where the WF$_6$ and H$_2$S were used as precursor in ADL of WS$_2$ film at 300°C. The growth initiation stage of film requires a catalyst, otherwise without reaction initiator the ALD growth does not occur in ALD window of 250 to 300°C. Interestingly, it was observed that Zn played the role of catalyst for the growth of WS$_2$ layer. Either a single pulse of DiEthylZinc (DEZ) and H$_2$S assisted the rapid nucleation of WS$_2$ film over a wide range of substrates [43, 44]. Zn plays the role of reducing agent for WF$_6$ favoring the adsorption of WF$_6$ on the surface of substrates. The initial growth rate of 1

Å/cycle was recorded which was reduced after 50 ALD cycles of WS_2 deposition (Figure 2.6a). By the employment of single ZnS pulse, the growth rate was retrieved. It was found that the deposition of a monolayer ZnS film also facilitated the growth of WS_2 film. However, in this case the GPC is modulated by the availability of free electrons in the substrates [44, 45]. The deposited WS_2 film has a crystalline hexagonal structure. The electron microscopic observations showed the formation of ZnF_2 in as-deposited WS_2 film (Figure 2.6b). In particular, the growth of ZnF_2 clusters on the Au substrates at the interfaces can also affect the uniformity of films. The excess usage of Zn with DEZ catalyst caused extreme damages at the interfaces. To

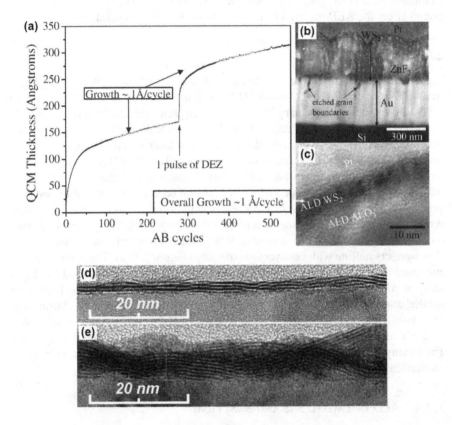

FIGURE 2.6 (a) Thickness of WS_2 film as the function of ALD cycle number. (b) The cross-sectional SEM image of WS_2 film using DEZ catalyst. (c) The HRTEM image of WS_2 grown on Al_2O_3 substrate. (d, e) The cross sectional HRTEM images of 2D WS_2 film by using WF_6, H_2 plasma and H_2S on crystalline Al_2O_3 substrate and deposited at (d) 300°C and (e) 450°C. (a, b, and c) reprinted with permission from Scharf, T., Prasad, S. V., Dugg, M. T., Kotula, P., Goeke, R., & Grubbs, R. (2006). Growth, structure, and tribological behavior of atomic layer-deposited tungsten disulphide solid lubricant coatings with applications to MEMS. *Acta Materialia, 54*(18), 4731–4743. (d and e) reprinted with permission from Delabie, A., Caymax, M., Groven, B., Heyne, M., Haesevoets, K., Meersschaut, J., ... Thean, A. (2015). Low temperature deposition of 2D WS_2 layers from WF_6 and H_2S precursors: Impact of reducing agents. *Chemical Communications, 51*(86), 15692–15695.

overcome this challenge, a separating Al_2O_3 film was employed at interface between substrate and WS_2 layer (Figure 2.6 c). In another research, the effects of reducing agents on the growth of ALD WS_2 layer were investigated. The partial reduction of WF_6 was required to form the WS_2 from deposited WF_6. By the release of SiF_4, the WF_6 was reduced by Si into its metallic tungsten. The metallic tungsten then can be either oxidized or sulfurized by the following H_2S pulse in the reactor. The phase transformation from WF_6 to WS_2 is controlled by the self-limiting catalyst effect of Si substrate. It was found that maximum of 6 nm Si layer can react with the halide compounds limiting the growth of upper WS_2 layer. The plasma exposure can guarantee the deposition of stoichiometric WS_2 films. The ALD parameters by high extend affect the ALD growth of the WS_2 film on basal plane orientation.

2.3.3 ALD OF 2D WSe₂

The ALD was also employed to fabricate 2D tungsten selenide (WSe_2). Two main strategies were employed to deposit 2D WSe_2 films. The self-limiting deposition of WSe_2 film was achieved by using WCl_6 and DiEthyl selenide (DESe) at the high ALD temperature of 600–800°C [28]. The WCl_6 and chloride were removed from the chamber by using DESe. The film thickness dependence on ALD temperature was observed. Respectively, five, three and one crystalized layers of WSe_2 were deposited on silicon oxide wafers (8 cm²) at 600°C, 700°C, and 800°C. After 100 ALD cycles, the growth of WSe_2 was stopped which was attributed to the growth saturation. The WSe_2 growth saturation of the film was enabled to the chemical inertness of TMDC surfaces and the physical adsorption of tungsten precursor at the high temperature processing (Figure 2.7a). The deposited WSe_2 films with mono and few layer thickness were crystalline with honeycomb structure (Figure 2.7b,c). The same film was employed as a p-type semiconductor in the FET instrument and showed the capability of ALD deposited WSe_2 film for application in electronic devices. In another parallel attempt, the thermal ALD of continuous WSe_2 film on S/SiO_2 substrate was successfully carried out at 390°C [47]. WCl_5 and H_2Se were employed to deposit a highly uniform film (Figure 2.7d) with well-crystalline structure (Figure 2.7e). The quality and crystallinity of the ALD WSe_2 film were fairly comparable to the mechanical exfoliated and CVD WSe_2 films [47].

2.3.4 ALD OF LAYERED SnS AND SnS₂ FILMS

Sn and SnS_2 are the most investigated layered materials after the TMDCs. The first ALD case of SnS has been developed by using tin (II) 2,4-pentanedionate ($Sn(acac)_2$) precursor as the Sn source and H_2S as the sulfur source. The ALD window was selected within the temperature range of 125 to 225°C. The growth rate of 0.23 Å/cycle was recorded for this ALD process. There are other tin precursors which are employed to react with H_2S gas in ALD chamber, including bis(N,N'-diisopropylacetamidinato) tin (II) ($Sn(amd)_2$) [49, 50] and N^2,N^3-di-tert-butylbutane-2,3-diamine tin (II) [49]. These precursors are effectively used to deposit crystalline SnS film at ALD temperature less than 200°C. To deposit SnS_2 and SnS_3 thin layer films, the precursors containing the Sn atoms with the oxidation state under +II are utilized

ALD of 2D Semiconductors

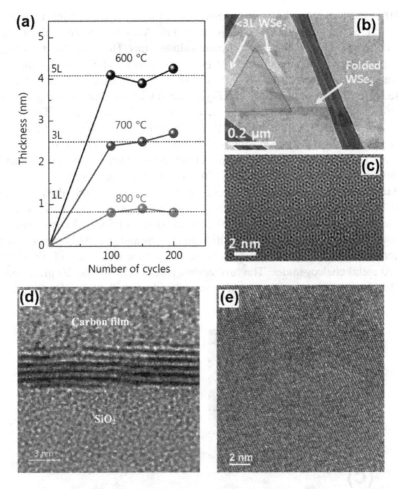

FIGURE 2.7 (a) The effect of a number of ALD cycles and ALD temperatures on the thickness of WSe$_2$ film during the self-limiting ALD of WSe$_2$ film by using WCl$_6$ and diethyl selenide (DESe) as the precursor. (b) TEM image of WSe$_2$ film and its corresponding (c) HRTEM image of the surface of Few-layer WSe$_2$ film. (d) Cross-sectional TEM image and (e) HRTEM image of the surface of ALD WSe$_2$ film deposited by using WCl$_5$ and H$_2$Se precursors. (a, b, and c) Reprinted with permission from Park, K., Kim, Y., Song, J. G., Jin Kim, S., Wan Lee, C., Hee Ryu, G., ... Kim, H. (2016). Uniform, large-area self-limiting layer synthesis of tungsten diselenide. *2D Materials*, *3*(1), 014004. (d) and (e) Reprinted with permission from Browning, R., Kuperman, N., Solanki, R., Kanzyuba, V., & Rouvimov, S. (2016). Large area growth of layered WSe$_2$ films. *Semiconductor Science and Technology*, *31*(9), 095002.

for ALD process [50]. For example, layered SnS$_2$ films were deposited by using the tetrakis (dimethylamino) tin (TDMASn) [51, 52] and tin acetate (Sn(OAc)$_4$) [53] as the Sn (IV) metal precursors. The ALD fabrication at the temperature of 140–150°C with TDMASn and H$_2$S developed smooth and uniformly deposited polycrystalline hexagonal SnS$_2$ films with the growth rate of 0.64–0.8 Å/cycle [51]. Postannealing at 300°C and H_2S are required to develop crystalline tin disulfide film (Figure 2.8a) with S-Sn-S triatomic planar arrangement. The ALD temperature above 160°C resulted in the development of thin monosulfide films. However, the as-deposited film always contains a combination of both mixed SnS and SnS_2 compounds. The postannealing treatment at 300°C in H_2 atmosphere assisted the formation of a single phased orthorhombic 2D SnS_2 films (Figure 2.8b) with interlayer spacing of 0.6 nm (Figure 2.8c) [54]. In another case, $Sn(OAc)_4$ was used for deposition of few layers of SnS_2 films at the low temperature of 150°C. The small growth rate of 0.17 Å/cycle was measured during the deposition of stoichiometric amorphous SnS_2 film. The post-thermal annealing at 250°C and under the controlled atmosphere of H_2S and N_2 resulted in the crystallization of SnS_2 film. As an interesting approach, few-layered crystalline films were employed to directly deposit crystalline SnS_2 films. Generally, in ALD deposition of few-layered SnS_x films with Sn precursors and H_2S gas, the chemical composition, structure and the stoichiometry of SnS_x ALD structures can be modulated depending on the deposition conditions and post-ALD treatments [53].

Generally, ALD is proven to be vital technique for deposition of ultrathin 2D layered metal chalcogenides. The further progress in ALD of the 2D metal chalcogenides should cover a wide range of technical and scientific topics. It is required to

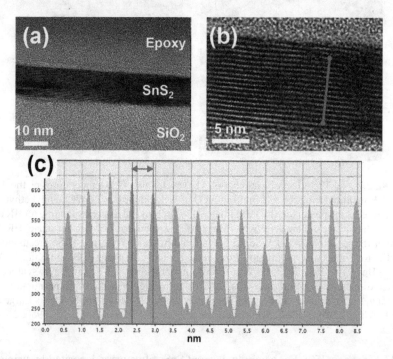

FIGURE 2.8 (a) TEM interface layer of SnS_2 film and its (b) high-resolution cross-sectional TEM image. (c) Interlayer spacing along the marker line in (b). Reprinted with permission from Ham, G., Shin, S., Park, J., Lee, J., Choi, H., Lee, S., & Jeon, H. (2016). Engineering the crystallinity of tin disulfide deposited at low temperatures. *RSC Advances*, 6(59), 54069–54075.

ALD of 2D Semiconductors 59

enhance the existing knowledge about the chemistry of precursors, the ALD reaction mechanisms, and the ALD parameters. Furthermore, *in-situ* characterization techniques of the thin films should be developed to facilitate monitoring of film growth during ALD process. The presence of both amorphous and crystalline structures characterized in as-deposited ALD films, whereas the control of structural properties of materials is facilitated via controlling the ALD parameters. Usually, further thermal annealing is required to improve the crystallinity of 2D TMDC films.

2.3.5 ALD OF LAYERED METAL DICHALCOGENIDE HETEROSTRUCTURES

ALD, as a precise thin-film deposition technique, opens up further opportunity for fabrication of VdW heterostructures. There are some technological challenges for development of VdW hetero-interfaces. The basal plane of 2D materials is inert, thus the lack of chemical anchors for nucleation of the second 2D films may induce difficulties for development of VdW hetero-interfaces. In this case, the nucleation delay, island growth and unsuccessful growth observed. Thus, it is necessary to facilitate the initial adsorption of precursors on the surface of 2D substrate films. For example, layered Sb_2Te_3 was successfully ALD deposited on graphene substrate by using $SbCl_3$ and $(Me_3Si)_2Te$ as precursors [55]. To activate the surface of CVD grown graphene films, $(Me_3Si)_2Te$ was previously physically adsorbed on the graphene. The successful low temperature (70°C) deposition of 24 nm thick crystalline Sb_2Te_3 layer was finally achieved with crystalline plane parallel to the graphene substrates. The defects-free sharp interface between graphene and Sb_2Te_3 film is depicted in the HRTEM image of Figure 2.9a [55]. It was found that the deposition of few-layered Sb_2Te_3 on the 2D graphene films improved the crystallinity of layered metal chalcogenide films compared with that of the same 2D films (Sb_2Te_3) deposited on the Si wafer. Therefore, the oxidation stability of metal chalcogenide was improved.

The 2D heterostructured Sb_2Te_3/Bi_2Te_3 metal chalcogenides were also fabricated entirely by ALD. Since the capability of Bi_2Te_3 for coverage of Si wafer is better than the Sb_2Te_3, the bismuth telluride was selected to be deposited over the silicon wafer by ALD technique at 165–170°C. The lack of dangling bonds on the surface of Bi_2Te_3 should be compensated to facilitate the deposition of Sb_2Te_3 upper layer. Considering the low ALD temperature of Bi_2Te_3 (at 65–70°C), the employment of an initial long $SbCl_3$ pulse in ALD chamber was taken as a capable strategy to saturate the surface of Bi_2Te_3 and complete the surface reactions for the next deposition step [56]. By using this technique, three stacking layers of Bi_2Te_3/Sb_2Te_3 are successfully deposited with highest level of crystallinity with local epitaxy between 2D polycrystalline metal telluride films of the heterostructures (Figure 2.9b) [57]. Another example of the heterostructured metal chalcogenides deposition is related to the growth of multiple WS_2/SnS layered semiconductor heterojunctions on the Si/SiO_2 substrates by alternating ALD of WS_2 and SnS at the ALD temperature of 390°C. The investigation of hetero-interfaces revealed that the stacked layers of WS_2/SnS have orientation with an angle of 15° between their c axis, which was related to their different crystalline structures and lattice mismatch (Figure 2.9c, d). This misaligned characteristic of WS_2/SnS hetero-interfaces caused a noticeable decrease of the hole mobility of SnS film [58], which was a considerable promising

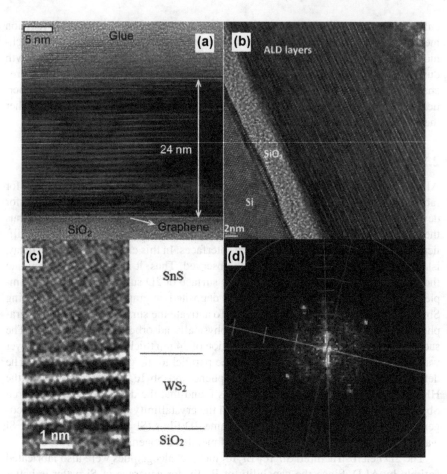

FIGURE 2.9 (a) Cross-sectional image from the interface between ALD grown Sb_2Te_3 on graphene/SiO_2 substrate. (b) Cross-sectional image of heterostructured Sb_2Te_3/Bi_2Te_3 stacked layers. (c) HRTEM image of SnS/WS$_2$ hetero-interfaces and (d) its corresponding fast Fourier transform (FFT) image. Reproduced with permission from Hao, W., Marichy, C., & Journet, C. (2019). Atomic layer deposition of stable 2D materials. *2D Materials*, 6(1), 012001.

property for electronic application of 2D heterostructures. Table 2.2 provides a brief overview of the ALD techniques employed to develop 2D metal chalcogenides and their heterostructures [32].

2.4 ALD OF 2D METAL OXIDE FILMS

Within the past few years, 2D metal oxides are found as promising materials for many engineering applications. Few-layered and ultrathin metal oxides are competitive rivals to their semiconductor and insulator counterparts. The concept of 2D oxide structures can be assigned to ultrathin oxide films with the thickness of few nanometers (several fundamental layers). In this context, the characterization and

ALD of 2D Semiconductors

TABLE 2.2
Review of the Reported ALD Deposition of Ultrathin TMDC Films

Material	Reactants	ALD temperature	Post-treatment	Crystallinity
MoS_2	$MoCl_5/H_2S$	500°C (3L) 700°C (2L) 900°C (2L)	None	Polycrystalline
MoS_2	$MoCl_5/H_2S$	300°C	800°C for 30 min	Well-crystallized, hexagonal
MoS_2	$MoCl_5/H_2S$	330°C–450°C	None	Well-crystallized, hexagonal
MoS_2	$MoCl_5/H_2S$	450°C	None	Well-crystallized, hexagonal
MoS_2	$MoCl_5/H_2S$	375°C, 475°C	600°C–900°C for	Small crystallites
MoS_2	$Mo(NMe_2)_4/$ $HS(CH_2)_2SH$	50°C	450°C for 30 min	In H_2: Nanocrystalline
MoS_2	$Mo(CO)_6/H_2S$	155°C–170°C	None	Amorphous
MoS_2	$Mo(CO)_6/H_2S$ plasma	175°C–225°C	None	Polycrystalline 2H-MoS_2
MoS_2	$Mo(CO)_6/$ozone	165°C	Sulfurization S atm	Polycrystalline
MoS_2	$(NtBu)_2(NMe_2)_2Mo/$ O_3	300°C	Sulfurization H_2S	2H-phase
MoS_2	$Mo(CO)_6/O_2$ plasma	200°C	2 step sulfurization	Polycrystalline
MoS_2	$(NtBu)_2(NMe_2)_2Mo/$ O_2 plasma	150°C	Sulfurization H_2S	Function of the process
MoS_2	Cl radical ads/Ar^+ desorp	–	None	2H-MoS_2, well-crystallized
WS_2	WF_6/H_2 plasma/H_2S	300°C–450°C	None	Nanocrystalline
WS_2	WF_6/H_2S +Si layer catalyst	300°C–450°C	None	Nanocrystalline
WS_2	WC_{l5}/H_2S	390°C	None	2H-phase
WS_2	$W(CO)_6/H_2S$	175°C–205°C	None	Amorphous
	$WH_2(iPrCp)_2/O_2$ plasma	300°C	Sulfurization Ar/ H_2S	Hexagonal
WSe_2	WCl_5/H_2S	390°C	None	Well-crystallized, hexagonal
WSe_2	$WCl_6/DESe$	600°C (5L), 700°C (3L), 800°C (1L)	None	Well-crystallized, hexagonal
Bi_2S_3	$Bi(thd)_3/H_2O$	300°C	Sulfurization 500°C–700°C,	Orthorhombic
Bi_2S_3	$Bi(thd)_3/H_2S$	125°C–300°C	None	Polycrystalline orthorhombic
Bi_2Se_3	$BiCl_3/(Et_3Si)_2Se$	165°C	None	Crystalline rhombohedral with amount of orthorhombic

(Continued)

TABLE 2.2 (CONTINUED)
Review of the Reported ALD Deposition of Ultrathin TMDC Films

Material	Reactants	ALD temperature	Post-treatment	Crystallinity
Bi_2Te_3	$BiCl_3/(Et_3Si)_2Te$	160°C–250°C	None	Crystalline rhombohedral
Bi_2Te_3	$Bi(NMe_2)_3/(Et_3Si)_2Te$	70°C, 120°C	None	Polycrystalline rhombohedral
FeSx	$Fe(amd)_2/H_2S$	80°C–200°C	None	Well-crystallized
GaS	$Ga_2(NMe_2)_6/H_2S$	125°C–225°C	None	Amorphous
GeS	N,N_3-di-tert-butylbutane$_{2,3}$-diamine Ge (II)/H_2S	50°C–75°C	None	Amorphous
InSe	$InCl_3/H_2Se$ (8% in Ar)	310°C–380°C	None	Hexagonal (γ-InSe)
In_2Se_3	$InCl_3/(Et_3Si)_2Se$	295°C	None	Crystalline
γ-MnS	$Mn(EtCp)_2/H2S$	100°C–225°C	None	T < 150°C: (γ-MnS) T > 150°C: (α-MnS)/γ-MnS
NiS	$Ni(thd)_2/H_2S$	175°C–350°C	None	Polycrystalline β-NiS
Sb_2Se_3	$SbCl_3/H_2Se$	270°C–320°C	None	Orthorhombic
Sb_2Te_3	$SbCl_3/(Et_3Si)_2Te$	60°C–140°C	None	Crystalline rhombohedral
SnS	$Sn(acac)_2/H_2S$	80°C–225°C 80°C–160°C 390°C	None	Long pulse: cubic Short pulse: orthorhombic $T \geqslant 300$°C: Orthorhombic
TiS$_x$	$TiCl_4/H_2S$	400°C–500°C	None	Hexagonal on soda Amorphous on Rh
$Bi_2Te_3/$ Sb_2Te_3	$BiCl_3/(Me_3Si)_2Te$ & $SbCl_3/(Me_3Si)_2Te$	165°C–170°C	Long $SbCl_3$ pulse	Polycrystalline
InSe/ Sb_2Se_3	$InCl_3/H_2Se$ & $SbCl_3/$ H_2Se	310°C	None	–
WS_2/SnS	WCl_5/H_2S & $Sn(acac)_2/H_2S$	300°C	None	Hexagonal/Orthorhombic

Source: Reproduced with permission from [32].

exploiting of the 2D oxide materials properties for the practical applications are considered as valuable technical knowledge along with the progressive advancement in the device miniaturization. Apart from pure scientific exploration of 2D semiconductor oxides, novel technologies nowadays use the privilege of quasi-ultrathin oxide structures as the main components of *state-of-the-art* nanodevices. Consequently, the promising applications of 2D oxide nanostructures are highly envisaged. The ultrathin oxide materials have already found several applications in solid oxide fuel cells, catalyst films, corrosion protection layers, gas sensors, spintronic devices, and UV and visible light sensors. Specifically, 2D oxide semiconductors are considered as one of the main components of metal oxide semiconductor field effect transistors (MOSFET) and the other novel nanoelectronic instruments. The solar energy cells, plasmonic devices, data storage applications

ALD of 2D Semiconductors

biocompatibility, and biosensing features and lately supercapacitance properties and electrochemical sensing are among the recently announced and discovered applications of 2D oxide semiconductors [32, 59]. The fabrication of large-area 2D films is extremely challenging since the structural stability and molecular integrity of 2D films can be deteriorated easily. Accordingly, one of the main challenges is the versatile and conformal deposition of these quasi-2D oxide nanostructures on the convenient substrates. Considering the capabilities of ALD, this deposition technique is found highly promising method for development of 2D oxide films over the large-area substrates. The following section will focus on the ALD of metal oxide materials, and particular attention is given to molybdenum, titanium, aluminum and tungsten oxide 2D films.

2.4.1 ALD OF 2D MoO_3

Molybdenum trioxide (MoO_3) exhibits unique catalytic, electrical and optical properties and also several promising functionalities originated from excellent characteristics of MoO_3. The bis(tert-butylimido) bis(dimethylamido) molybdenum (($tBuN)_2(NMe_2)_2Mo$) is the most common precursor for MoO_3 deposition, since it provides good volatility and thermal stability [60–62]. The temperature window in ALD process of molybdenum oxide with O_2 plasma and (($tBuN)_2(NMe_2)_2Mo$) precursor is in the range of 50–350°C. The as-deposited films at temperature ranges below 250°C are amorphous; thus, a postdeposition annealing process is required in order to transition from an amorphous to crystalline state. While the as-deposited films are substoichiometric, the stoichiometry can be adjusted by the modulation of plasma parameters [63]. $Mo(CO)_6$ is another chemical precursor mostly employed with O_2 plasma to deposit MoO_3 films in the narrow ALD window of 152–172°C [64, 65]. It has been observed that the employment of combination of various oxygen precursors resulted in the linear growth rate of 0.75 Å/cycle. The as-deposited amorphous films were highly uniform with the low level of surface roughness. The oxygen deficiency was detected at the hetero-interfaces between substrate and MoO_3 film. The improvement of crystallinity was achieved after postannealing at temperature above 500–600°C. In another strategy, the ozone plasma was used on the surface of ALD MoO_3 film deposited on a 300 mm Si wafer [66]. Mo film was deposited by using Si_2H_6 and MoF_6 precursors at 200°C.

In another case, the wafer-scale deposition of MoO_3 film over 4-Inch Au wafer is reported [67]. $C_{12}H_{30}N_4Mo$ as molybdenum precursor and O_2 plasma as oxygen source were used at two ALD temperatures of 150 and 250°C, respectively. The schematic interpretation of ALD process is demonstrated in Figure 2.10. The initial growth stage includes the chemisorption of $C_{12}H_{30}N_4Mo$ molecules by the active sites on the Au. Furthermore, the pre-exposure of bare Au surface by O_2 plasma assists the formation of OH bonding on the Au surface, as depicted in Figure 2.10a [67]. Then, the introduction of $C_{12}H_{30}N_4Mo$ molecules into the reaction chamber is accompanied by the chemisorption of precursor molecules. In this reaction, $C_{12}H_{30}N_4Mo$ molecules chemically interact with OH groups on the surface of substarte. the ligands exchange occurs within the designed ALD window, as shown in Figure 2.10b, c. Then, the process continues by the purge of remained products of the

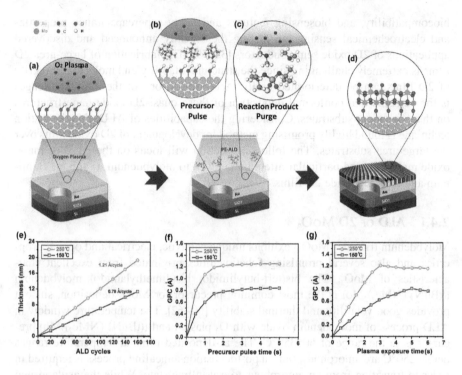

FIGURE 2.10 (a) Graphical scheme of deposition of ultrathin MoO$_3$ nanofilms by PE-ALD. (b) The injection of precursor into ALD chamber and (c) the reaction between precursor and Au surface (d) and the complete deposition of a monolayer film over Au substrate. (e) The thickness of PE-ALD MoO$_3$ film over Au substrate vs. the ALD cycle number based on the usage of (NtBu)$_2$(NMe$_2$)$_2$Mo precursor and O$_2$ plasma at 150°C and 250°C, with the precursor dosing time of 2 s and plasma exposure time of 5 s. (f) The saturation curve of (NtBu)$_2$(NMe$_2$)$_2$Mo precursor. The GPC is demonstrated as the function of precursor dosing time. (g) The saturation curve for O$_2$ plasma showing the GPC as the function of plasma exposure time. Reprinted and reproduced with permission from Xu, H., Karbalaei Akbari, M., Hai, Z., Wei, Z., Hyde, L., Verpoort, F., ... Zhuiykov, S. (2018). Ultra-thin MoO$_3$ film goes wafer-scaled nano-architectonics by atomic layer deposition. *Materials and Design*, *149*, 135–144.

first ALD cycle to the outside of reaction chamber by using Ar gas. After purging, the O$_2$ plasma is again conducted into ALD film. The plasma exposure is continued to complete the oxidation process. The Figure 2.10d graphically shows the development of Au-O-Mo bonding at the first ALD cycle. The development of Mo-O-Mo bonding is expected after completion of oxidation process [67]. Finally, a 4.6 nm thick MoO$_3$ film was deposited over the wafer substrate.

The *in-situ* ellipsometry map was employed to monitor the thickness of MoO$_3$ film during ALD process. The measurements confirmed the increase of GPC from 0.78 Å/cycle to 1.21 Å/cycle, by increasing the ALD temperature from 150 to 250°C (Figure 2.10e). By using the O$_2$ gas instead of O$_2$ plasma in ALD chamber, the growth rate did not change considerably. It confirmed that the thermal CVD growth mechanism did not take place during ALD process. The saturation curves of $C_{12}H_{30}N_4Mo$ precursor and O_2 plasma exposure time for the different ALD temperatures are respectively demonstrated in Figure 2.10f, g. The GPC changed according to the precursor pulse time indicating the soft saturation process. The growth rate is sensitive to precursor dosage, and it usually stabilizes at the highest value when the full monolayer coverage of surface is achieved. Regarding the obtained results, the 2 s pulse time was the saturation time for $C_{12}H_{30}N_4Mo$ precursor over Au substrate, which did not change considerably at various ALD temperatures. On the contrary, the O_2 pulse exposure saturation durations for deposition at 150 and 250°C were respectively 5 s and 3 s. It confirms that the oxidation of $C_{12}H_{30}N_4Mo$ was improved at the higher ALD deposition temperatures [67].

2.4.2 ALD OF 2D WO$_3$

The uniform and ultrathin WO$_3$film is one of the most interesting transition metal oxide semiconductors, which was successfully employed for significant number of optical, energy and environmental applications. Several different precursors have been utilized to deposit WO$_x$ film over the different substrates. One of the most widely used precursors is tungsten hexacarbonyl W(CO)$_6$. The narrow ALD window between 195 and 205°C with a GPC of 0.2 Å/cycle was noted when the ozone was used as the oxygen source in the ALD process [68, 69]. The deposited film was already amorphous, which can be later annealed to improve its crystallinity and be converted to monoclinic WO$_3$ [69].

The bis(ter-butylimido) bis(dimethylamino) tungsten(VI) $(^tBuN)_2W(NMe_2)_2$ as tungsten precursor and H$_2$O as oxygen source were also used. This W source presents a low vapor pressure suitable for ALD technique. Ultrathin 2D films with a thickness of less than 1.0 nm were deposited on wafer-scale Si/SiO$_2$ substrate. To confirm the ALD growth of oxide film, the *in-situ* ellipsometry was employed during deposition process. In experiments, the dependence of growth rate on the precursor pulse times, substrate temperature and cycle numbers were investigated. As presented in Figure 2.11a, b, the growth rate was sensitive to the precursor doses. As expected, the growth rate was stabilized to a maximum thickness per cycle when precursor dosing achieved full monolayer coverage. Growth rate saturation was observed after 50 ms for H$_2$O and after 2 s for $(^tBuN)_2W(NMe_2)_2$, indicating self-limiting growth characteristics of the ALD. In addition, it was established that the growth was also highly sensitive to the substrate temperature. The first deposition at operating temperature of ~ 300°C showed the deposition of the limited thickness of 2D WO$_3$ film. Consequently, the optimum growth condition yielded stable WO$_3$ growth (Figure 2.11c, d). Figure 2.11e demonstrates the cross-sectional view of the ultrathin WO$_3$ film deposited on Si/SiO$_2$. Scanning transmission electron microscopy (STEM) image displays the uniform development of the ALD WO$_3$ film across a large area of the Si/SiO$_2$ wafer. The film was semi-amorphous without clearly identified crystalline structure, as shown in Figure 2.11f. The effect of the postannealing process on the microstructure of ALD developed WO$_3$ films was investigated. It was

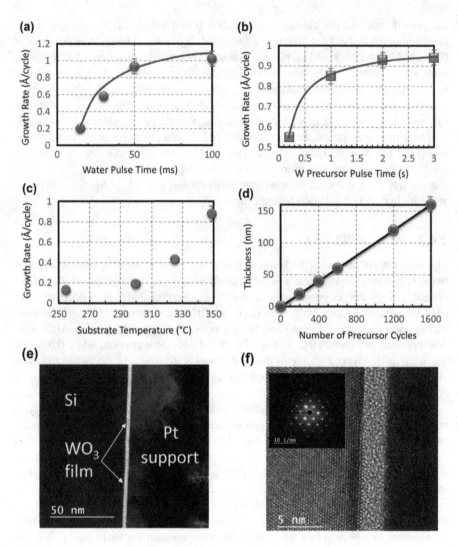

FIGURE 2.11 The experimental ALD data for deposition of WO$_3$ film. The growth rate versus precursors pulse time for (a) H$_2$O and (b) (^{t}BuN)$_2$W(NMe$_2$)$_2$ precursors. (c) The initial growth rate versus temperature, and (d) the thickness variation of WO$_3$ film per ALD cycle number. (e) The cross-sectional STEM image of the ~3.3 nm thick ALD 2D WO$_3$ film deposited on Si/SiO$_2$ wafer, (f) HRTEM image of the interface between Si/SiO$_2$ and WO$_3$ film. Inset shows the SAED pattern indicating crystalline structure of SiO$_2$. Reprinted with permission from Zhuiykov, S., Hyde, L., Hai, Z., Akbari, M. K., Kats, E., Detavernier, C., ... Xu, H. (2017). Atomic layer deposition-enabled single layer of tungsten trioxide across a large area. *Applied Materials Today, 6,* 44–53.

found that the postannealing process can affect the crystallinity, stoichiometry and electrical properties of 2D WO_3 films. From the structural point of view, when post-annealing was performed at low temperatures (T < 300°C), the 2D ALD film kept its mechanical stability on the substrate. However, by the increase of postannealing temperature to higher values (T > 300°C) the nucleation and the growth of granular WO_3 nanostructures were observed. Further increase of the annealing temperature or annealing time led to a deterioration of integrity of 2D films over the substrate, which finally resulted in the growth of coarse granular WO_3 nanostructures.

2.4.3 ALD of 2D TiO_2

The wafer-scale synthesis and conformal growth of 2D TiO_2 film on Si/SiO_2 sub-strate were achieved by ALD using tetrakis (dimethylamino) titanium (TDMAT) precursor and H_2O as an oxidation agent [71]. For precise fabrication of the extremely thin 2D TiO_2 films, where their thickness must be below ~0.5 nm, the optimum operating temperature of 250°C was selected. To achieve a complete coverage of monolayer TiO_2, two individual ALD super-cycles were designed and implemented [71]. The first supercycle consisted of 10 consecutive cycles of pulse/purge TDMAT to assure the reliable coverage of surface by precursors. Afterward, 10 pulse/purge stages of H_2O were performed to assure the completion of oxidation [71]. The XPS and FTIR studies revealed the mechanism of ALD process. In the XPS studies, the changes of binding energies in both Si 2p and Ti 2p peaks after the deposition of 2D TiO_2 could be elucidated by the development of Si–O–Ti bonds at the interface between native SiO_2 and deposited 2D TiO_2 film. As titanium was introduced into the Si–O bonds during ALD process, the binding energy of Si^{4+} decreased slightly from its bulk SiO_2, which was related to this fact that the silicon was more electronegative than titanium. The evidence of Oxygen Bridge bonding was also demonstrated by characterization of the vibrational mode of Ti–O–Ti bonds in the FTIR spectrum of monolayer TiO_2 film and also by investigation of XPS spectrum of monolayer film [71]. The detection of FITR characteristics of Ti^{4+} confirmed the existence of Ti(IV)O oxide in crystalline state [71]. It was observed that the postannealing process has improved the crystallinity and electrical con-ductivity of 2D TiO_2 film. The 2D TiO_2 films annealed at 150°C showed widen-ing of their bandgap to 3.37 eV compared to the reported values for bulk rutile (3.03 eV) and anatase (3.20 eV) [71]. TiO_2 nanosheets were also fabricated by using $TiCl_4$ and deionized (DI) water as the precursors. In particular, free standing TiO_2 nanosheets with various thicknesses were deposited by ALD on the surface of dis-solvable sacrificial polymer layers [72, 73].

2.4.4 ALD of 2D Al_2O_3

ALD of monolayer alumina film over Cu and Cu_2O/Cu substrates was carried out by using trimethylaluminum (TMA) as Al precursor and O_2 as oxidant agent [74]. The interaction of TMA with Cu surface faces several challenges. Regarding the DFT calculations the adsorption and dissociation of TMA on surface of pure Cu

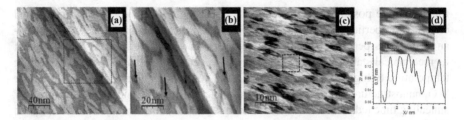

FIGURE 2.12 (a) The low- and (b) high-magnification STM images of the Cu$_2$O/Cu(111) surface exposed to first cycle and (c) second cycle of TMA, and (d) Zoom-in region of the highlighted section of image and the line profile along the solid line indicated in the image. Reproduced from Gharachorlou, A., Detwiler, M. D., & Gu, X. K. (2015). Trimethyl aluminum and oxygen atomic layer deposition on hydroxyl-free Cu(111). *ACS Applied Materials and Interfaces*, *7*, 16428–16439.

(111) are endothermic, confirming that the interaction of TMA by pure Cu surface is thermodynamically unfavorable in the absence of hydroxyl groups. XPS measurements and high-resolution electron energy loss spectroscopy did not show any characteristic vibration of Al atoms on pure Cu surface. On the other hand, the first-principle calculations demonstrated that TMA is capable of reacting with a copper oxide surface in the absence of hydroxyl species. The adsorption of Cu by Cu$_2$O/Cu surface is limited by the initial amount of oxygen in Cu$_2$O lattice structure. This adsorption phenomenon resulted in the reduction of some surface Cu^{1+} to metallic copper (Cu^0) and the formation of a copper aluminate compounds. The XPS studies showed that the Al:O atomic percentage ratio at the first deposited layer was approximately 0.46 while the stoichiometric Al$_2$O$_3$ would yield an Al:O ratio of 0.66. It confirmed the development of substoichiometric 2D alumina layers. The further XPS measurements represented the formation of copper aluminate, most likely CuAlO$_2$ on the basis of the Al:O ratio of 0.5. The STM image of surface of TMA-exposed Cu$_2$O/Cu film demonstrated the growth of 2D islands of CuAlO$_2$ with an average height of approximately 0.19 nm with flat and uniform surfaces which was close to Cu–O and Al–O bond length, as depicted in Figure 2.12a, b. After the second ALD cycle, TMA continued to react with surface Al–O, forming stoichiometric Al$_2$O$_3$. The observed morphological changes after consecutive ALD cycles proposed the development of second alumina islands with thickness of 0.17 nm as shown by STM image in Figure 2.12c, d [74]. It was confirmed that TMA readily reacted with oxide surfaces even at the absence of coadsorbed hydroxyls. For ALD applications on an air-exposed Cu surface, large domains of oxides might still exist which can facilitate the selective ALD of metal surfaces. This is of great importance to thin-film applications such as microelectronics and catalysis where only few ALD cycles are desirable. Table 2.3 has summarized the ALD techniques employed to develop 2D metal oxide with the different structural characteristics.

TABLE 2.3

Review of the Reported ALD Deposition of Ultrathin 2D Metal Oxide Films

Material	Reactants	ALD temperature	Post-treatment	Crystallinity
MoO_3	$(tBuN)_2(NMe_2)_2Mo/O_2$-PE-ALD	50°C–350°C	None	Amorphous films at < 200°C Polycrystalline at > 250°C
MoO_3	$(tBuN)_2(NMe_2)_2Mo/O_2$-PE-ALD	150°C	Sulfidation to MoS_2	2H-MoS_2 after annealing at T > 900°C
MoO_3	$(tBuN)_2(NMe_2)_2Mo/O_2$-ALD	300°C-	Sulfidation to MoS_2	2H-MoS_2 after annealing at T > 900°C
MoO_3	$Mo(CO)_6/O_3$	152°C–175°C	Annealing in air at 500°C–600°C	Amorphous as-deposited α- and β-MoO_3 phases after annealing at 500°C
MoO_3	$Mo(CO)_6/H2O$	152°C–172°C	Annealing in air at 500°C and 600°C	Amorphous as-deposited α- and β-MoO_3 phases after annealing at 500°C Phase-pure, highly oriented α-MoO_3 at 600°C
MoO_3	$MoO_2(R_2amd)_2$ (R = Cy; iPr)/O_3	150°C–225°C	None	Amorphous
MoO_3	MOTSMA/O_3	250°C–300°C	None	–
MoO_3	$CoCp_2$, Co(thd)$_2$ or $Mo(CO)_6/O_3$, H_2O or ($O_3 + H_2O$)	167°C	Annealing in air at 600°C for 6 min or 10 min	Amorphous as-deposited Crystallize into β-$CoMoO_4$ under annealing
WO_3	$W(CO)_6/H_2O_2$	180°C–200°C	*Ex-situ* annealing	Amorphous
WO_3	$W(CO)_6/O_3$	195°C–205°C	Annealing at 600°C–1000°C in O_2 or N_2	Partially crystalline as-deposited Crystallinity enhanced after annealing

(*continued*)

TABLE 2.3 (CONTINUED)

Review of the Reported ALD Deposition of Ultrathin 2D Metal Oxide Films

Material	Reactants	ALD temperature	Post-treatment	Crystallinity
WO_3	$W(CO)_6/H_2O$	150°C–320°C	Annealing at O_2	Amorphous layer completely crystallizes into polycrystalline film under postannealing
WO_3	$(tBuN)_2 W(NMe_2)_2/H_2O$	250°C–350°C	Annealing at 550°C under of O_2	Crystalline
	$(tBuN)_2 W(NMe_2)_2/H2O$	300°C–350°C	Annealing at 200°C–400°C	Amorphous as-deposited
WO_3	$WO_2(tBuamd)_2/H2O$	120°C–270°C	Annealing at 500°C	Crystallize as WO_3 nanowires
WO_3	WF_6/H_2O	30°C–180°C	None	Amorphous
W_2O_3	$W_2(NMe_2)_6/H2O$	140°C–200°C	None	Amorphous
Al_2O_3	TMA/H_2O	180°C	None	Amorphous
AlO_x	TMA/H_2O	90°C	Annealing in air 450°C–500°C	Amorphous
TiO_2	$TiCl_4/H_2O$ or O_2	30°C–180°C	None	Amorphous
TiO_2	$TDMAT/H_2O$	150°C	–	Amorphous
TiO_2	$TDMAT/H_2O$	250°C	Annealing at 150°C–1100°C in air	As-deposited amorphous, Annealed crystalline: $T > 280°C$ anatase $T > 400°C$ rutile
TiO_2	$TiCl_4/H_2O$	100°C	Annealing in air at 450°C – 500°C	Amorphous
ZnO	DEZ/H_2O	200°C	None	Single layer ZnO on graphene presents graphene-like structure instead of wurtzite structure

Source: Reproduced with permission from [32].

2.5 CONCLUSION

This chapter has highlighted the outstanding ALD capabilities for the development of ultrathin 2D nanostructured semiconductors. Compared with other chemical vapor deposition techniques, the ALD features for designing, characterization, and tailoring of these 2D nanostructures make broad appeals. ALD techniques have been successfully employed to fabricate various types of 2D materials, including insulators, semiconductors, and metal films. Tacking the advantages of ALD mechanism, thickness controllability, and conformity, the ALD technique seems to be one of the most promising techniques for the wafer-scale deposition of 2D semiconductors for the large-scale production. With the constant rising of 2D materials, it is expected that ALD and its family will take position in the 2D materials related fields in near future. ALD of 2D materials will rely on the development of precursors, improvement of the quality of as-grown films and controlling the surface reactivity. Most of the as-deposited ALD films do not have a fully crystallized structure; thus, postannealing is often required to improve the crystallinity of the film to reach to desired structures. There are some fundamental challenges that should be considered for further development of ALD systems for the wafer-scale fabrication of 2D semiconductors:

i. The chemistry of precursors should be well understood. The ALD mechanisms, the thermodynamic of reactions, the adsorption of chemicals on the different substrates and the effect of technological parameters on nucleation and growth of semiconductor films should be investigated. Novel precursors must be synthesized in regards to the requirements for new applications.

ii. The in-situ characterization techniques are highly desired to monitor the growth mechanisms of 2D films. Therefore, it is highly required to either develop new techniques or improve the present characterization systems.

iii. The substrate is one of the most important components in the ALD deposition of 2D films. The type of substrate, its crystalline structure, physical and chemical properties fundamentally affect the nucleation and growth of 2D films. Surface chemistry represents a key point for the advancement of ALD of 2D materials. The lack of reactivity of substrates can basically hinder the nucleation and growth of ALD films. Moreover, the basal planes in 2D films might strongly affect the growth and successful fabrication of VdW heterostructures, and also affects the ability of the second layer to nucleate on the first one. Furthermore, the surface activation mechanisms of the former 2D films should be further investigated.

iv. The crystalline orientation of 2D films is governed by the lattice mismatch between substrate and upper 2D film. Consequently, it is necessary to investigate new substrates and analyze their surface characteristics. The control of stack orientation in heterostructured films would allow the fabrication of novel materials with advanced properties, which have not yet been explored. This provides new pathways for 2D electronics and their capabilities, where the properties and functionalities of 2D-based instruments such as field effect transistors, memory capacitors, resistive switching memory cells, and microelectromechanical systems are fundamentally affected by the structural characteristics of 2D films and their structural relation with their substrates. For example, the relative rotation angle between two-heterostructured 2D films directly influences the structural orientation and elastic deformation of as-grown film. The control and tuning of the structural characteristics of 2D films permit tailoring the electronic properties of 2D films.

v. Innovative ALD techniques including PE-ALD and hot-wire assisted ALD (HWALD) can be employed to facilitate the growth of ultrathin films over challenging substrates. Especially, two-step processes as well as enhanced or HWALD enable low temperature deposition over sensitive/organic substrates. Nonetheless, it should be taken into account that the PE-ALD and other assisted ALD techniques are currently limited to the flat surfaces, since the methods suffer from the lack of homogeneity on high-aspect ratio substrates.

vi. The crystallinity and stoichiometry issues are the other important parameters in ALD of 2D films that remained as challenging factor. While most of the ALD chalcogenide films are crystalline, 2D oxide films and graphene are always amorphous and need postannealing treatment at high temperatures. However, the postannealing has its own drawbacks. The high temperature annealing can deteriorate the integrity of 2D films, cause agglomerated particles, change the stoichiometry of films, affect the vacancy number in 2D films (especially 2D oxides, annealed in atmospheric condition). Postannealing treatment can also affect the substrate structure of 2D films. For example, it can alter the atomic arrangement in hetero-interfaces, change the crystalline structure of substrate, or even deteriorate the mechanical stability of substrate. This is a serious challenge when the deposition of transparent 2D oxides on the transparent polymeric flexible substrates is targeted. Furthermore, ALD 2D oxide films are always suffering from the lack of control over their stoichiometry, thus it is highly required to improve the ALD parameters to attain the desired stoichiometry.

vii. ALD allows the conformal deposition of 2D films on various substrates (polymers, biological substrates, metal). Especially, it is a highly capable technique for the high-aspect ratio deposition of films for development of the advanced biomedical materials, as ALD allows coatings, functionalization, and modification of heat-sensitive substrates such as polymers, biological and organic components or even complex structures, like woven fabrics, fibers, and bundles. ALD is capable of playing a prominent role in the fabrication of bio-microelectronic systems (bio-MEMS). The bio-MEMS devices, similar neurosurgical tools, implants, and catheter-blood sensors are required to be protected from the fluidic environment that they are working on it. Apart from the challenges of the high temperature ALD process and also the problems of postannealing treatment for polymers and biological substrates, for some applications it is necessary to scale down the ALD temperature to the temperature lower than 100°C. In this context, the emergence of alternative ALD techniques (PE-ALD, HWALD), as well as the improvement or introduction of novel precursors, can open up new opportunities in biomicroelectronics and biomedical sensors.

REFERENCES

1. Kim, H. G., & Lee, H. B. R. (2017). Atomic layer deposition on 2D materials. *Chemistry of Materials*, *29*(9), 3809–3826.
2. Zhuiykov, S., Kawaguchi, T., Hai, Z., Karbalaei Akbari, M., & Heynderickx, P. M. (2017). Interfacial engineering of two-dimensional nano-structured materials by atomic layer deposition. *Applied Surface Science*, *392*, 231–243.
3. Lee, Y. H., Zhang, X. Q., Zhang, W., Chang, M. T., Lin, C. T., Chang, K. D., ... Lin, T. W. (2012). Synthesis of large area MoS_2 atomic layers with chemical vapor deposition. *Advanced Materials*, *24*(17), 2320–2325.
4. Schoiswohi, J., Surnev, S., & Netzar, F. P. (2005). Reaction on inverse model catalyst surface: Atomic view by STM. *Topics in Catalysis*, *36*(1–4), 91–105.
5. Detavernier, C., Dendooven, J., Sree, S. P., Ludwig, K. F., & Martens, J. A. (2011). Tailoring nanoporous materials by atomic layer deposition. *Chemical Society Reviews*, *40*(11), 5242–5253.
6. Warren, R., Sammoura, F., Tounsi, F., Sanghadasa, M., & Lin, L. (2015). Highly active ruthenium oxide coating via ALD and electrochemical activation in supercapacitor applications. *Journal of Materials Chemistry A*, *3*(30), 15568–15575.
7. Zhu, C., Yang, P., Chao, D., Wang, X., Zhang, X., Chen, S., ... Fan, H. J. (2015). All metal nitrides solid state asymmetric supercapacitors. *Advanced Materials*, *27*(31), 4566–4571.
8. Dasgupta, N. P., Lee, H., Bent, V., & Weiss, P. S. (2016). Recent advances in atomic layer deposition. *Chemistry of Materials*, *28*(7), 1943–1947.
9. Ponraj, J. S., Attolini, G., & Bosi, M. (2013). Review on atomic layer deposition and application of oxide thin films. *Critical Reviews in Solid State and Materials Sciences*, *38*(3), 203–233.
10. Niinisto, L., Paivasaari, J., Niinisto, J., Putkonen, M., & Nieminen, M. (2004). Advanced electronic and optoelectronic materials by atomic layer deposition: An overview with special emphasis on recent progress in processing of high-K dielectrics and other oxide materials. *Physical Status Solidi. Part A*, *201*, 1443–1452.
11. George, S. M. (2010). Atomic layer deposition: An overview. *Chemical Reviews*, *110*(1), 111–131.
12. Suntola, T. (1992). Atomic layer epitaxy. *Thin Solid Films*, *216*(1), 84–89.
13. Jõgi, I., Pärs, M., Aarik, J., Aidla, A., Laan, M., Sundqvist, J., ... Kukli, K. (2008). Conformity and structure of titanium oxide films grown by atomic layer deposition on silicon substrates. *Thin Solid Films*, *516*(15), 4855–4862.
14. Du, X., & George, S. M. (2008). Thickness dependence of sensor response for CO gas sensing by tin oxide films grown using atomic layer deposition. *Sensors and Actuators. Part B*, *135*(1), 152–160.
15. Malik, M. A., & O'Brien, P. (2005). Organometallic and metallo-organic precursors for nanoparticles. In R. A. Fischer (Ed.), *Precursor chemistry of advanced materials: CVD, ALD and nanoparticles* (pp. 125–145). Springer-Verlag, Berlin, Heidelberg.
16. Potts, S. E., & Kessel, W. M. M. (2013). Energy-enhanced atomic layer deposition for more process and precursor versatility. *Coordination Chemistry Reviews*, *257*(23–24), 3254–3270.
17. Liang, X. H., Zhou, Y., Li, J., & Weimer, A. W. (2011). Reaction mechanism studies for platinum nanoparticle grown by atomic layer deposition. *Journal of Nanoparticle Research*, *13*(9), 3781–3788.

18. Laskela, M., & Ritala, M. (2002). Atomic layer deposition (ALD) from precursors to thin film structures. *Thin Solid Films*, *409*(1), 138–146.

19. Philip, A., Thomas, S., & Kumar, K. R. (2014). Calculation of growth per cycle (GPC) of atomic layer deposited aluminum oxide nanolayers and dependence of GPC on surface OH concentration. *Pram. Journal de Physique*, *82*, 563–569.

20. Liang, X., George, S. M., Weimer, A. W., Li, N., Blackson, J. H., Harris, J. D., & Li, P. (2007). Synthesis of a novel porous polymer/ceramic composite material by low-temperature atomic layer deposition. *Chemistry of Materials*, *19*(22), 5388–5394.

21. Hatanpää, T., Ritala, M., & Leskelä, M. (2013). Precursors as enablers of ALD technology: Contributions from University of Helsinki. *Coordination Chemistry Reviews*, *257*(23–24), 3297–3322.

22. Lee, S. W., Choi, B. J., Eom, T., Han, J. H., Kim, S. K., Song, S. J., … Hwang, C. S. (2013). Influences of metal, non-metal precursors, and substrates on atomic layer deposition processes for the growth of selected functional electronic materials. *Coordination Chemistry Reviews*, *257*(23–24), 3154–3176.

23. Burton, B. B., Lavoie, A. R., & George, S. M. (2008). Tantalum nitride atomic layer deposition using (tert-butylimido)tris (diethylamido) tantalum and hydrazine. *Journal of the Electrochemical Society*, *155*(7), D508–D516.

24. Suntola, T., & Antson, J. (1974). Method for producing compound thin films. US Patent US4058430A.

25. Dasgupta, N. P., Meng, X., Elam, J., W., Martinson, A., & B. F. (2015). Atomic layer deposition of metal sulfide material. *Accounts of Chemical Research*, *48*(2), 341–348.

26. Lin, T., Kang, B. T., Jeon, M. H., Huffman, C., Jeon, J., Lee, S., … Kim, K. (2015). Controlled layer-by-layer etching of MoS_2. *ACS Applied Materials and Interfaces*, *7*(29), 15892–15897.

27. Kim, Y., Song, J. G., Park, Y. J., Ryu, G. H., Lee, S. J., Kim, J. S., … Kim, H. (2016). Self-limiting layer synthesis of transition metal dichalcogenides. *Scientific Reports*, *6*, 18754.

28. Park, K., Kim, Y., Song, J. G., Jin Kim, S., Wan Lee, C., Hee Ryu, G., … Kim, H. (2016). Uniform, large-area self-limiting layer synthesis of tungsten diselenide. *2D Materials*, *3*(1), 014004.

29. Liu, H. (2017). Recent progress in atomic layer deposition of multifunctional oxides and two-dimensional transition metal dichalcogenides. *Journal of Engineering Materials and Technology*, *4*, 1640010.

30. Van Bui, H., Grillo, F., & van Ommen, J. R. (2017). Atomic and molecular layer deposition: Off the beaten track. *Chemical Communications*, *53*(1), 45–71.

31. Karbalaei Akbari, M., Hai, Z., Depuydt, S., Kats, E., Hu, J., & Zhuiykov, S. (2017). Highly sensitive, fast-responding, and stable photodetector based on ALD-developed monolayer TiO_2. *IEEE Transactions on Nanotechnology*, *16*(5), 880, 87.

32. Hao, W., Marichy, C., & Journet, C. (2019). Atomic layer deposition of stable 2D materials. *2D Materials*, *6*(1), 012001.

33. Wang, H., Lu, Z., Xu, S., Kong, D., Cha, J. J., Zheng, G., … Cui, Y. (2013). Electrochemical tuning of vertically aligned MoS_2 nanofilms and its application in improving hydrogen evolution reaction. *Proceedings of the National Academy of Sciences of the United States of America*, *110*(49), 19701–19706.

34. Martella, C., Melloni, P., Cinquanta, E., Cianci, E., Alia, M., Longo, M., … Molle, A. (2016). Engineering the growth of MoS_2 via atomic layer deposition of molybdenum oxide film precursor. *Advanced Electronic Materials*, *2*(10), 1600330.

35. Keller, B. D., Bertuch, A., Provine, A., Sundaram, G., Ferralis, N., & Grossman, J. C. (2017). Process control of atomic layer deposition molybdenum oxide nucleation and sulfidation to large-area MoS_2 monolayers. *Chemistry of Materials*, *29*(5), 2024–2032.

ALD of 2D Semiconductors

36. Tan, L. K., Liu, B., Teng, H., Guo, S., Low, H. Y., Tan, H. R., ... Loh, K. P. (2014). Atomic layer deposition of a MoS₂ film. *Nanoscale, 6*(18), 10584–10588.

37. Valdivia, A., Tweet, D. J., & Conley, J. F. (2016). Atomic layer deposition of two dimensional MoS₂ on 150 mm substrates. *Journal of Vacuum Science and Technology. Part A, 34*(2), 021515.

38. Kim, Y., Song, J. G., Park, Y. J., Ryu, G. H., Lee, S. J., Kim, J. S., ... Kim, H. (2016). Self-limiting layer synthesis of transition metal dichalcogenides. *Scientific Reports, 6,* 18754.

39. Nandi, D. K., Sen, U. K., Choudhury, D., Mitra, S., & Sarkar, S. K. (2014). Atomic layer deposited MoS₂ as a carbon and binder free anode in Li-ion battery. *Electrochimica Acta, 146,* 706–713.

40. Oh, S., Kim, J. B., Song, J. T., Oh, J., & Kim, S.-H. (2017). Atomic layer deposited molybdenum disulfide on Si photocathodes for highly efficient photoelectrochemical water reduction reaction. *Journal of Materials Chemistry A, 5*(7), 3304–3310.

41. Jin, Z., Shin, S., Kwon, D., H., Han, S.-J., & Min, Y.-S. (2014). Novel chemical route for atomic layer deposition of MoS₂ thin film on SiO₂/Si substrate. *Nanoscale, 6,* 14453–14458.

42. Jurca, T., Moody, M. J., Henning, A. Emery, J. D., Wang, B., Tan, J. M., ... Marks, T. J. (2017). Low-temperature atomic layer deposition of MoS₂ films. *Angewandte Chemie, 129,* 5073–5077.

43. Cadot, S., Renault, O., Frégnaux, M., Rouchon, D., Nolot, E., Szeto, K., ... Quadrelli, E. A. (2017). A novel 2-step ALD route to ultra-thin MoS₂ films on SiO₂ through a surface organometallic intermediate. *Nanoscale, 9*(2), 538–546.

44. Scharf, T., Prasad, S. V., Dugg, M. T., Kotula, P., Goeke, R., & Grubbs, R. (2006). Growth, structure, and tribological behavior of atomic layer-deposited tungsten disulphide solid lubricant coatings with applications to MEMS. *Acta Materialia, 54*(18), 4731–4743.

45. Scharf, T. W., Diercks, D. R., Gorman, B. P., Prasad, S. V., & Dugger, M. T. (2009). Atomic layer deposition of tungsten disulphide solid lubricant nanocomposite coatings on rolling element bearings. *Tribology Transactions, 52*(3), 284–292.

46. Delabie, A., Caymax, M., Groven, B., Heyne, M., Haesevoets, K., Meersschaut, J., ... Thean, A. (2015). Low temperature deposition of 2D WS₂ layers from WF₆ and H₂S precursors: Impact of reducing agents. *Chemical Communications, 51*(86), 15692–15695.

47. Browning, R., Kuperman, N., Solanki, R., Kanzyuba, V., & Rouvimov, S. (2016). Large area growth of layered WSe₂ films. *Semiconductor Science and Technology, 31*(9), 095002.

48. Kim, J. Y., & George, S. M. (2010). Tin monosulfide thin films grown by atomic layer deposition using tin 2,4-pentanedionate and hydrogen sulfide. *Journal of Physical Chemistry C, 114*(41), 17597–17603.

49. Kim, S. B., Sinsermsuksakul, P., Hock, A. S., Pike, R. D., & Gordon, R. G. (2014). Synthesis of N-heterocyclic stannylene (Sn(II)) and germylene (Ge(II)) and a Sn(II) amidinate and their application as precursors for atomic layer deposition. *Chemistry of Materials, 26*(10), 3065–3073.

50. Sinsermsuksakul, P., Heo, J., Noh, W., Hock, A. S., & Gordon, R. G. (2011). Atomic layer deposition of tin monosulfide thin films. *Advanced Energy Materials, 1*(6), 1116–1125.

51. Ham, G., Shin, S., Park, J., Choi, H., Kim, J., Lee, Y. A., ... Jeon, H. (2013). Tuning the electronic structure of tin sulfides grown by atomic layer deposition. *ACS Applied Materials and Interfaces, 5*(18), 8889–8896.

52. Jang, B., Yeo, S., Kim, H., Shin, B., & Kim, S.-H. (2017). Fabrication of single-phase SnS film by H₂ annealing of amorphous SnSx prepared by atomic layer deposition. *Journal of Vacuum Science and Technology. Part A, 35*(3), 031506.

53. Mattinen, M., King, P. J., Khriachtchev, L., Meinander, K., Gibbon, J. T., Dhanak, V. R., ... Leskelä, M. (2018). Low-temperature wafer-scale deposition of continuous 2D SnS_2 films. *Small*, *14*(21), 1800547.

54. Ham, G., Shin, S., Park, J., Lee, J., Choi, H., Lee, S., & Jeon, H. (2016). Engineering the crystallinity of tin disulfide deposited at low temperatures. *RSC Advances*, *6*(59), 54069–54075.

55. Zheng, L., Cheng, X., Cao, D., Wang, Q., Wang, Z., Xia, C., ... Shen, D. (2015). Direct growth of Sb_2Te_3 on graphene by atomic layer deposition. *RSC Advances*, *5*(50), 40007–40011.

56. Zhang, K., Nminibapiel, D., Tangirala, M., Baumgart, H.,. & Kochergin, V. (2013). Fabrication of Sb_2Te_3 and Bi_2Te_3 multilayer composite films by atomic layer deposition. *ECS Transactions*, *50*(13), 3–9.

57. Nminibapiel, D., Zhang, K., Tangirala, M., Baumgart, H., Chakravadhanula, V. S., Kubel, C., & Kochergin, V. (2013). Microstructure analysis of ALD Bi_2Te_3/Sb_2Te_3 thermoelectric nanolaminates. *ECS Transactions*, *58*(10), 59–66.

58. Browning, R., Plachinda, P., Padigi, P., Solanki, R., & Rouvimov, S. (2016). Growth of multiple WS_2/SnS layered semiconductor heterojunctions. *Nanoscale*, *8*(4), 2143–2148.

59. Karbalaei Akbari, M., & Zhuiykov, S. (2018). Tailoring two-dimensional semiconductor oxides by atomic layer deposition. In: Walia, S., (Ed.), *Low power semiconductor devices and processes for emerging applications in communications, computing, and sensing* (pp. 117–156). CRC Press, Taylor and Francis Group, Boca Raton, FL.

60. Bivour, M., Macco, B., Temmler, J., Kessels, W. M. M., & Hermle, M. (2016). Atomic layer deposited molybdenum oxide for the hole-selective contact of silicon solar cells. *Energy Procedia*, *92*, 443–449.

61. Seghete, D., Rayner, G. B., Cavanagh, A. S., Anderson, V. R., & George, S. M. (2011) Molybdenum atomic layer deposition using MoF_6 and Si_2H_6 as the reactants. *Chemistry of Materials*, *23*, 1668–1678.

62. Macco, B., Vos, M. F. J., Thissen, N. F. W., Bol, A. A., & Kessels, W. M. M. (2015). Low-temperature atomic layer deposition of MoOx for silicon heterojunction solar cells. *Physica Status Solidi*, *9*(7), 393–396.

63. Ziegler, J., Mews, M., Kaufmann, K., Schneider, T., Sprafke, A. N., Korte, L., & Wehrspohn, R. B. (2015). Plasma-enhanced atomic-layer-deposited MoO_x emitters for silicon heterojunction solar cells. *Applied Physics. Part A*, *120*(3), 811–816.

64. Nandi, D. K., & Sarkar, S. K. (2014). Atomic layer deposition of molybdenum oxide for solar cell application. *Applied Mechanics and Materials*, *492*, 375–379.

65. Diskus, M., Nilsen, O., & Fjellvåg, H. (2011). Growth of thin films of molybdenum oxide by atomic layer deposition. *Journal of Materials Chemistry*, *21*(3), 705–710.

66. Tseng, Y.-C., Mane, A. U., Elam, J. W., & Darling, S. B. (2012). Ultrathin molybdenum oxide anode buffer layer for organic photovoltaic cells formed using atomic layer deposition. *Solar Energy Materials and Solar Cells*, *99*, 235–239.

67. Xu, H., Karbalaei Akbari, M., Hai, Z., Wei, Z., Hyde, L., Verpoort, F., ... Zhuiykov, S. (2018). Ultra-thin MoO_3 film goes wafer-scaled nano-architectonics by atomic layer deposition. *Materials and Design*, *149*, 135–144.

68. Nandi, D. K., & Sarkar, S. K. (2014). Atomic layer deposition of tungsten oxide for solar cell application. *Energy Procedia*, *54*, 782–788.

69. Malm, J., Sajavaara, T., & Karppinen, M. (2012). Atomic layer deposition of WO_3 thin films using $W(CO)_6$ and O_3 precursors. *Chemical Vapor Deposition*, *18*(7–9), 245–248.

70. Zhuiykov, S., Hyde, L., Hai, Z., Akbari, M. K., Kats, E., Detavernier, C., ... Xu, H. (2017). Atomic layer deposition-enabled single layer of tungsten trioxide across a large area. *Applied Materials Today*, *6*, 44–53.

71. Zhuiykov, S., Karbalaei Akbari, M., Hai, Z., Xue, C., Xu, H., & Hyde, L. (2017). Wafer-scale fabrication of conformal atomic-layered TiO_2 by atomic layer deposition using

tetrakis (dimethylamino) titanium and H_2O precursors. *Materials and Design, 120,* 99–108.

72. Lee, K., Kim, D. H., & Parsons, G. N. (2014). Free-floating synthetic nanosheets by atomic layer deposition. *ACS Applied Materials and Interfaces, 6*(14), 10981–10985.

73. Lee, K., Losego, M. D., Kim, D. H., & Parsons, G. N. (2014). High performance photo-catalytic metal oxide synthetic bicomponent nanosheets formed by atomic layer deposition. *Materials Horizons, 1*(4), 419.

74. Gharachorlou, A., Detwiler, M. D., & Gu, X. K. (2015). Trimethyl aluminum and oxygen atomic layer deposition on hydroxyl-free Cu(111). *ACS Applied Materials and Interfaces, 7,* 16428–16439.

3 Self-Limiting Two-Dimensional Surface Oxides of Liquid Metals

3.1 INTRODUCTION

Atomically thin semiconductors are among the fastest-growing categories of modern advanced materials. The emergence of large variety of two-dimensional (2D) semiconductors has led to the tremendous changes and novel paradigm in the design of electronic devices [1–3]. 2D layered crystals are one of the main players of this advancement in electronic technology [4, 5]. The synthesis methods of 2D crystals have been restricted either to the exfoliation of intrinsically layered crystalline materials by mechanical exfoliation or limited to the high-vacuum deposition techniques [6–8]. However, the diversity of novel 2D materials for advanced applications is limited by conventional synthesis methods of 2D films. The mechanical exfoliation is not technologically a precise technique since the size distribution and thickness of produced nanosheets are not adequately or accurately controlled [9]. These high-vacuum techniques are also expensive and furthermore the library and diversity of deposited 2D films are restricted [10]. As a continuous research process, it is always required to introduce new types of 2D materials and find novel synthesis methods to obtain advanced 2D materials with superior properties. Thus, the synthesis of 2D semiconductors remains a fundamental challenge.

Room-temperature liquid metals have shown several numbers of interesting properties originated from their unique electron-rich metallic cores and the interface of liquid metals with their surrounding environment [11, 12]. Unlike molecular and ionic liquids, liquid metals are rarely used as reaction solvents. One of the main components of liquid metal alloys is their self-limiting surface oxide films (skin). This skin oxide film of liquid metals is one of the most perfect planar materials with atomic scale thickness [13, 14]. The crystalline nature of surface oxide of liquid metals and their alloys is the host of several distinguished properties, thus providing new platforms for the development of high-quality atomically thin materials for advanced electronic applications [15]. While considerable efforts have been devoted to the development of our knowledge about 2D surface oxide of liquid alloys, the field is still in its infancy. With respect to the atomic structure, elemental composition, fluidity and thermodynamic of liquid metals, and furthermore by considering the nucleation and growth characteristics of surface oxide films countless parameters are engaged to determine the properties of extracted 2D surface oxide films [16].

Consequently, this chapter will review the fundamental concepts of the atomic scale 2D surface oxide of liquid metal alloys. The metallurgical and thermodynamic aspects of liquid metal alloys are explained. Specific attention will be devoted to the individual properties of these 2D materials, the methods of their synthesis, characterization, and functionalization. Moreover, the electronic properties of 2D oxide films will be discussed and emerging applications will be reviewed in each individual case. The surface skins of liquid metal semiconductors are not restricted to the oxide semiconductors. 2D post-transition metal chalcogenide and metal nitride semiconductors were also synthesized by using the surface oxide of liquid metals. In this case, the deposition and wafer-scale patterning of 2D semiconductor films will be individually reviewed.

3.2 THE FUNDAMENTAL PROPERTIES OF LIQUID METALS

3.2.1 MONOPHASIC AND BIPHASIC LIQUID METALS AND ALLOYS

While traditional solvent systems and molecular and ionic liquid follow the law of definite proportion, this law does not cover atomic relations in liquid metals [17–19]. Instead, concentration range and phase structure are used to define the characteristics of liquid metals and are, in fact, the solvent mixture. Monophasic liquid alloys contain homogenously distributed elements without any segregation. In contrast, the biphasic liquid metals contain solid domains inside the liquid matrix. The low-melting monophasic metal alloys are composed of post-transition metals (Ga, In, Sn, Pb, Al, and Bi) and the elements of group 12 (Zn, Cd, and Hg). These elements can be combined together to produce a group of liquid alloys with a low melting temperature.

A biphasic liquid metal (mixture of solid and liquid) can be formed by either deviation of alloy composition from eutectic point or via adding solute elements above the solubility limit into the liquid metal solvent [20]. The colloidal suspension of solid nano and micro-particles can nucleate inside of liquid metal alloy. By considering the size of precipitations, the biphasic liquid metal can be stable and homogenous. The biphasic liquid metal can be a eutectic alloy, which does not melt congruently, thus it has a gradual melting process accompanied by the composition changing as the function of temperature [20, 21]. The metallic bonds are the main characteristics of metallic materials facilitating the low directional dependency of atomic metallic bonds to each other in both solid and liquid states. As the results, liquid metals have high boiling points. However, the melting temperature is depended on the crystalline structure of solid phase and the degree of the delocalization (ionization) of the valence electrons of metallic structure [22, 23]. Considering the diversity of crystalline structures of metallic elements, the melting temperature can vary considerably in metallic materials.

Among different post-transition metals, gallium (Ga) shows the lowest melting temperature with both covalent and metallic bonds in solid state. The characterization techniques displayed that the interatomic distance between a Ga atom and its closest neighbor is 2.7 ~ 2.9 Å, which is a significant distance between the gallium dimers [24, 25]. It causes strong structural anisotropy and creation of weak atomic bonding between Ga dimmers. Thus, the gallium crystals can break up at the lower temperatures compared with that of other metals [26–28]. The alloying of Ga with some of the other elements may lead to further decrease of melting temperature of liquid alloy. One of the most famous examples is the Ga–In alloy, when the melting temperature decreases to 16°C at the eutectic point of the In–Ga alloy with atomic concentration of 14.2% indium [29]. This information can be observed in the phase diagram of the In–Ga alloy in Figure 3.1a [30].

The incorporation of third metallic elements in metal alloys can even lead to a further decrease of the melting temperature of ternary alloy (Table 3.1). One of the famous example is galinstan, which is the ternary alloy of Ga, In, and Sn [31]. It is worth mentioning that the melting point of galinstan is often given as −19.1°C, which refers to the freezing temperature which differs from the melting point due to supercooling of liquid alloy. The mechanism behind the low melting point of the Ga–In–Sn alloy is attributed to two principal effects. The first one is related to the altered average interatomic distance by incorporation of In and Sn atoms into Ga alloy [30]. The second reason returns to the difference among localized charge densities of alloying metallic atoms. It was found that the free-electron charge density surrounding In atoms is tangibly lower than that of Ga and Sn atoms in a galinstan alloy [30]. Therefore, In atoms are not strongly attached to each other and behave similar to lubricant among the other alloying elements in atomic scales [30]. The schematic model for atomic distribution inside of eutectic Ga–In–Sn alloy was produced via density functional theory (DFT) calculations and was depicted in Figure 3.1b. The other factor is related to the decreased melting temperature of alloys due to formation of eutectic composition in Ga alloys. To have considerable solubility, the atomic diameter of the solute and solvent should not differ by more than 15%, their crystalline structure must be the same and both elements are expected to have similar electronegativity and the same valency [30]. For example, since the saturation limit of iron in Ga is less than 1%, the low-melting eutectic alloy cannot be formed in a Ga–Fe system [32].

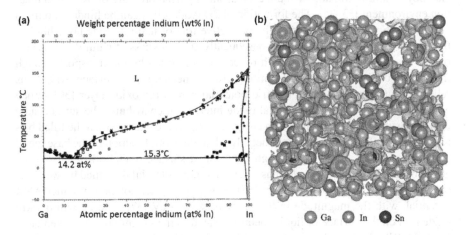

FIGURE 3.1 (a) The phase diagram of the Ga–In eutectic alloy. (b) The proposed atomic model for galinstan alloy (Ga–In–Sn). Reprinted with permission from Daeneke, T., Khoshmanesh, K., Mahmood, N., de Castro, I. A., Esrafilzadeh, D., Barrow, S. J., ... Kalantar-Zadeh, K. (2018). Liquid metals: Fundamentals and applications in chemistry. *Chemical Society Reviews*, 47(11), 4073–4111.

TABLE 3.1
Compositions and Melting Points of Several Ternary, and Quinary Alloys

Alloyed elements	Element A (at %)	Element B (at %)	Element C (at %)	Element D (at %)	Element E (at %)	Melting point (°C)
Ga/In (EGaIn)	85.8	14.2	0	0	0	15.4
Ga/In/Sn (galinstan)	78.3	14.9	6.8	0	0	13.2
In/Sn/Bi (Field's alloy)	60.1	18.8	21.1	0	0	62.0
Bi/Pb/Sn (Newton's alloy)	40.7	16.4	42.9	0	0	98.0
Bi/Pb/Sn (Rose's alloy)	43.1	23.5	33.4	0	0	95.0
Bi/Pb/Sn (D'Arcet's alloy)	43.0	24.3	32.7	0	0	95.0
Bi/Pb/Sn (Onion's alloy)	43.3	26.2	30.5	0	0	92.0
Bi/Pb/Sn/Cd (Wood's alloy)	41.5	20.9	18.3	19.3	0	70.0
Bi/Pb/Sn/Cd (Lipowitz's alloy)	42.1	22.9	19.3	15.7	0	74.0
Bi/Pb/Sn/Cd/In (French's alloy)	31.5	17.1	14.4	11.7	25.3	46.9

Source: Reprinted with permission from [30].

3.2.2 CHARACTERISTICS OF THE LIQUID METAL INTERFACE

Similar to solid metals, the surface of liquid metals and alloys is also covered by ultrathin metal oxide film. The liquid metals also react to the atmospheric oxygen even at very low oxygen pressure forming in a self-limiting metal oxidation reaction [33]. The surface oxide layers are extraordinary planar and are among the most perfect naturally developed 2D materials. Upon exposure to the atmosphere, a thin layer of oxide forms instantaneously on the surface of liquid metals such as Ga and Ga alloys, indium, tin, and bismuth even at small partial pressure of oxygen in reaction environment [33–35]. The formation of 2D oxide films can be hindered only in an ultra-high-vacuum condition. The protective role of natural oxide layer hinders the further oxidation process, and consequently, the thickness of natural 2D surface oxides changes slightly. The growth process is controlled by ion transport through the surface oxide films. The self-limiting growth mechanism of surface oxide films relies on the formation of an electric potential across the oxide layer [30]. When an electron is adsorbed on the metal oxide interface it contributes the surface state energy of surface oxide film with an energy amount more positive than the filled balance band of metal oxide and more negative than the work function of the base metal alloy. By electron tunneling through the created surface oxide, adsorbed oxygen atoms on the surface is reduced, thus an electrostatic potential is formed between the oxide-air and oxide-metal interfaces [36]. This electrostatic potential is called Mott potential, with the magnitude in the range of several eV [37]. The growth of skin oxide film on the surface of liquid metal is related to the magnitude of created electric field. When the thickness of oxide film is ultrathin, the strength of electric field is sufficiently strong to drive ion diffusion through the oxide thin film and facilitate the growth of oxide film. By the increase of film thickness, the electric field cannot efficiently drive the ionic species and the growth of thin film is stopped, thus, the rest of liquid metal alloy is passivated from further oxidation [36, 37]. There are different

2D Surface Oxides of Liquid Metals

parameters that determine the final thickness of self-limiting surface oxide films, including the work function of the liquid metal, the solubility of metal and oxygen ions in the oxide film and the oxidation process temperature. For example, the thickness of self-limited oxide layer for Ga is around 0.3 to 0.5 nm [38]. It was found that the exposure of Ga to the oxygen with a pressure of $180*10^{-6}$ Torr for one second is sufficient enough to grow a complete surface oxide layer and higher oxygen pressure and temperature (up to 300°C) do not considerably affect the thickness of oxide films [38, 39]. The developed natural oxide layer is amorphous and barely crystalline [38, 39]. The other surface oxide films of liquid alloys have different growth characteristics. In other sections of the present chapter, the growth properties of naturally occurred surface oxide films of liquid alloys will be discussed separately.

For liquid alloys with different alloying elements, the surface oxide is mostly composed of one particular oxide film. The formation of this oxide film leads to the highest reduction in Gibbs free energy (ΔG_f) [40]. For instance, the surface oxide of eutectic In–Sn alloys is predominantly indium oxide, which contains 6% doped tin oxide. Therefore, it is possible to extract 2D mixed indium tin oxide (ITO) from the In–Sn alloy with numerous possibilities in transparent electronic applications [41]. Since the ΔG_f of Ga_2O_3 is lower than that of In_2O_3 and SnO_2 in oxygen atmosphere, the surface oxide of galinstan (Ga–In–Sn alloy) alloy is mostly composed of Ga_2O_3 [42]. The direct exposure of the surface of galinstan to other atmosphere leads to development of the different types of 2D surface compounds. For example, when the HCl is exposed to galinstan surface, the rapid reaction of gallium oxide with hydrochloric acid leads to the removal of gallium oxide from the surface and formation of metal chlorides and H_2O [43]. Interestingly, in the case of galinstan alloy, the ΔG_f of $InCl_4$ is higher than that of $SnCl_2$ and $GaCl_3$ [43]. As a result, the surface of galinstan alloy in HCl atmosphere is mostly composed of indium chloride. The transformation process is explained by the following equations [43]:

$$2Ga_{(galinstan)} + \frac{1}{2}O_2 \rightarrow Ga_2O_3 \tag{3.1}$$

$$Ga_2O_3 + 6HCl \rightarrow 2GaCl_3 + 3H_2O \tag{3.2}$$

$$GaCl_3 + In \rightarrow InCl_3 + Ga \tag{3.3}$$

Generally, the aforementioned observation has confirmed that the reactivity of metal surface is governed by the well-established thermodynamic rules, where the compound attributed to the highest level of decrement of free energy grows at the interface between liquid metal and the surrounding environment. By controlling the metallurgical and thermodynamic parameters, it is possible to develop novel synthetic approaches toward the development of novel low dimensional inorganic nanostructures. The next section will specifically focus on different synthesis cases of 2D surface oxides of liquid metals.

3.3 SYNTHESIS AND APPLICATIONS OF 2D SURFACE OXIDES OF LIQUID METALS

The interface between the liquid metal and the surrounding environment presents a unique platform for synthesis and growth of 2D materials. The outer layer of a

84   Ultrathin 2D Semiconductors

liquid metal is one of the highest quality natural exteriors with smooth surface and a defect and stress fee interface for growth of the 2D ultrathin materials. Due to non-polar nature of liquid metals, the attraction forces between the liquid metal and the created 2D surface compound films are weak and localized. Thus, the isolation of these ultrathin surface films is feasible. The process typically includes the natural self-limiting growth of ultrathin film on the surface of liquid alloy which is accompanied by the mechanical separation and exfoliation of 2D films by applying mechanical exfoliation methods including sonication, touching of liquid metal with the appropriated selected substrate and separation of substrate from the liquid metal, mechanical rolling of the liquid metal and alloy over the smooth substrate and separation and extraction of 2D films through density differences and gradients between the synthesized compounds. The next section will individually introduce different synthesis methods of 2D ultrathin films of surface components of liquid alloys and will explain about their applications.

3.3.1 SCREEN PRINTING OF 2D SEMICONDUCTORS

3.3.1.1 Screen Printing of Ga$_2$O$_3$ and GaS 2D Films

There are several methods for the synthesis of high-quality ultrathin 2D films that rely on rolling liquid metals over the appropriately selected substrate [44]. This rolling method is known as the screen-printing technique and is capable of synthesis of high-quality 2D films without or with post-synthesis treatments. In one practical case the 2D Ga$_2$O$_3$ films were synthesized by rolling low-melting gallium alloy over Si/SiO$_2$ wafer [44]. To facilitate the substrate surface patterning, the selective area deposition technique has been developed where the photolithography technique was accompanied by the surface functionalization of substrate by perfluorinated silanes [44]. The functionalization agent (silane) in this technique effectively decreased the van der Waals force between the natural surface oxide of Ga alloy and the substrate preventing the exfoliation and delamination of the surface oxide film. By using this approach, a patterned Ga$_2$O$_3$ were deposited on non-functionalized Si/SiO$_2$ substrate with precise controllability [44]. Due to rapid oxidation of Ga liquid metal at atmospheric conditions, continuous deposition was enabled which is quite similar to the screen-printing technique depicted in Figure 3.2. For the same deposited film, another post-treatment process was employed to transform the Ga$_2$O$_3$ film into Ga sulfide by direct sulfurization process [44]. In this method, the deposited Ga$_2$O$_3$ 2D ultrathin film was first transformed into Ga$_2$Cl$_3$ by passing the hydrochloric acid vapor over the surface of Ga$_2$O$_3$ film patterned on Si/SiO$_2$ substrate at 45°C. The following heat treatment of the samples at 300°C transformed the Ga$_2$Cl$_3$ into GaS 2D films. One of the most important advantages of this technique is the low-temperature chalcogenization step (300°C). In regular conditions, this requires high processing temperatures (900°C) [45]. The low-temperature fabrication techniques are highly desired for semiconductor processing instruments, and low-temperature processing is a basic technical requirement for the fabrication of transparent and flexible electronic devices.

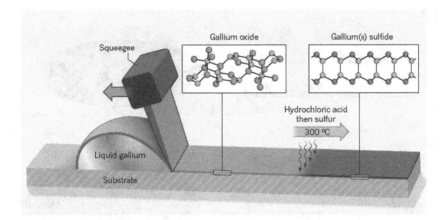

FIGURE 3.2 The screen-printing of 2D semiconductor films. The liquid gallium alloy is scraped across the Si/SiO₂ substrate by employing a squeegee. The surface oxide of the galinstan alloy is attached to the surface of substrate while the instantaneous oxidation of gallium leads to the formation of Ga₂O₃ film. The following low-temperature chalcogenization step results in the formation of 2D gallium sulfide. Reprinted with permission from Kim, Y., & Hone, J. (2017). Screen printing of 2D semiconductors. *Nature*, *544*(7649), 167–168.

3.3.1.2 Wafer-Scale Screen Printing of GaS 2D Films

Wafer-scale deposition of ultrathin GaS film was developed over Si/SiO₂ substrate with patterned tungsten electrode [46]. This technique has facilitated the development of optical photodetectors with high capabilities. In this method, the tungsten was initially deposited on the Si/SiO₂ film and the electrodes were prepared by photolithography. Then, the surface of the wafer was immersed into FDTES to functionalize the surface of Si/SiO₂ electrode with a monolayer of fluorocarbon. After the removal of photoresist materials, the patterned substrate was employed for the gallium alloy printing. With the screen-patterning method, Ga was rolled on the surface of Si/SiO₂ substrate to produce ultrathin 2D patterned films. Before rolling, Ga was placed into HCl at 60°C to remove the natural oxide from the surface. Surface of the printing paper was placed in direct contact with HCl vapor to halogenize Ga₂O₃ and produce GaCl₃. Then the GaCl₃ again was placed in direct contact with the sulfur gas to turn it into GaS films. The process was graphically demonstrated in Figure 3.3. The two-step sulfurization process helps to reduce the sulfurization temperature of Ga₂O₃ film down to 300°C, since the direct conversion of chemical inert Ga₂O₃ into GaS requires a high-temperature process (900°C) [46].

Both Raman spectroscopy and XRD characterization confirmed the development of stoichiometric GaS films. Atomic force microscopy measurements certified that the 2D GaS film has a thickness of 1.5 nm, which is equal to the thickness of two fundamental layers of GaS (Figure 3.4a). In addition, the transmission electron microscopy (TEM) observation established that the developed 2D crystalline film has lattice spacing of 3.1 Å (Figure 3.4 b). The investigation of bandgap of 2D GaS films revealed the value of ~ 2.5 eV (Figure 3.4c). Moreover, the photoluminescence

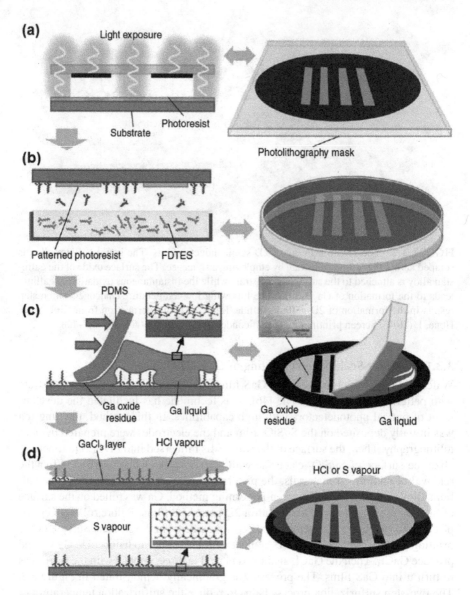

FIGURE 3.3 The graphical scheme of wafer-scale development of GaS-patterned 2D films on the Si/SiO$_2$ substrate. (a) The lithography process for patterning the Si/SiO$_2$ substrate and production of lithography mask. (b) The coverage of the exposed area to the FDTES. (c) Coverage of patterned substrate with gallium using PDMS. (d) The two-step process to halogenize and chalcogenize ultrathin films on Si/SiO$_2$ substrate. Reprinted with permission from Carey, B. J., Ou, J. Z., Clark, R. M., Berean, K. J., Zavabeti, A., Chesman, A. S., ... Kalantar-Zadeh, K. (2017). Wafer-scale two-dimensional semiconductors from printed oxide skin of liquid metals. *Nature Communications, 8*, 14482.

2D Surface Oxides of Liquid Metals

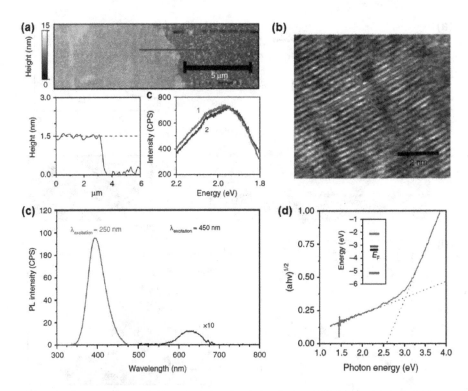

FIGURE 3.4 (a) The morphology and characteristic of printed 2D films with their thickness profile measured by AFM microscopy. (b) The high-resolution TEM image of GaS flake, which shows the lattice spacing of 3.1 Å. (c) The PL spectra (d) and bandgap measurements of 2D film of GaS which demonstrate the contributions of interband transition (red line) and also the deep trap recombination (black line). Reprinted with permission from Carey, B. J., Ou, J. Z., Clark, R. M., Berean, K. J., Zavabeti, A., Chesman, A. S., ... Kalantar-Zadeh, K. (2017). Wafer-scale two-dimensional semiconductors from printed oxide skin of liquid metals. *Nature Communications*, 8, 14482.

studies also confirmed that the synthesized 2D films are capable for sensing of the light at the middle and the end of the visible light spectrum range (Figure 3.4d). All of the characterization studies endorsed and confirmed the *n*-type nature of synthesized GaS 2D film. The *p*-type conductivity of GaS 2D film was gained by exposing the synthesized film to NO_2 gas [46]. Thus, combined screen-printing and two-step annealing approach is a strong technique for synthesizing 2D films with a variety of electronic properties.

An optical sensor was made based on the 2D GaS films on patterned Si/SiO_2 film by employing tungsten electrodes [46]. Tungsten electrode was employed due to several reasons. The liquid Ga does not amalgamate with tungsten (Figure 3.5a). The other privilege is the formation of WS_2 during the sulfurization process, which can act as the hole injector at heterointerfaces between WS_2 and GaS 2D films (Figure 3.5b). The WS_2 layer helps the electron to overcome the Schottky barrier at the interface between GaS and tungsten electrode to improve the capability of instruments for

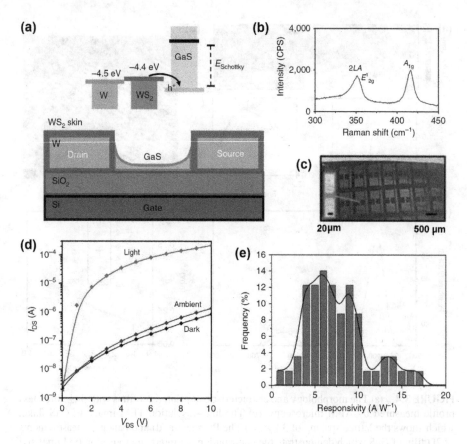

FIGURE 3.5 (a) The graphical scheme of backgated GaS phototransistor with WS$_2$/W electrodes. (b) The growth of WS$_2$ film was characterized at the heterointerfaces between tungsten and GaS 2D film. (c) The optical image of series of phototransistor devices based on GaS 2D films. (d) The I/V characteristics of device when the V$_g$ = 0. (e) The distribution of values of photo responsivity of phototransistors. Reprinted with permission from Carey, B. J., Ou, J. Z., Clark, R. M., Berean, K. J., Zavabeti, A., Chesman, A. S., ... Kalantar-Zadeh, K. (2017). Wafer-scale two-dimensional semiconductors from printed oxide skin of liquid metals. *Nature Communications*, 8, 14482.

high mobility carrier properties. The fabricated phototransistor devices (Figure 3.5c) demonstrated a high *on/off* ratio of 170 (Figure 3.5d) and medium photoresponsivity of 6.4 A/W under the solar simulator (Figure 3.5e). Thus, the fabricated phototransistors showed high capability and uniformity when several cases of measurements confirmed the average success rate of 89.4% for the fabricated devices [46].

3.3.2 Reactive Environment for Synthesis of 2D Semiconductor Films

As mentioned earlier, Ga$_2$O$_3$ is the prevalent metal oxide on the surface of galinstan (Ga–In–Sn) alloy. However, if more reactive elements are incorporated into galinstan alloy, the other type of oxide films are expected to form on the surface of liquid alloys due to changing the thermodynamic condition of surfaces. In this case, the oxide dominates the surface, which causes the highest reduction of the Gibbs free energy [47]. From the thermodynamic point of view, it is estimated that all lanthanide oxide, a tangible number of transition metals, and post-transition metal oxides can be synthesized by using the liquid metal printing approach [47]. One of the best known examples is the incorporation of less than 1 wt. % of elemental aluminum, gadolinium, and hafnium in a galinstan alloy, leading to the formation of Al_2O_3, Gd_2O_3, and HfO_2 (Figure 3.6a,b) [47]. These oxides fulfilled the requirement of the representing lower ΔG_f than that of Ga_2O_3. However, the incorporation of copper and

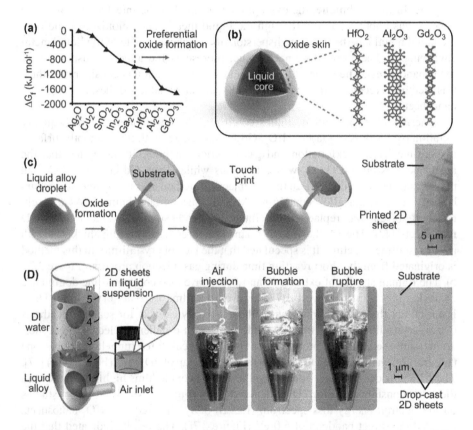

FIGURE 3.6 (a) ΔG_f of some selected metal oxide films. (b) The graphical scheme of the cross-sectional part of a liquid galinstan droplet with Hf, Gd, and Al alloying elements. (c) The graphical scheme showing the vdW exfoliation of 2D oxide films and the optical image of exfoliated film. (d) Schematic representing the principles of gas injection method. The optical image represents the image of 2D oxide nanosheets extracted from the aqueous suspension. Reprinted with permission from Zavabeti, A., Ou, J. Z., Carey, B. J., Syed, N., Orrell-Trigg, R., Mayes, E. L. H., ... Daeneke, T. (2017). A liquid metal reaction environment for the room-temperature synthesis of atomically thin metal oxides. *Science*, *358*(6361), 332–335.

silver did not satisfy the thermodynamic conditions. Therefore, the surface oxides of galinstan/Cu and galinstan/Ag alloys were mostly composed of Ga_2O_3 [47].

These transition (HfO_2), post-transition (Al_2O_3), and rare metal (Gd_2O_3) oxides cannot be produced by exfoliation methods, since they have nonstrained crystalline structure [47]. Two methods were proposed to extract these nanostructured 2D films. The first method is the mechanical exfoliation of surface oxide of liquid metal by touching the surface of liquid metal to an appropriately selected substrate. Due to the nature of liquid metal and the absence of microscopic forces between the liquid metal and its natural surface oxide, the vdW exfoliation of 2D films is facilitated. For example, high-quality delaminated films depicted in Figure 3.6c were produced by this technique. The second method relies on the exfoliation of surface oxide films by using a gas injection method inside of the liquid metal droplets. In this technique, the oxide layer is formed at the interface between the injected gas bubbles passing through the liquid metal. The deionized liquid over the liquid metal lets the uniform dispersion of nanosheets inside of the liquid metal environment. Therefore, the final product represents the aqueous suspension of oxide nanosheets. The scalability of the liquid metal reaction is considerable; thus, this method is recognized as a reliable approach for scalable synthesis of 2D oxide nanosheets.

Materials characterization studies confirmed the development of high-quality ultrathin oxide films of Ga_2O_3, HfO_2, Gd_2O_3, and Al_2O_3 with ultra-smooth surfaces in both mechanical exfoliation and gas injection methods. It was found that the mechanical exfoliated Ga_2O_3 was amorphous while HfO_2, Gd_2O_3, and Al_2O_3 had polycrystalline structure [47]. In the gas injection method, the strong reaction occurs between a galinstan alloy containing Al and aqueous environment. As a result, the inert solvent was replaced to facilitate the synthesis of Al_2O_3 and Gd_2O_3 2D nanosheets [47]. The synthesized 2D nanostructures had ultrathin thickness with poor crystalline structure. It is speculated that the lack of crystallinity in this method is originated from the short reaction time during gas injection process. (Figure 3.7a, b). The Raman studies of extracted nanosheets have also confirmed the dominance of oxide films with the lowest ΔG_f. However, the films were not crystalline. Thus, it turned out that the gas injection method is highly capable for scalable synthesis of non-crystalline pinhole-free 2D oxide films with excellent dielectric properties. To measure the electrical properties of HfO_2, the peak force tunneling microscope (Figure 3.7c) was employed to produce the current map of 2D films (Figure 3.7d) [47]. The *I-V* electrical characteristics of synthesized 2D oxide films highlighted the high dielectric constant of 39 for 2D oxide nanosheets (Figure 3.7e) [47]. Further studies on the electron energy loss spectrum (EELS) of synthesized 2D HfO_2 nanosheets revealed the direct bandgap of 6.0 eV (Figure 3.7f). The results indicated that the reactive environment synthesis approach is a capable method for large-scale synthesis of the chemically manipulated nanosheets, which are not naturally present in the environment. Consequently, considerable opportunities are expected to be found by the synthesis of rare 2D oxide with unexpected electronic, magnetic, optical, and catalytic properties.

2D Surface Oxides of Liquid Metals 91

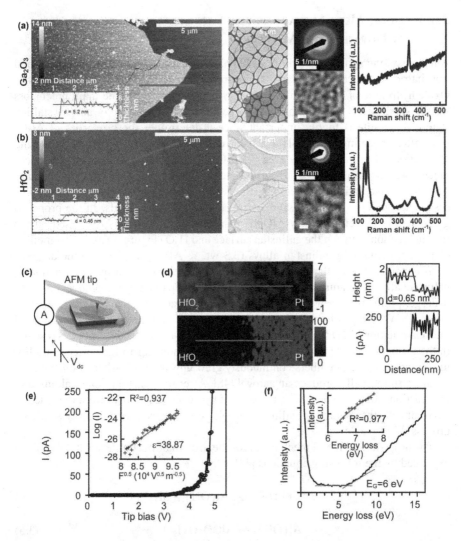

FIGURE 3.7 The extracted (a) Ga$_2$O$_3$ and (b) HfO$_2$ 2D films from aqueous suspension in gas injection technique, accompanied by the TEM image and Raman spectrum. (c) The schematic of peak force tunneling AFM for the measurement of dielectric properties of HfO$_2$ 2D films. (d) The AFM measurement of HfO$_2$ 2D film height and current maps of the film. (e) The *I-V* curve of 2D HfO$_2$ layer. The inset is produced based on Schottky emission model to determine the dielectric constant. (f) The graph of low-loss energy EELS spectrum, which gives the bandgap of 2D HfO$_2$ film. The inset shows the direct nature of the bandgap. Reprinted with permission from Zavabeti, A., Ou, J. Z., Carey, B. J., Syed, N., Orrell-Trigg, R., Mayes, E. L. H., … Daeneke, T. (2017). A liquid metal reaction environment for the room-temperature synthesis of atomically thin metal oxides. *Science*, *358*(6361), 332–335.

3.3.3 Liquid Metal Media for Green Synthesis of Ultrathin Flux Membranes

Among existing synthesis methods of 2D nanostructures, a green synthesis method was introduced recently for large-scale production of γ-AlOOH nanosheets [48]. Oxide hydroxide of aluminum (boehmite γ-AlOOH) is one of the main precursors for the synthesis of aluminum oxides. The control of morphology and crystalline state of boehmite is highly important since it determines the properties of Al_2O_3 films after annealing. γ-AlOOH is a stable compound in environment with filtering applications. The high surface area γ-AlOOH can be synthesized by several techniques, which are mostly based on the techniques that require high-temperature, power-intensive and long-processing times [48]. Furthermore, the production process is accompanied by the usage and release of toxic chemicals. In the proposed green synthesis method, 2D γ-AlOOH nanosheets were synthesized based on the instant reaction between the galinstan surface and H_2O (Figure 3.8a). In this method aluminum containing galinstan alloys (3.3 wt. % Al) are exposed to the aqueous environment either in the form of liquid or vapor phases. The instant reaction of galinstan alloy with aqueous environment was accompanied by the production of the hydrogel products. When the galinstan alloy was placed adjacent to the distilled water (DIW), the gas bubbles were formed on the liquid metal surface and caused the delamination of 2D films from the surface. Interestingly, water vapor gas resulted in the formation of 1D nano-pillar structures perpendicular to the surface of a liquid galinstan alloy. The fibers continuously grew upward as clearly demonstrated in Figure 3.8b to finally produce an aerogel [48]. After the consumption of aluminum, the reaction stopped. Then, after removal of hydrogel and aerogel from the reaction environment, the remaining galinstan droplet was recycled for the other synthesis process (Figure 3.8c).

The proposed chemical mechanism for the growth of boehmite γ-AlOOH can be explained by the following equations [48]:

$$2Al + 6H_2O \rightarrow 2Al\left(OH\right)_3 + 3H_2 \tag{3.4}$$

$$Al\left(OH\right)_3 \rightarrow AlOOH + H_2O \tag{3.5}$$

$$2Al + 4H_2O \rightarrow 2AlOOH + 3H_2 \tag{3.6}$$

$$2AlOOH \rightarrow Al_2O_3 + H_2O \tag{3.7}$$

Equation 3.4 explains the formation of hydrogel products which later turn into AlOOH after heating at 170°C for 4 hours (Equation 3.5). However, it is possible to synthesize AlOOH with a single step-growth, where the AlOOH aerogel is prepared by the exposure of the galinstan alloy surface to the H_2O vapor (Eq. 3.6). To convert the Boehmite into Al_2O_3, the following annealing at 550°C is required (Eq. 3.7). The synthesized AlOOH 2D nanosheets are found as crystalline layered structures (Figure 3.9a) with a thickness of 1.2 nm (Figure 3.9b). The HRTEM and XRD studies (Figure 3.9c) have also confirmed that the γ-AlOOH had orthorhombic

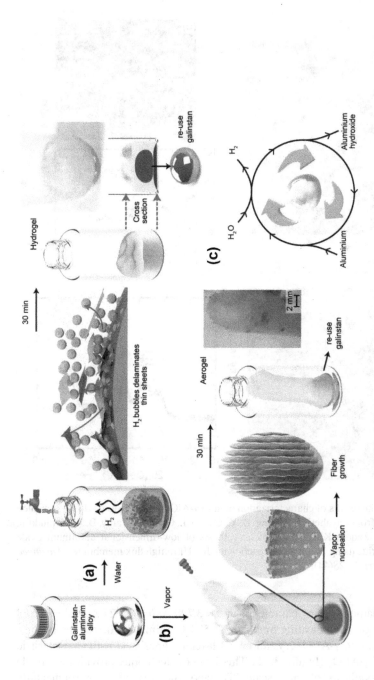

FIGURE 3.8 The schematic of green synthesis of 2D and 1D nanostructured materials from the reaction between galinstan-Al alloy with (a) distilled water (DIW) and with (b) water vapor. In section (a) the severe reaction of water with galinstan-Al alloy is accompanied by the generation of gas bubbles, which results in the delamination of 2D nanosheets and the formation of a hydrogel. In (b) the process is accompanied by the growth and emergence of fibrous structures, and the final product is an aerogel. (c) After the consumption of Al in galinstan, the reaction is stopped. Reprinted with permission from Zavabeti, A., Zhang, B. Y., Castro, I., Ou, J. Z., Carey, B. J., Mohiuddin, M., ... Kalantar-Zadeh, K. (2018). Green synthesis of low-dimensional aluminum oxide hydroxide and oxide using liquid metal reaction media: Ultrahigh flux membranes. *Advanced Functional Materials*, 28(44), 1804057.

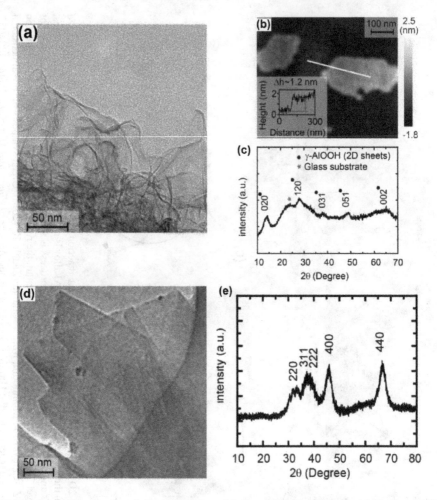

FIGURE 3.9 The results of characterization of (a-c) γ-AlOOH and (d-e) γ-Al$_2$O$_3$. Reprinted with permission from Zavabeti, A., Zhang, B. Y., Castro, I., Ou, J. Z., Carey, B. J., Mohiuddin, M., ... Kalantar-Zadeh, K. (2018). Green synthesis of low-dimensional aluminum oxide hydroxide and oxide using liquid metal reaction media: Ultrahigh flux membranes. *Advanced Functional Materials*, *28*(44), 1804057.

structure with lattice spacing parameters of $a = 3.700$ Å, $b = 12.227$ Å, and $c = 2.868$ Å [48]. The following annealing treatment at 550°C ultimately altered the boehmite and turned it into γ-Al$_2$O$_3$ crystalline 2D-layered nanosheets with characteristic XRD peaks of γ-Al$_2$O$_3$ (Figure 3.9d). This low-cost technique provides a reliable approach for synthesis of large-scale 2D nanostructures with environmentally friendly characteristics. This approach shows high efficiency with a yield of almost 100% for large-scale synthesis of Al$_2$O$_3$ 2D nanosheets with reduced reaction and synthesis time at low processing temperature [48]. The structural characteristics of produced nanosheets are quite similar to their perfect crystals, giving the materials a high mechanical stiffness and considerable Young's modulus. The synthesized 2D nanosheets were later employed as the membrane, which supported a tangible flux for the separation of heavy metals and oil from the polluted water. The ultrahigh ion filtration mechanism of γ-AlOOH ($8.4*10^4$ L m^{-2} h^{-1} bar^{-1}) was attributed to the capability of highly wrinkled and distorted γ-AlOOH films, which allowed the facile permeation of water molecules among the layered 2D structures [48]. The other filtration mechanism of 2D γ-AlOOH membranes was attributed to the formation of layered double hydroxides (LDH) with divalent metal cations in water. Another mechanism was based on the electrostatic attraction of the metal cations into the surface hydroxide groups of the surface [48].

3.3.4 THE vdW EXFOLIATION AND PRINTING METHODS OF 2D SEMICONDUCTORS

3.3.4.1 Semiconducting SnO Monolayers

The mechanical delamination or van der Waals exfoliation is one of the most employed techniques for preparation of 2D oxide films separated from the surface of liquid metal alloys. Elemental tin (Sn) has a high conductivity in electrical connections in electrical and electronic instruments and also shows a low-melting point. It was confirmed that tin oxide layer has protective effects against the corrosion. Tin oxide naturally covers the surface of tin alloy under atmospheric condition [49]. The surface oxide of tin is almost composed of SnO [49]. SnO_2 is the other type of tin oxide, which only occurs when the surface of molten tin is exposed to the large quantity of oxygen gas, which does not naturally occur in atmospheric conditions [49]. Thus, the natural surface oxide of tin alloy can be a source of thin SnO layers. Here the precise control of oxygen exposure is the main technical factor to determine the composition of surface oxide film of tin alloy. This factor is highly important when we know that the SnO_2 is an n-type semiconductor, and SnO is a p-type semiconductor with a variety of electrical properties. For example, the tin oxide can be employed as the ultrathin material for field effect transistor [49]. It can also be employed as the ultrathin invertor material originated from bipolar properties of SnO. Considering the stability and sensitivity of well-established SnO films, it has tremendous capability for application in chemical-based FET sensors and catalyst films. Transparency is the other characteristic of SnO films that originated from the wide band gap of this semiconductor film.

To produce ultrathin SnO film, the tin alloy was melted at 300°C, and after removal of thick oxide layer, the surface of the liquid alloy was subjected to the atmospheric conditions or within a continuous purge of oxygen gas inside of glovebox [49]. The mechanical delamination of the surface oxide film was performed by touching the Si/SiO$_2$ substrate on the surface of liquid tin alloy. Due to the vdW force between the substrate and surface oxide film, the tin oxide film is exfoliated from the liquid Sn. The development of microscopic attachment force between the oxide ultrathin films and liquid alloy allows the lifting and delamination of the oxide films, leaving the fresh liquid metal behind (Figure 3.10a) [49]. The AFM studies revealed that the thickness of ultrathin films at the ambient atmosphere was approximately

96 Ultrathin 2D Semiconductors

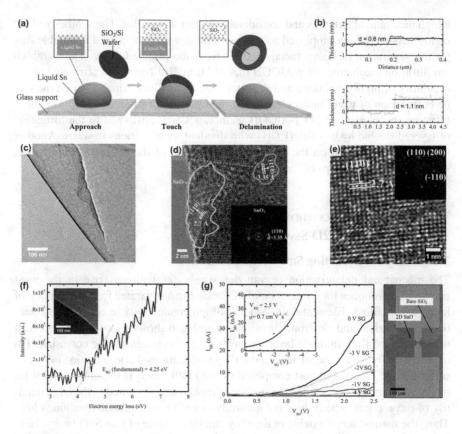

FIGURE 3.10 (a) The graphical representation of vdW exfoliation method. The surface of molten Sn alloy over the glass substrate is covered with ultrathin SnO film. When a Si/SiO$_2$ substrate is attached to the molten tin alloy, the created vdW force between the tin oxide and SiO$_2$ surface helped the delamination and following attachment of 2D oxide film to the SiO$_2$ substrate. (b) The thickness measurement of SnO film by AFM showed the exfoliation of monolayer (0.6 nm) and bilayer (1.2 nm) tin oxide film. (c) The TEM image of tin oxide film developed in a controlled atmosphere. (d) The HRTEM of tin oxide film developed in ambient atmosphere with mixed SnO and SnO$_2$ structure. (e) The HRTEM image of SnO film developed in a controlled atmosphere. (f) The band gap of monolayer SnO film measured by electron energy loss spectroscopy. (g) The *I-V* characteristics of fabricated SnO FET device at various gate voltages. The inset demonstrates the I_{SD}-V_{SG} graph of FET device at the constant voltage of 2.5 V. The optical image of FET device based on 2D SnO film is depicted in (g). Reprinted with permission from Daeneke, T., Atkin, P., Orrell-Trigg, R., Zavabeti, A., Ahmed, T., Walia, S., ... Kalantar-Zadeh, K. (2017). Wafer-scale synthesis of semiconducting SnO monolayers from interfacial oxide layers of metallic liquid tin. *ACS Nano*, *11*(11), 10974–10983.

0.6 nm thick [49]. Considering the interlayer spacing of crystalline SnO (0.484 nm), 0.6 nm thick-film can be considered as a monolayer SnO film and 1.1 nm thick-film is therefore a bilayer SnO 2D film (Figure 3.10b). It was found that the samples synthesized in glovebox with the oxygen concentration of 10–100 ppm oxygen gas had ultra-fine and smooth surface, compared with that of 2D films formed in the ambient atmosphere [49]. Further characterization of synthesized tin oxide films with HRTEM revealed the formation of ultra-fine tin oxide films over the surface of the liquid Sn in the glovebox atmosphere (Figure 3.10c). The measurement of interlayer spacing showed that the nanosheets formed under ambient atmosphere were composed of two distinct materials with interlayer spacing of 2.7 Å and the other 3.35 Å, which were respectively attributed to the crystal lattice of SnO and SnO_2. The 2.7 Å lattice spacing was related to the (110) plane of SnO, while the 3.35 Å lattice spacing was attributed to the (110) plane of SnO_2, respectively. However, the HRTEM observations have also established that the sample formed under the reduced oxygen condition was mostly composed of SnO film (Figure 3.10e) [49]. The measurements of the bandgap of the monolayer SnO film by EELS in the low-loss region showed the direct bandgap nature of monolayer SnO film with a magnitude of 4.25 eV [49].

The few-layer SnO-based FET devices were fabricated based on the mechanically delaminated SnO film from the surface of molten tin alloy. The p-type characteristics of SnO-based developed FET unit was confirmed with the mobility of 0.7 $cm^2V^{-1}s^{-1}$ (Figure 3.10g) which was much higher than the similar FET devices based on the pulsed layer deposited bilayer SnO film with the mobility of 0.05 $cm^2V^{-1}s^{-1}$ [50]. The optical image of the FET unit based on the few-layer SnO film is depicted in Figure 3.10g. It is estimated that the developed technique is highly suitable for the synthesis of low-melting temperature metal and alloys similar indium, gallium, bismuth, lead and their alloys like galinstan and Field's metal.

3.3.4.2 Semiconducting 2D Gallium Phosphate Nanosheets

Gallium phosphate is a semiconductor material with trigonal structure with the cell parameters of $a = 4.87$ Å, $c = 11.05$ Å, and $\gamma = 120°$ (Figure 3.11 a, b) [51]. The high-temperature thermal stability of gallium phosphate (up to 930°C) placed this semiconductor material in an outstanding technical position compared with the similar materials like α-quartz. It was found that the $GaPO_4$ has recognizable piezoelectric properties. The piezoelectricity facilitates the mutual conversion of electrical energy or pulses to mechanical forces or oscillations. The advent of 2D materials with piezoelectric characteristics is highly interesting since it opens up new opportunities for development of miniature power and electric instruments. In these devices the atomic scale mechanical displacement, vibration and bending can be turned into electrical pulses and then facilitate the harvesting of kinetic energies of oscillation of piezoelectric materials. The loss of centrosymmetry is the main structural specifications of 2D materials with piezoelectric properties [51]. The doped graphene, h-BN and many transition metal chalcogenides (TMDCs) with odd numbers of layers are capable of showing piezoelectric properties. These theoretical investigations also confirmed the piezoelectric characteristics of transition metal oxides, aluminum nitride, GeS, and $SnSe_2$. Most of the piezoelectric functions of 2D materials are observed at the low temperatures [52–57]. The stable piezoelectric performance of

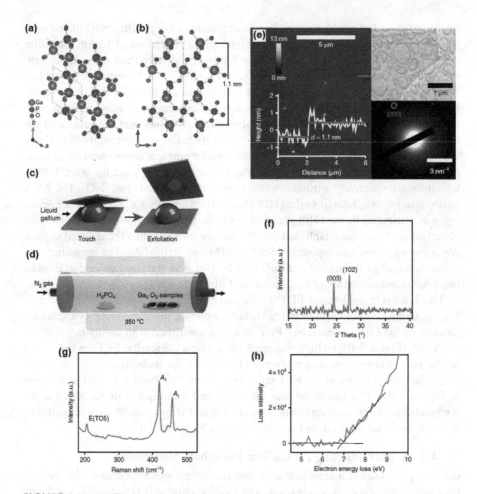

FIGURE 3.11 (a) The top view and (b) side view of GaPO$_4$ demonstrating the out-of-plane structure of GaPO$_4$. (c) The schematic of the vdW 2D printing of Ga$_2$O$_3$ surface oxide of gallium alloy. (d) The set-up for the secondary phosphatization treatment. (e) The AFM thickness profile and TEM image, (f) XRD, (g) Raman spectra and (h) bandgap measurement of GaPO$_4$ developed films. Reprinted with permission from Syed, N., Zavabeti, A., Ou, J. Z., Mohiuddin, M., Pillai, N., Carey, B. J., ... Kalantar-Zadeh, K. (2018). Printing two-dimensional gallium phosphate out of liquid metal. *Nature Communications*, 9(1), 3618.

2D materials at the high temperatures now is highly required for technical applications, since the performance of 2D piezoelectric materials are failed due to the structural changes at the high working temperatures. Furthermore, the ultra-uniform deposition of homogenous 2D piezoelectric materials over the large-area substrates is highly challenging. Thus, this field requires a tremendous attention for further technological developments.

The archetypal piezoelectric GaPO$_4$ does not naturally available in crystalline scarified structure [51], thus the conventional mechanical exfoliation techniques are not capable of synthesis of 2D crystalline gallium phosphate films. Therefore, it is necessary to develop other techniques to facilitate the synthesis of 2D $GaPO_4$ nanostructures. The surface oxide of liquid metals can be the appropriate platform for the growth of 2D oxide films. The naturally formed metal oxide films on the surface of liquid metals (Figure 3.11c) can be later extracted by the vdW exfoliation and be used in the secondary post-treatment to synthesize novel 2D structures which are not naturally available in 2D layered forms. With the same strategy, $GaPO_2$ 2D nanostructures are fabricated. In this method, the Ga_2O_3 surface oxide of gallium alloy was extracted. To this aim the fresh surface of liquid Ga is touched by the Si/SiO_2 substrate, and then the exfoliated Ga_2O_3 was moved into a tube furnace for the secondary phosphatization treatment [51]. At this stage, the synthesized ultrathin gallium oxide 2D nanosheets were subjected to the chemical vapor containing phosphor. To this aim, phosphoric acid (H_3PO_4) powders were heated up to 350°C inside of the tube furnace. The N_2 gas was used to transport the vapor phases toward the Ga_2O_3 film over the Si/SiO_2 substrate at 300 ~ 350°C (Figure 3.11d). The prepared samples were later kept inside of the glovebox containing N_2 gas. The following AFM measurements confirmed the development of ultrathin $GaPO_4$ film with the typical step-height of 1.1 nm, which is in agreement with dimensions of unit cell of the trigonal gallium phosphate ($GaPO_4$) in the c crystalline direction (Figure 3.11b, e) [51]. The HRTEM studies of the selected area electron diffraction (SAED) pattern reaffirmed the growth of crystalline structure with lattice spacing of 0.21 nm corresponding to the d-s pacing of (200) plane of the trigonal $GaPO_4$ nanosheets (Figure 3.11e) [51]. The crystalline structure of 2D gallium phosphates was further investigated by XRD. The dominance of (003) and (102) planes of trigonal α-$GaPO_4$ was vividly observed in the XRD results of 2D films (Figure 3.11f). The Raman spectroscopy helped to characterize the characteristics peaks of PO_4 tetrahedra at ~420.5 and ~456.6 cm^{-1}. Those peaks are attributed to the A1 vibration mode related to the internal bending of PO_4 structure (Figure 3.11g). The lack of Raman characteristic peaks of Ga_2O_3 endorsed the successful conversion of Ga_2O_3 to $GaPO_4$ film after phosphatization treatment. The wide band gap of 6.9 eV was measured for the developed $GaPO_4$ film by EELS method (Figure 3.11h), which has shown an excellent agreement with the other reported values for the band gap of $GaPO_4$ films [51]. The experimental evidences presented the strong out-of-the-plane piezoelectric properties for the $GaPO_3$ 2D nanostructures which was relatively around 8.5 pm V^{-1} [51]. This is a considerable value for such an ultrathin 2D film. It demonstrates the capacity of 2D $GaPO_4$ film for the sensing and energy harvesting targets. The similar behavior of the free standing 2D $GaPO_4$ film has highlighted the capability of this novel synthesized material for applications in miniature electromechanical systems for employment in high-temperature processing situations.

3.3.4.3 Semiconducting GaN and InN Nanosheets

The synthesis of large-area ultrathin GaN and InN film was reported by using an ammonolysis method [58]. In this technique the postammonolysis treatment of the surface oxide of liquid metals facilitated the development of high-quality wafer-scale ultrathin films. The technique includes the synthesis of centimeter scaled 2D GaN and InN films by multiple stage synthesis approach [58]. The technique is based on squeeze printing of a gallium droplet between two Si/SiO_2 substrates (Figure 3.12a).

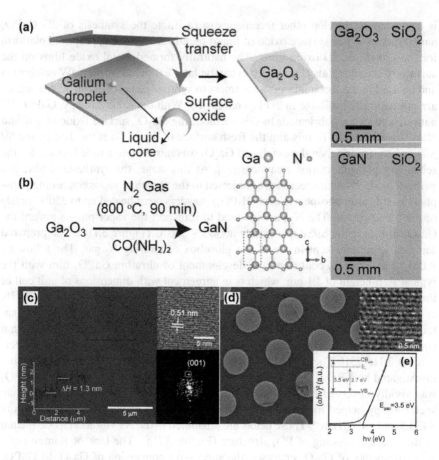

FIGURE 3.12 (a) The schematic illustration of squeezing of liquid metal galinstan between SiO$_2$ substrates and (b) following post-ammonolysis treatment for phase transform of Ga$_2$O$_3$ to GaN film. (c) High-resolution AFM image of the surface of GaN film. (d) HRTEM image of GaN film and (e) the bandgap measurement of GaN film. Reprinted with permission from Syed, N., Zavabeti, A., Messalea, K. A., Della Gaspera, E., Elbourne, A., Jannat, A., ... Daeneke, T. (2019). Wafer-sized ultrathin gallium and indium nitride nanosheets through the ammonolysis of liquid metal derived oxides. *Journal of the American Chemical Society*, *141*(1), 104–108.

After the squeeze printing, the instantaneous Ga liquid oxidation was accompanied by the facile transfer of Ga$_2$O$_3$ surface oxide film to the substrate by vdW adhesion. The process has resulted in the formation of ultrathin Ga$_2$O$_3$ films with an approximate thickness of 1.4 nm [58]. The following ammonolysis reaction occurred at 800°C in the presence of urea. It was found that lower reaction temperatures did not facilitate the conversion of Ga$_2$O$_3$ into the wurtzite GaN films (Figure 3.12b) [58]. The characterization techniques also revealed important information about the synthesized GaN films. The thickness of the film after ammonolysis process was found to be 1.3 nm which is corresponded to the to three wurtzite GaN unit cells (Figure 3.12c). The more precise AFM technique showed that the GaN crystals with 5.18 nm lattice constant

2D Surface Oxides of Liquid Metals

has grown along the (001) crystalline direction with high crystalline characteristics [58]. The same ammonolysis reaction was used to convert the Ga_2O_3 films into GaN over the TEM grid. The Ga_2O_3 was directly moved onto the TEM gird (Si_3N_4) using the vdW exfoliation and then was subjected to the ammonolysis reactions. The high crystallinity of developed GaN film was confirmed similar to the wurtzite GaN with the same growth direction along the (001) crystalline plate and interlayer spacing of 0.28 Å [58]. XRD test also confirmed the development of such a thin crystalline structure over the substrate after deposition technique. The X-ray photoelectron spectroscopy affirmed that the substantial amount of nitrogen was replaced by the oxygen atoms in GaN structure which is due to the similar bond length of Ga–O and Ga–N. It was targeted to decrease the level of oxygen in 2D GaN films. However, the successful removal of oxygen from the 2D GaN films was not achieved even during the synthesis process at the high temperatures [58]. It also did not alter the optical properties of the 2D GaN film. The bandgap measurement shows $E_g = 3.5$ eV for 2D GaN film (Figure 3.12e). The substantial alteration of optical properties is expected since the surface substitution has significant effects on the electronic properties and band structure of 2D GaN films. The measured band gap of 3.5 eV was much lower than the previously reported band gap of graphene encapsulated GaN nanosheets. The reason for the considerable decrease of the bandgap was attributed to the upward shift of the valence band of GaN film due to the hybridization of the O_{2p}–N_{2p} orbitals [58].

The credibility of method was confirmed when the liquid indium was used as the body material for the vdW exfoliation of In_2O_3. The following intermediate bromination step was required to transform In_2O_3 to $InBr_3$. To this aim, the screen-printed In_2O_3 nanosheet was subjected to the generated gases from the HBr source for 20 seconds. The ammonolysis process was later successfully carried out. In this stage the urea was employed as the NH_3 source at the temperature of 630°C [58]. The characterization results confirmed the homogenous deposition of InN films. The high-resolution AFM image depicted the growth of crystalline 2D ultrathin wurtzite InN nanostructures with the crystalline lattice constant of 5.5 Å [58]. The XPS studies of developed InN 2D films confirmed the stoichiometric developed InN film [58]. The measured band gap of synthesized 2D InN was about 3.5 eV [58]. The two-step synthesis technique was found as reliable approach toward the synthesis of advanced 2D nanostructures by using the surface oxide of liquid metals. The developed 2D nanostructures have clearly demonstrated novel electronic and optoelectronic characteristics which are not expected and observed in the similar 2D nanostructures, synthesized via high-vacuum deposition-based methods.

3.3.4.4 SnO/In$_2$O$_3$ 2D van der Waals Heterostructure

In modern systems and units based on vdW, heterostructures from stacking of various 2D-layered crystals are widely adopted and employed for the novel technological high-performance electronic and optoelectronic applications. The *p-n* heterojunctions are now the building blocks of the semiconductor devices where the hetero-interfaces between two semiconductor materials with the different types of charge carriers are used to fabricate the different types of electronic and photonic instruments. Specifically, the 2D heterostructured-based devices can be developed by several different approaches and techniques including mechanical exfoliation from the host material, the chemical vapor deposition, atomic layer deposition, molecular beam epitaxy and pulsed laser deposition [59, 60]. The innovative techniques are always introduced to fabricate various novel 2D heterostructured materials [59, 60]. For instance, the liquid metal-based synthesis of 2D materials and the following printing framework offer the possibility of synthesis of 2D heterostructured materials. The technique was employed to prepare 2D SnO/In_2O_3 van der Waals heterostructures [61]. The heterostructured 2D electronic devices were fabricated by using liquid metal vdW transfer method to develop large-area heterostructured thin metal oxide films of p-SnO/n-In_2O_3. In this method, Si/SiO_2 substrate was touched respectively to the surface of liquid tin (1st) and indium metals (2nd) to produce large-scale 2D vdW heterostructures (Figure 3.13a) [61]. The SEM images of the heterostructures are presented in Figure 3.13b. The HRTEM observations have confirmed that the presence of two crystalline structures with different characteristics developed over the Si/SiO_2 substrate (Figure 3.14c). The lattice spacing of 0.27 nm corresponds to the (321) plane of crystalline In_2O_3, while the other crystalline structure with 0.298 Å is attributed to the (101) plane of SnO. The AFM image as related to the same heterostructures is also shown in Figure 3.13d. The heterostructured film had the thickness of 5.5 nm composed of 2D SnO film (1.0 nm) and 2D In_2O_3 (4.5 nm), which was equal to four unit cells of indium oxide. The Raman characterization technique also revealed the development of SnO/In_2O_3 van der Waals heterostructures with ultra-fine homogeneity (Figure 3.13f). The absorption spectra of individual and heterostructured 2D materials provided valuable information about the optical properties of 2D heterostructured films. The maximum optical absorption peak of SnO/In_2O_3 heterostructured film located at 251 nm, which was a number between 246 nm (SnO absorption peak) and 256 nm (In_2O_3 absorption peak) [61]. It was found that the strong interlayer coupling of charge carriers affected the optical absorption of 2D heterostructured film. The band gap of SnO 2D film was found to be 4.8 eV, and the value of 3.65 eV was also measured as the bandgap of In_2O_3 2D film [61]. However, the band gap of 2D SnO_2/In_2O_3 vdW heterostructures was narrowed to 2.30 eV [61]. The schematic of band alignment at 2D SnO_2/In_2O_3 vdW heterointerfaces was demonstrated at Figure 3.13h. The effective bandgap at p-n junction was expected to be 2.30 eV, which was close to the calculated value of band gap from Tauc plot. The narrow bandgap assisted the charge separation at heterointerfaces, thus the photoexcited electron-holes can be separated easily and the photoexcited electrons can feasibly migrate from the valence band of SnO to the conduction band of In_2O_3 film. The specific energy band alignment at the heterointerfaces of vdW SnO_2/In_2O_3 demonstrated quite impressive capabilities and excellent photodetectivity of 5 * 10^9 J with the photocurrent ratio of 24 and the outstanding photoresponsivity of 1047 A/W [61]. Thus, the liquid metal printing technique was found as the valuable potential approach for fabrication of heterostructured 2D materials, which do not naturally exist as the layered structures in nature.

3.3.5 Sonochemical-Assisted Functionalization of Surface Oxide of Galinstan

The employment of ultrathin layered functional materials and the precise control of their surface characteristics offer a remarkable platform for the development of

2D Surface Oxides of Liquid Metals

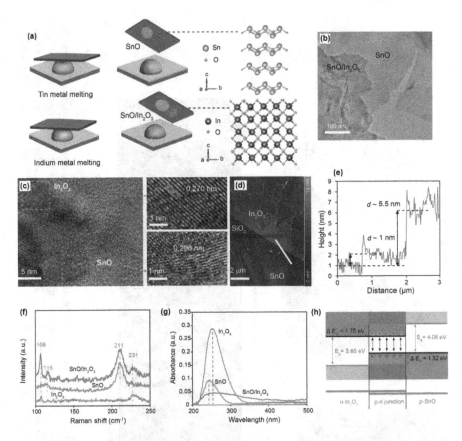

FIGURE 3.13 (a) Schematic illustration of fabrication of 2D vdW heterostructured SnO/In$_2$O$_3$ heterostructured film. (b) HRTEM image of heterostructured SnO/In$_2$O$_3$ films. (c) The HRTEM image of atomic structure of heterostructured films shows the high crystalline structure of SnO and In$_2$O$_3$ corresponding to their lattice spacing distance. (d) AFM of heterostructured SnO/In$_2$O$_3$ film with (e) the profile thickness of the films. (f) Raman spectra of SnO, In$_2$O$_3$, and SnO/In$_2$O$_3$ 2D films. (g) Optical absorption of SnO, In$_2$O$_3$, and SnO/In$_2$O$_3$ 2D films. (h) Band alignment of SnO/In$_2$O$_3$ 2D heterostructures. Reprinted with permission from Alsaif, M., Kuriakose, S., Walia, S., Syed, N., Jannat, A., Zhang, B. Y., ... Zavabeti, A. (2019). 2D SnO/In$_2$O$_3$ van der Waals heterostructure photodetector based on printed oxide skin of liquid metals. *Advanced Materials Interfaces*, 6(7), 1900007.

advanced electronic and optoelectronic technologies. Herein, the wide bandgap and self-limiting surface oxides of galinstan were employed as templates for the next sonochemical-assisted reactions to finally synthesize the ultrathin functionalized Ga$_2$O$_3$ nanosheets with controllable electronic and optoelectronic properties. To functionalize the surface oxide of galinstan liquid metal alloy, Ga$_2$O$_3$ nanosheets were extracted from the sonochemically dispersed Ga$_2$O$_3$ nanoparticles inside of aqueous AgNO$_3$ and SeCl$_4$ solutions [62]. Galinstan fluid was flowing freely under mechanical loads during sonication process. The ultrasound wave creates strong shear forces which can simply delaminate the surface oxide film of the galinstan.

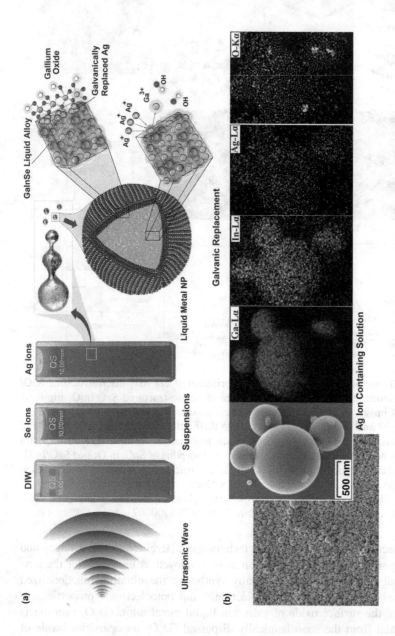

FIGURE 3.14 (a) Schematic of sonochemical-assisted synthesis of galinstan nanoparticle in Ag and Se containing aqueous solutions. (b) The distribution of alloying elements on the surface of galinstan nanoparticles synthesized in Ag ion-containing aqueous solutions Reprinted with permission from Karbalaei Akbari, M., Hai, Z., Wei, Z., Ramachandran, R. K., Detavernier, C., Patel, M., … Zhuiykov, S. (2019). Sonochemical functionalization of the low dimensional surface oxide of galinstan for heterostructured optoelectronic applications. *Journal of Materials Chemistry C, 7*(19), 5584–5595.

2D Surface Oxides of Liquid Metals

The spherical morphology of galinstan droplet was due to the high surface energy of fresh galinstan alloy (Figure 3.14a). The Se ion and Ag ion-containing aqueous solutions were employed to functionalize the surface oxide of galinstan nanoparticles during ultra-sonication process. The galvanic replacement of Ag with all of the metallic components of galinstan is thermodynamically possible. This is because the standard reduction potential of the Sn^{2+}/Sn^0, In^{3+}/In^0, and Ga^{3+}/Ga reactions are less than that of Ag^+/Ag [which are respectively −0.138, −0.340, −0.529, and 0.799 V vs. standard hydrogen electrode (SHE)]. Ga with its lowest reduction potential is the main candidate for replacement with Ag^+ ions inside of solution. The elemental analysis demonstrated the uniform distribution of alloying elements on the surface of liquid nanoparticles (Figure 3.14b). While the galvanic replacement is the most possible scenario for the functionalization of Ag^+ ions, it cannot explain the presence of Se on the surface of galinstan nanoparticles. The observation of Se containing nanostructures and compounds was attributed to the strong interaction between galinstan alloy and selenium in aqueous solution [62].

Another example depicted in Figure 3.15a is the gallium hydroxide 2D nanosheets, which were extracted from their aqueous suspension [62]. The slurry of nanoparticle suspension contains ultrathin nanosheet films with various lateral size distributions (Figure 3.15b, c). Nanosheets were separated by centrifugation method and then were extracted by using hydrographical printing or by passing the suspension through a paper sieve (Figure 3.15d). The gallium hydroxide nanosheets were then dehydrated via subsequent drying process at 400°C to transfer the possible GaOOH nanosheets into Ga_2O_3. The chemical analysis of dehydrated nanosheets evidently showed the distribution of alloying elements on the surface of 2D nanosheets extracted from DIW (Figure 3.15e), DIW-AgNO$_3$ (Figure 3.15f), and DIW-SeCl$_4$ solutions (Figure 3.15g). The presence of Se and Ag alloying elements and nanostructures on the surface of Ga_2O_3 nanosheets was attributed to the sonochemical-assisted nucleation and growth of nanostructured Ag and Se on the surface of Ga_2O_3 nanosheets which acted as the platform for nucleation of nanostructures [62]. The Raman spectra of Ag functionalized Ga_2O_3 nanosheets demonstrated the Raman characteristics of gallium oxides (Figure 3.15h). It was found that the nucleated silver nanoparticles had surface enhanced Raman scattering (SERS), which was originated from plasmonic properties of silver nanostructures. It was also observed that the SERS properties of Ag can intensify the characteristic peaks of the Ga_2O_3 nanosheets [62]. The characterization of Raman peaks of Ga_2O_3 nanosheets functionalized in Se ion-containing solutions showed the nucleation of crystalline Se nanostructures and Ga_xSe_y compounds on the Ga_2O_3 nanosheets. The photoluminescence effect was characterized by Ga_2O_3 (Se) nanosheets (Figure 3.15i) [62]. The XRD spectra of Ga_2O_3 (Se) nanosheets revealed the diffraction peaks of trigonal Se. It was observed that the nanosheets act as the platforms for the nucleation of amorphous selenium, which was accompanied by the presence of α-Ga_2O_3 in the structure of nanosheets. In addition, the bandgap measurements provided valuable data on the optical characteristics and structural composition of the samples. The bandgap of the Ga_2O_3 nanosheets was 4.82 eV (Figure 3.15j), which was fairly consistent with the previously reported bandgap of Ga_2O_3. However, in the Ga_2O_3 (Ag) nanosheets, two individual bandgap values were measured. The 4.82 eV was assigned to the bandgap of Ga_2O_3 while the 3.44 eV was

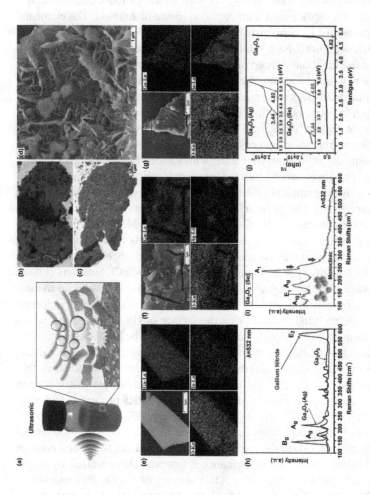

FIGURE 3.15 (a) Schematic of ultrasonic assisted separation and functionalization of surface oxide of galinstan. (b), (c) The optical micrograph and (d) SEM image of extracted 2D nanosheets. (e) EDX elemental map analysis of the nanosheets synthesized in water, (f) Ag containing aqueous solutions and (g) Se containing aqueous solutions. (h) Raman spectra of Ga_2O_3, Ga_2O_3 (Ag), and (i) Ga_2O_3 (Se) nanosheets. (j) The bandgap measurement of the corresponding nanosheets. Reprinted with permission from Karbalaei Akbari, M., Hai, Z., Wei, Z., Ramachandran, R. K., Detavernier, C., Patel, M., ... Zhuiykov, S. (2019). Sonochemical functionalization of the low dimensional surface oxide of galinstan for heterostructured optoelectronic applications. *Journal of Materials Chemistry C*, 7(19), 5584–5595.

attributed to the GaN compounds [62]. The presence of GaN was also observed in the Raman spectra of the nanosheets. In the case of Ga_2O_3 (Se) nanosheets, the calculated bandgap of 4.65 eV was lower than that of the Ga_2O_3 nanosheets. In addition, the bandgap of 1.44 eV was the characteristic peak of the Ga_xSe_x compounds [62]. Generally, it was shown that the sonochemical functionalization of surface oxide of galinstan was a capable and cheap technique for large-scale synthesis of composite 2D films with novel electronic and optoelectronic properties [62].

CONCLUSIONS

The naturally occurring surface oxides of liquid metals are remarkable candidates for synthesis and extraction of 2D nanostructures, which are not available as layered structures in nature. The liquid metal is always covered with ultrathin 2D films in the ambient atmosphere. Because of the lack of a strong attachment bonding between the ultrathin surface oxide films and internal liquid core, delamination of surface oxide films from their host alloy is possible. One of the main strategies is separation of surface oxide films by using the vdW attraction forces between the surface oxide films and an appropriately selected substrate. The extracted ultrathin nanosheets were found as the high-quality and defects-free 2D films. The surface oxide of gallium (Ga_2O_3), tin (SnO_x) and indium (In_2O_3) are the most famous extracted 2D surface oxide of the low-melting alloys. The synthesized nanosheets were characterized in both amorphous and crystalline structures. The crystalline nature of surface oxide of liquid metal and alloys has indeed provided new platform for development of high-quality atomically thin materials for the advanced electronic applications. The post-treatment of surface oxide of nanosheets ensures an excellent background for the following synthesis of novel non-oxide 2D semiconductors with unexpected outstanding properties. Several types of 2D films, including $GaPO_4$, GaN, and InN, were synthesized by post-treatment of Ga_2O_3 and In_2O_3 surface oxide films of liquid gallium and indium. The same techniques were employed to fabricate 2D vdW heterostructures.

REFERENCES

1. Tan, C. L., Cao, X. H., Wu, X. J., He, Q., Yang, J., Zhang, X., ... Zhang, H. (2017). Recent advances in ultrathin two-dimensional nanomaterials. *Chemical Reviews*, *117*(9), 6225–6231.
2. Novoselov, K. S., Jiang, F., Schedin, D., Booth, T. J., Khotkevich, V. V., Morozov, S. V., & Geim, A. K. (2005). Two-dimensional atomic crystals. *Proceedings of the National Academy of Sciences of the United States of America*, *102*(30), 10451–10453.
3. Mak, K. F., Lee, C., Hone, J., Shan, J., & Heinz, T. F. (2010). Atomically thin MoS_2: A new direct-gap semiconductor. *Physical Review Letters*, *105*(13), 136805.
4. Zhou, F., Chen, J., Tao, X., Wang, X., & Chai, Y. (2019). 2D materials based optoelectronic memory: Convergence of electronic memory and optical sensor. *Research*, 9490413.
5. Akinwandeh, D., Huyghebaert, C., Wang, C. H., Serna, M. I., Goossens, S., Li, L. J., ... Koppens, F. H. L. (2019). Graphene and two-dimensional materials for silicon technology. *Nature*, *573*(7775), 507–518.

6. Sun, Z., Liao, T., Dou, Y., Hwang, S. M., Park, M. S., Jiang, L., ... Dou, S. X. (2014). Generalized self-assembly of scalable two-dimensional transition metal oxide nanosheets. *Nature Communications, 5*, 3813.

7. Cai, Z., Liu, B., Zou, X., & Cheng, H. M. (2018). Chemical vapor deposition growth and applications of two-dimensional materials and their heterojunctions. *Chemical Reviews, 118*(13), 6091–6033.

8. Hao, W., Marichy, C., & Juornet, C. (2019). Atomic layer deposition of stable 2D materials. *2D Materials, 6*(1), 012001.

9. Tao, H., Zhang, Y., Gao, Y., Sun, Z., Yan, C., & Texter, J. (2017). Scalable exfoliation and dispersion of two-dimensional materials-an update. *Physical Chemistry Chemical Physics, 19*(2), 921–960.

10. Yu, J., Li, J., Zhang, W., & Chang, H. (2015). Synthesis of high quality two-dimensional materials via chemical vapor deposition. *Chemical Science, 6*(12), 6705–6716.

11. De Castro, I. A., Chrimes, A. F., Zavabeti, A., Berean, K. J., Carey, B. J., Zhuang, J., ... Daeneke, T. (2017). A gallium-based magnetocaloric liquid metal ferrofluid. *Nano Letters, 17*(12), 7831–7838.

12. Kanatzidis, M. G., Pottgen, R., & Jeitschko, W. (2005). The metal flux: A preparative tool for the exploration of intermetallic compounds. *Angewandte Chemie International Edition, 44*(43), 6996–6923.

13. Kalantar-Zadeh, K., Tang, J., Daeneke, T., O'Mullane, A. P., Stewart, L. A., Liu, J., ... Dickey, M. D. (2019). Emergence of liquid metals in nanotechnology. *ACS Nano, 13*(7), 7388–7395.

14. Dickey, M. D. (2014). Emerging applications of liquid metals featuring surface oxides. *ACS Applied Materials and Interfaces, 6*(21), 18369–18379.

15. Tabatabai, A., Fassler, A., Usiak, C., & Majidi, C. (2013). Liquid-phase gallium-indium alloy electronics with microcontact printing. *Langmuir, 29*(20), 6194–6200.

16. Wang, Q., Yu, Y., & Liu, J. (2018). Preparation, characteristics and application of the functional liquid metal materials. *Advanced Engineering Materials, 20*, 1700781.

17. Deming, H. G. (1942). The laws of definite composition and definite proportions. *Journal of Chemical Education, 19*(7), 336.

18. Paranjpe, G. R., & Joshi, R. M. (1931). The problem of liquid sodium-amalgams. *Journal of Physical Chemistry. Part A, 36*(9), 2474–2482.

19. Richards, T. W., Frevert, H. L., & Teeter, C. E. (1928). A thermodynamical contribution to the study of the system cadmium-mercury. *Journal of the American Chemical Society, 50*(5), 1293–1302.

20. Hernandez, M. J. Q., Pero-Sanz, J. A., & Verdeja, L. F. (2017). *Solidification and solid-state transformations of metals and alloys.* Saint Louis: Elsevier Science.

21. Wu, Y., Deng, Z., Peng, Z., Zheng, R., Liu, S., Xing, S., ... Liu, L. (2019). A novel strategy for preparing stretchable and reliable biphasic liquid metal. *Advanced Functional Materials, 29*(36), 1903840.

22. Jensen, W. B. (2003). The place of zinc, cadmium, and mercury in the periodic table. *Journal of Chemical Education, 80*(8), 952.

23. Norrby, L. J. (1991). Why is mercury liquid? Or, why do relativistic effects not get into chemistry textbooks? *Journal of Chemical Education, 68*(2), 110–113.

24. Heine, V. (1968). Crystal structure of gallium metal. *Journal of Physics – Condensed Matter, 1*(1), 222.

25. Zuger, O., & Durig, U. (1992). Atomic structure of the α-Ga(001) surface investigated by scanning tunneling microscopy: Direct evidence for the existence of Ga_2 molecules in solid gallium. *Physical Review. B: Condensed Matter and Materials Physics, 46*(11), 7319–7321.

26. Flower, S. C., & Saunders, G. A. (1990). The elastic behaviour of indium under pressure and with temperature up to the melting point. *Philosophical Magazine Part B, 62*(3), 311–328.

27. Cucka, P., & Barrett, C. S. (1962). The crystal structure of Bi and of solid solutions of Pb, Sn, Sb and Te in Bi. *Acta Crystallographica, 15*(9), 865–872.
28. Tolmacheva, Z. I., & Eremenko, V. N. (1963). Relationship between some properties of the transition metals, characterizing their interatomic bond strength, and their electronic structure. *Powder Metallurgy and Metal Ceramics, 2*(5), 369–376.
29. Anderson, T. J., & Ansara, I. (1991). The Ga-In (gallium-indium) system. *Journal of Phase Equilibrium, 12*(1), 64–72.
30. Daeneke, T., Khoshmanesh, K., Mahmood, N., de Castro, I. A., Esrafilzadeh, D., Barrow, S. J., ... Kalantar-Zadeh, K. (2018). Liquid metals: Fundamentals and applications in chemistry. *Chemical Society Reviews, 47*(11), 4073–4111.
31. Bo, G., Ren, L., Xu, X., Du, Y., & Dou, S. (2018). Recent progress on liquid metals and their applications. *Advances in Physics: X, 3*(1), 412–442.
32. Ueno, T., Summers, E., & Higuchi, T. (2007). Machining of iron-gallium alloy for microactuator. *Sensors and Actuators. Part A, 137*(1), 134–140.
33. Xu, Q., Oudalov, N., Guo, Q., Jaeger, H. M., & Brown, E. (2012). Effect of oxidation on the mechanical properties of liquid gallium and eutectic gallium-indium. *Physics of Fluids, 24*(6), 063101.
34. Preuss, A., Adolphi, B., & Wegener, T. (1995). The kinetic of the oxidation of InSn48. *Analytical and Bioanalytical Chemistry, 353*(3–4), 399–402.
35. Leontie, L., Caraman, M., Alexec, M., & Harnagea, C. (2002). Structural and optical characteristics of bismuth oxide thin films. *Surface Science, 507*, 480–485.
36. Cabrera, N., & Mott, N. F. (1949). Theory of the oxidation of metals. *Reports on Progress in Physics, 12*(1), 163.
37. Gelderman, K., Lee, L., & Donne, S. W. (2007). Flat-band potential of a semiconductor: Using the Mott-Schottky equation. *Journal of Chemical Education, 84*(4), 685–688.
38. Regan, M. J., Tostmann, H., Pershan, P. S., Magnussen, O. M., DiMasi, E., Ocko, B. M., & Deutsch, M. (1997). X-ray study of the oxidation of liquid-gallium surfaces. *Physical Review. B: Condensed Matter and Materials Physics, 55*(16), 10786–10790.
39. Wang, Y., Li, Y., Zhang, J., Zhuang, J., Ren, L., & Du, Y. (2019). Native surface oxides featured liquid metals for printable self-powered photoelectrochemical device. *Frontiers in Chemistry, 7*, 356.
40. Caturla, M. J., Jiang, J. Z., Louis, E., & Molina, J. M. (2015). Some issues in liquid metals research. *Metals, 5*(4), 2128–2133.
41. Thirumoorthi, M., & Prakash, J. T. J. (2016). Structure, optical and electrical properties of indium tin oxide ultra-thin films prepared by jet nebulizer spray pyrolysis technique. *Journal of Asian Ceramic Societies, 4*(1), 124–132.
42. Sen, P., & Kim, C. J. (2009). Microscale liquid-metal switches-A review. *IEEE Transactions on Industrial Electronics, 56*(4), 1314–1330.
43. Zhang, W., Naidu, B. S., Ou, J. Z., O'Mullane, A. P., Chrimes, A. F., Carey, B. J., ... Kalantar-Zadeh, K. (2015). Liquid metal/metal oxide frameworks with incorporated Ga_2O_3 for photocatalysis. *ACS Applied Materials and Interfaces, 7*(3), 1943–1948.
44. Kim, Y., & Hone, J. (2017). Screen printing of 2D semiconductors. *Nature, 544*(7649), 167–168.
45. Downs, A. J. (1993). *Chemistry of aluminium, gallium, indium and thallium*, p. 162. Springer, New York.
46. Carey, B. J., Ou, J. Z., Clark, R. M., Berean, K. J., Zavabeti, A., Chesman, A. S., ... Kalantar-Zadeh, K. (2017). Wafer-scale two-dimensional semiconductors from printed oxide skin of liquid metals. *Nature Communications, 8*, 14482.
47. Zavabeti, A., Ou, J. Z., Carey, B. J., Syed, N., Orrell-Trigg, R., Mayes, E. L. H., ... Daeneke, T. (2017). A liquid metal reaction environment for the room-temperature synthesis of atomically thin metal oxides. *Science, 358*(6361), 332–335.
48. Zavabeti, A., Zhang, B. Y., Castro, I., Ou, J. Z., Carey, B. J., Mohiuddin, M., ... Kalantar-Zadeh, K. (2018). Green synthesis of low-dimensional aluminum oxide

hydroxide and oxide using liquid metal reaction media: Ultrahigh flux membranes. *Advanced Functional Materials*, 28(44), 1804057.

49. Daeneke, T., Atkin, P., Orrell-Trigg, R., Zavabeti, A., Ahmed, T., Walia, S., ... Kalantar-Zadeh, K. (2017). Wafer-scale synthesis of semiconducting SnO monolayers from interfacial oxide layers of metallic liquid tin. *ACS Nano*, 11(11), 10974–10983.

50. Saji, K. J., Tian, K., Snure, M., & Tiwari, A. (2016). 2D tin monoxide- an unexplored p-type van der Waals semiconductor: Material characteristics and field effect transistors. *Advanced Electronic Materials*, 2(4), 1500453.

51. Syed, N., Zavabeti, A., Ou, J. Z., Mohiuddin, M., Pillai, N., Carey, B. J., ... Kalantar-Zadeh, K. (2018). Printing two-dimensional gallium phosphate out of liquid metal. *Nature Communications*, 9(1), 3618.

52. Wu, W., Wang, L., Li, Y., Zhang, F., Lin, L., Niu, S., ... Wang, Z. L. (2014). Piezoelectricity of single-atomic layer MoS_2 for energy conversion and piezotronics. *Nature*, 514(7523), 470–474.

53. Duerloo, K. A. N., Ong, M. T., & Reed, E. J. (2012). Intrinsic piezoelectricity in two dimensional materials. *Journal of Physical Chemistry Letters*, 3(19), 2871–2876.

54. Zelisko, M., Hanlumyuang, Y., Yang, S., Liu, Y., Lei, C., Li, J., ... Sharma, P. (2014). Anomalous piezoelectricity in two-dimensional graphene nitride nanosheets. *Nature Communications*, 5, 4284.

55. Blonsky, M. N., Zhuang, H. L., Singh, A. K., & Hennig, R. G. (2015). Ab initio prediction of piezoelectricity in two-dimensional materials. *ACS Nano*, 9(10), 9885–9891.

56. Fei, R., Li, W., Li, J., & Yang, L. (2015). Giant piezoelectricity of monolayer group IV monochalcogenides: SnSe, SnS, GeSe, and GeS. *Applied Physics Letters*, 107(17), 173104.

57. Michel, K., & Verberck, B. (2011). Phonon dispersions and piezoelectricity in bulk and multilayers of hexagonal boron nitride. *Physical Review. Part B*, 83(11), 115328.

58. Syed, N., Zavabeti, A., Messalea, K. A., Della Gaspera, E., Elbourne, A., Jannat, A., ... Daeneke, T. (2019). Wafer-sized ultrathin gallium and indium nitride nanosheets through the ammonolysis of liquid metal derived oxides. *Journal of the American Chemical Society*, 141(1), 104–108.

59. Wang, J., Li, Z., Chen, H., Deng, G., & Niu, X. (2019). Recent advanced in 2D lateral heterostructures. *Nano-Micro Letters*, 11(1), 48.

60. Liu, Y., Zhang, S., He, J., Wang, Z. M., & Liu, Z. (2018). Recent progress in the fabrication, properties and devices of heterostructures based on 2D materials. *Nano-Micro Letters*, 11(1), 13.

61. Alsaif, M., Kuriakose, S., Walia, S., Syed, N., Jannat, A., Zhang, B. Y., ... Zavabeti, A. (2019). 2D SnO/In_2O_3 van der Waals heterostructure photodetector based on printed oxide skin of liquid metals. *Advanced Materials Interfaces*, 6(7), 1900007.

62. Karbalaei Akbari, M., Hai, Z., Wei, Z., Ramachandran, R. K., Detavernier, C., Patel, M., ... Zhuiykov, S. (2019). Sonochemical functionalization of the low dimensional surface oxide of galinstan for heterostructured optoelectronic applications. *Journal of Materials Chemistry C*, 7(19), 5584–5595.

4 Hetero-Interfaces in 2D-Based Semiconductor Devices

4.1 INTRODUCTION TO 2D HETERO-INTERFACES

Over the past few years, a great deal of research has been dedicated to the development of 2D materials for a wide range of applications in electronic, optic, magnetic, spintronic and memory devices [1–3]. Accordingly, the idea of development of heterostructured 2D materials is introduced. 2D heterostructured materials provide a platform for exploration of novel physical properties which are not observed in individual 2D materials. Recent advancement in controllable and scalable fabrications of high-quality 2D materials offer the opportunity to design the heterostructured 2D materials with extraordinary performances in various areas including tunneling transistors, photodetectors and spintronic devices [4–6]. These 2D heterostructures can be divided according to the interaction force at their hetero-interfaces. Specifically, theoretical and experimental studies have concentrated on the scientific and technological aspects of development and application of both vertical and lateral 2D heterostructures.

Vertical 2D heterostructures are composed of individual 2D layers without the creation of junctions at the atomic levels. In vertical 2D heterostructures, the isolated atomic component can be assembled to form the new layer materials stacked in a precisely selected sequence. The different layers in the vertical heterostructures are generally combined by van der Waals (vdW) interaction [7, 8].

Another type of 2D heterostructured material is *lateral heterojunctions*. In this group of 2D heterostructures, different 2D atomic panels are stitched together in the 2D atomic level hetero-interfaces. Considering the level of lattice mismatch, the lateral hetero-interfaces can be homogenous or heterogeneous. The chemical bonding at the edges of the different 2D panels considerably affects the physical and chemical properties of 2D heterostructured materials. These hetero-interfaces substantially alter the charge transfer mechanisms, thus unique properties are expected from the lateral heterojunction, which is not found in the vertical one [9–12].

The vdW interactions between 2D layered vertical heterostructures are quite weaker than the chemical bonding formed between the lateral hetero-interfaces. Due to the different nature of interaction forces of the 2D lateral and vertical heterostructures at their hetero-interfaces, the synthesis methods of them have distinct differences [13, 14]. As a result of the weak vdW interactions between 2D layers, the mechanical exfoliation is the most prominent technique for fabrication of the vertical

2D heterostructures. The crystalline structures in mechanically exfoliated 2D films with their high-quality specifications have become the main candidate for application in research fields. However, two main challenges have restricted the applications and reliable performance of 2D vertical heterostructured devices [15–17]:

  i. Contamination of hetero-interfaces caused by synthesis method.
  ii. Stacking orientation issues.

Lateral heterojunctions are proposed to tackle the difficulties of employment of 2D vertical structures. The nature of hetero-interfaces in lateral 2D heterojunctions has enhanced the capabilities of 2D lateral based devices. The direct growth techniques can facilitate the development of seamlessly stitched panels of hetero-interfaces and creation of sharp interfaces. The microstructural features of lateral 2D heterostructures induce many attractive properties. The quality and orientation of the inner interfaces in the lateral 2D heterostructures can be controlled precisely. Furthermore, the larger tunability of band offset in the lateral heterostructures is facilitated. Therefore, the developed lateral 2D hetero-interfaces offer tunable electronic characteristics. These characteristics make this specific design of 2D heterostructures a potential option for the fabrication of novel electronic instruments. Different from the vertical stacking 2D heterostructures, many devices were fabricated with the lateral heterostructure configuration and they have clearly demonstrated superior performance and unique properties. The successful control of the properties of hetero-interfaces is highly important to exploit the highest capability of the 2D heterostructured devices.

From another point of view, the fabricated heterostructures can be selected according to the intrinsic electronic properties of their 2D components. In this view, the heterostructures can be divided into semiconductor/semiconductor, metal/semiconductor, and insulator/semiconductor heterostructured materials. One of the most well-known metal/semiconductor junctions is the lateral interface between graphene and another 2D material. Furthermore, the electronic and optoelectronic performance of 2D based instruments are related by high extend to the electrical contacts that make the connection between 2D materials with external circuit system. Thus, from practical point of view, the understanding and engineering of electrical and physical phenomena at the metal/semiconductor 2D hetero-interfaces are highly important and it is fundamental step toward the industrialization of 2D-based electronic devices.

The present chapter summarizes the current progress achieved for 2D lateral heterostructures. The characteristics of homogenous and heterogeneous lateral heterojunctions will be discussed. The structural characteristics of 2D heterostructures including the atomic level sharpness at hetero-interfaces, and the mechanical strain due to structural mismatch at hetero-interfaces, doping of hetero-interfaces and other phenomena will also be reviewed. The chapter will continue by the investigation of electrical contact in the 2D based semiconductors and the related technical parameters. In the individual section the practical applications of heterostructured 2D based instruments will be introduced.

4.2 PROPERTIES OF 2D HETERO-INTERFACES

4.2.1 BAND ALIGNMENT

Band alignment is one of the most fundamental characteristics of the 2D hetero-structured materials. The charge transfer mechanism, illumination properties, and charge-trapping phenomena are all stem from the energy band structure at 2D hetero-interfaces. The band alignment can be theoretically calculated by using the XPS spectra and by measuring the valence band maximum (VBM) of materials. The band alignment of several 2D transition metal dichalcogenides (TMDCs) nanostructures is calculated and presented in Figure 4.1a [18]. The results interestingly confirmed that the energy of the conduction band minimum (CBM) and the valence band maximum (VBM) of monolayer TMDCs with MX_2 structure enhanced by the increase of atomic number of chalcogenide element of TMDCs 2D structure. Generally, the heterojunction of two different materials develops three types of band alignment [19]. These three types of band alignment are called *straddling gap, staggered gap* and *broken gap*, respectively (Figure 1.4b) [19]. The type-I band alignment or straddling gap facilitates the fast recombination of electrons and holes at hetero-interfaces, since the valence band maximum and conduction band minimum of one of 2D films is located within the band gap of the other 2D structure. Thus, the type-I is a desired structure for application in luminescent devices, such as light-emitting diodes (LEDs) [20, 21]. Type-II or staggered gaps can effectively contribute to the spatial separation of electrons and holes. Therefore, it facilitates the prolongation of the interlayer excitation lifetime since the VBM of heterostructure is attributed to one of 2D materials while the CBM is located in another 2D film. The type-II hetero-interfaces are the desired structure for the electron/holes separation, and, consequently, have several applications in optoelectronic related devices. In type-III hetero-interfaces, the CBM of one of the components of 2D heterostructure is lower than the VBM of the other one. Thus, the valence band of one of 2D films overlaps with the conduction band of another 2D film. This type of 2D films allows the band-to-band tunneling (BTBT) effect of carriers and enables the operation of the tunnel field effect transistors (TFET). The speed of electron and holes transportation in type-III heterostructures is tangibly faster compared with type-II hetero-interfaces. It results in the separation of a large number of electron/holes at the hetero-interfaces, promoting semi-metallic features for type-III hetero-interfaces. This phenomenon generates a strong built-in electric field at the type-III heterostructures making this type of hetero-interfaces as the ideal candidate for new-generation of the thermal photovoltaic cells.

Band bending at the hetero-interfaces has a considerable effect on the hetero-structural properties. For example, a notch on the accumulation side and a peak on the depletion side of the PN junction are formed, where the valence and conduction bands on the different sides are bent in the opposite directions (Figure 1.4c) [22]. Due to the lack of symmetry in ΔE_C and ΔE_V values, electrons and holes experience different potential barriers at the hetero-interfaces. The band structure of the hetero-interfaces can be affected by various factors. There are several main issues, which should be taken into consideration:

114   Ultrathin 2D Semiconductors

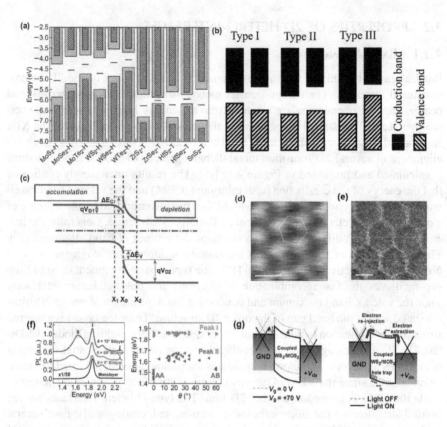

FIGURE 4.1 The band alignment at 2D hetero-interfaces. (a) The theoretically calculated location of VBM and CBM of TMDCS materials. (b) Different types of energy band alignment. (c) The formation of notch on the accumulation side and a peak on the depletion side. (d) and (e) show the STM measurement of hBN/graphene heterostructure stacked with the different rotation angle of 5°(d) and 0° (e). (f) PL spectra of MoS$_2$ monolayer and bilayer films with different twist angles. The direct transition at K-valley is assigned by peak I and the peak II is associated to the indirect bandgap. (g) The principle of photogating and band alignment at graphene-WSe$_2$/MoS$_2$ hetero-interfaces. (a) Reproduced with permission from Gong, C., Zhang, H. J., Wang, W., Colombo, L., Wallace, R. M., & Cho, K. (2015). Band alignment of two-dimensional transition metal dichalcogenides: Application in tunnel field effect transistors. *Applied Physics Letters*, *107*, 053513. (b) Reproduced with permission from Cai, Z., Liu, B., Zou, X., & Chen, H.-M. (2018). Chemical vapor deposition growth and applications of two dimensional materials and their heterostructures. *Chemical Reviews*, *118*(13), 6091–6133. (c, d, e) are reproduced with permission from Woods, C. R., Britnell, L., Eckmann, A., Ma, R. S., Lu, J. C., Guo, H. M., ... Novoselov, K. (2014). Commensurate–incommensurate transition in graphene on hexagonal boron nitride. *Nature Physics*, *10*(6), 451–456. (f) Reprinted with permission from Liu, K. H., Zhang, L. M., Cao, T., Jin, C., Qiu, D., Zhou, Q., ... Wang, F. (2014). Evolution of interlayer coupling in twisted molybdenum disulfide bilayers. *Nature Communications*, *5*, 4966 (g) Reprinted with permission from Tan, H. J., Xu, W. S., Sheng, Y. W., Lau, C. S., Fan, Y., Chen, Q., ... Warner, J. H. (2017). Lateral graphene-contacted vertically stacked WS$_2$/MoS$_2$ hybrid photodetectors with large gain. *Advanced Materials*, *29*(46), 1702917.

I. The quality of hetero-interfaces has tangible impacts on the hetero-interfaces phenomena. The presence of defects remained from the synthesis stage can cause the undesigned energy levels at hetero-interfaces. The Fermi level pinning effect is another phenomenon at the hetero-interfaces, which may be caused by intrinsic metal-induced gap states (MIGS) or extrinsic disorder induced gap states (DIGS) at the interface. As the evidence, it was found that the presence of metal like defects at MoS_2/metal heterostructures could hugely decrease the Schottky barrier height (SBH) that attributed to strong Fermi level pinning at the defects [23, 24]. On the other side, a weak Fermi level pinning facilitated the SBH modulation by electrical gating [25]. The confirmation of this phenomenon was observed at the spectra of scanning tunneling spectroscopy (STS) of the TMDCs flakes residing on the single layer (SLG) and bilayer graphene (BLG), after the voltage was shifted about 120 mv. This number is equal to the difference in the SLG and BLG work functions.

II. The lattice constant and stack orientation of 2D materials are the other important factors. The 2D materials with the same stack orientations and lattice constants develop hetero-interfaces with a periodic Moiré pattern, such as hetero-interfaces at graphene/hBN [22, 26] and SnS_2/MoS_2 [27]. Figure 4.1d and e show the STM results of rotation at graphene/hBN hetero-interfaces when it is more than 1° (Figure 4.1d) and when it is smaller than 1° (Figure 4.1e) [22]. The deformation of the lattice structure at heterojunction results in a much stronger vdW interaction at hetero-interfaces, which directly affects the band structure in graphene [41]. The investigation of photoluminescence (PL) spectra demonstrated the largest red shift for AA− stacked layer with the twist angle of 0°, while the red shift for AA- with twist angle of 60° is much smaller compared with that of previous case (Figure 4.1f) [28]. Worthwhile to mention that the constant red shift is observed for all of the other twist angles (Figure 4.1f).

III. The adjustment of the band alignments through the alteration of structural factors is sophisticatedly complicated. It is much more convenient to externally control the band alignment by using electric, magnetic and light fields [28]. The engineering of band alignment by employing the gate voltages is the most common strategy, which can facilitate the transition from type-I to type-II and type-III at the 2D hetero-interfaces. One of the well-known examples of this case is the electric field assisted gate modulation of graphene. It made the graphene a potential option for employment as the gate electrode and channel materials for the 2D based devices to reduce the contact resistance at hetero-interfaces. Furthermore, the band alignment can also be modulated by optical sources. In this concept, the heterostructure is a photogate assisted modulated system, instead of the voltaic electric field assisted systems. The actual example of a photo-gated 2D heterostructure is the WSe_2/MoS_2 photodetectors with graphene electrode (Figure 4.1g) [29]. The photo illumination of heterostructure causes the accumulation of holes at hetero-interfaces region, which acts as a positive gate at graphene electrode. It contributes the lowering of SBH at graphene/2D film hetero-interfaces by effective rising of the Fermi level of graphene electrode (Figure 4.1g) [29].

4.2.2 Charge Transport in Hetero-Interfaces

The charge transport mechanisms can be altered based on the developed energy band alignment at the hetero-interfaces. It further determines the mechanism of energy transfer between two components of hetero-interfaces. Parallel to the band alignment theories for the charge transfer, the single particle transportation is also another main mechanism of the charge transfer at hetero-interfaces, which widely affected the charge transportation by interlayer tunneling effect and charge-trapping phenomena. These mechanisms considerably contributed the low energy charge transfer phenomenon, and consequently, have been widely employed for fabrication of high-performance low consumption electronic devices. The body transport and electron/hole separation are the other main charge transport mechanisms enabling broad applications in optoelectronic field [30–34].

The band-to-band tunneling phenomenon can be realized by electrostatic gating to avoid the deprivation of band-edge sharpness resulted from the chemical doping. This charge transfer mechanism specifically happens at the hetero-interfaces with ultra-narrow channel length and atomically sharp structures [35]. Taking $SnSe_2/WSe_2$ as a practical example of 2D hetero-interfaces (Figure 4.2a), by applying a small V_G the CBM of WSe_2 is located at higher position than the VBM of $SnSe_2$. Thus, the electron tunneling from the $SnSe_2$ side to WSe_2 side is not possible. However, the charge diffusion across the interface can be realized by applying positive or negative bias voltages (Figure 4.2b) [36]. With keeping on enhancing the gate voltage, type-I band alignment is formed where the CBM of WSe_2 is set below the VBM of $SnSe_2$ (Figure 4.2c) [36]. In this condition, the tunneling window is opened at hetero-interfaces contributing to the transfer of charge carriers at 2D hetero-interfaces. In type-I heterostructures, the layer with a wide bandgap will cause spatial confinement of electrons and holes in the "well" layer with a narrow bandgap. This separation causes a strong PL in the tapping layer and a quenched PL spectrum in the other layer. Figure 4.2d depicts the PL spectra of $WSe_2/MoTe_2$ heterostructure [37]. In these graphs, the individual lines correspond to the PL intensity of the isolated WSe_2, isolated $MoTe_2$ region, and heterostacked $WSe_2/MoTe_2$ film, respectively.

The interfacial charge trapping at hetero-interfaces at PN junction is one of the main phenomena attributed to the band bending at type-II hetero-interfaces. In heterostructured p-MSB/WSe_2 electrogating switchable photodetector the p-MSB serves as a light absorber. The unidirectional carrier injection is facilitated at MSB/WSe_2 hetero-interfaces by the interfacial energy barrier and band bending at hetero-interfaces (Figure 4.2e,f) [38]. By applying a negative voltage at hetero-interfaces, the Fermi level of WSe_2 shifts downward and therefore increases the interfacial energy barrier at the hetero-interfaces. The direct effect of this phenomenon is the increase in the amount of trapped charges at hetero-interfaces. The methods for charge-trapping mechanisms are not merely restricted to band bending phenomenon, but also the other approaches can be employed to trap the charges at hetero-interfaces. To this aim a floating-gate layer can be used between the channel and control gate as the charge-trapping layer [39–44]. Figure 4.2g depicts the working mechanism of the memory unit based on the $MoS_2/$ graphene heterostructured 2D films. In this device, MoS_2 act as the channel layer while graphene plays the role of trapping layer. By applying a high drain bias voltage (–6 V

Hetero-Interfaces in 2D-Based Devices

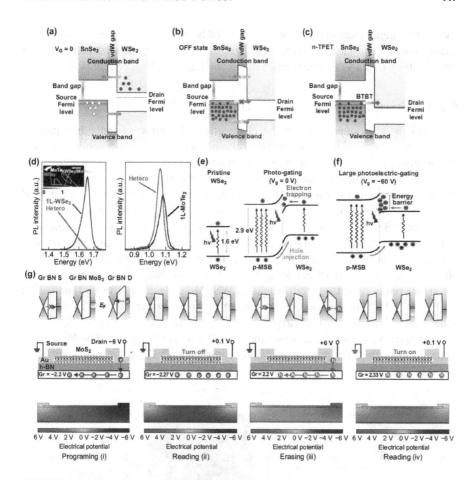

FIGURE 4.2 The single particle transport mechanisms. (a) The band alignment at 2D SnSe$_2$/WSe$_2$ heterostructures. (b) and (c) The electronic modulation of bandgap alignment. (d) The PL spectra of 1L-WSe$_2$ and 1L-MoTe$_2$ and stacked heterostructures. (e) and (f) depict the diagram of charge transfer in heterostructured 2D WSe$_2$-MoTe$_2$ stacked films at (e) V$_G$ = 0 V and (f) V$_G$ = −60 V. (g) The band alignment, charge transport and charge distribution in heterostructured MoS$_2$/h-BN/graphene. The charges are trapped by graphene floating gate. (a, b, c) Reproduce with permission from Yan, X., Liu, C. S., Li, C., Bao, W. Z., Ding, S., Zhang, D. W., & Zhou, P. (2017). Tunable SnSe$_2$/WSe$_2$ heterostructure tunneling field effect transistor. *Small*, 13(34), 1701478. (d) Reprinted with permission from Yamaoka, T., Lim, H. E., Koirala, S., Wang, X., Shinokita, K., Maruyama, M., ... Matsuda, K. (2018). Efficient photocarrier transfer and effective photoluminescence enhancement in type I monolayer MoTe$_2$/WSe$_2$ heterostructure. *Advanced Functional Materials*, 28(35), 1801021. (e, f) Reproduced with permission from Cai, Z., Cao, M., Jin, Z. P., Yi, K., Chen, X., & Wei, D. (2018). Large photoelectric-gating effect of two-dimensional van-der-Waals organic/tungsten diselenide heterointerface. *NPJ 2D Materials and Applications*, 2(1), 21. (g) Reprinted with permission from Vu, Q. A., Shin, Y. S., Kim, Y. R., Nguyen, V. L., Kang, W. T., Kim, H., ... Yu, W. J. (2016). Two-terminal floating-gate memory with van der Waals heterostructures for ultrahigh on/off ratio. *Nature Communications*, 7, 12725.

and +6 V) a high potential difference is generated between the instrument components (drain, source electrode, and graphene). Thus, the electron or holes are able to tunnel through the h-BN layer and be trapped in the floating gate (Figure 4.2g) [45].

4.2.3 Generation of Interlayer Excitons

The exciton is a bound state of the electrostatically attracted individual pair of electron/hole. This electrically neutral quasiparticle can exist in semiconductors and insulators. As an elementary excitation of condensed matter, the excitons can contribute to the energy transfer without the transport of the net electric charge. The hetero-interfaces can also contribute to the generation of excitons. For the band II alignment, the ultrafast separation of electron and holes is facilitated, since the CBM and VBM are in the different layers. The ultrafast electron/holes separation at the hetero-interfaces can be monitored by the pump–probe technique [46–48]. One of the examples of this phenomenon is the monitoring of ultrafast pumping of photoinduced signals from 2D MoS_2 film at the WS_2/MoS_2 heterostructure, where the signal rising time for heterostructured WS_2/MoS_2 device is shorter than the 50 fs (Figure 4.3a). The creation of excitons at tightly interlayer bounds of the WS_2/MoS_2 hetero-interfaces was clearly demonstrated by the measurement of PL spectra in monolayer MoS_2 and heterostructured WS_2/MoS_2 2D films [49]. The PL spectra at the MoS_2 monolayer created A-exciton resonances (1 & 2), but the signal from

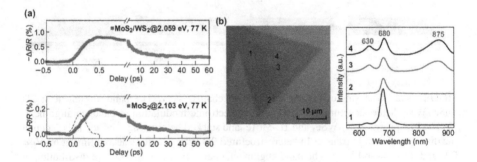

FIGURE 4.3 The generation of interlayer excitation. (a) The transient absorption measurements of WS_2 A-exciton resonance in the MoS_2/WS_2 heterostructure and B-exciton resonance in an isolated MoS_2 monolayer. The dynamic evolution signal is obtained by convoluting the instrument response function (dashed line). (b) The optical micrograph of monolayer MoS_2 and stacked MoS_2/WS_2 heterostructures. The PL spectra is produced from the 1-4 regions. The numbers 1 and 2 are assigned to the monolayer MoS_2 and 3 and 4 are attributed to heterostructured MoS_2/WS_2 films. The peak at 680 nm is attributed to the A-exciton resonance and two additional peaks at 630 and 875 nm are assigned to generation of interlayer excitation at hetero-interfaces. (a) Reprinted with permission from Hong, X., Kim, J., Shi, S. F., Zhang, Y., Jin, C., Sun, Y., ... Wang, F. (2014). Ultrafast charge transfer in atomically thin MoS2/WS2 heterostructures. *Nature Nanotechnology*, 9(9), 682–686. (b) Reprinted with permission from Gong, Y. J., Lin, J. H., Wang, X. L., Shi, G., Lei, S., Lin, Z., ... Ajayan, P. M. (2014). Vertical and in-plane heterostructures from WS_2/MoS_2 monolayers. *Nature Materials*, *13*(12), 1135–1142.

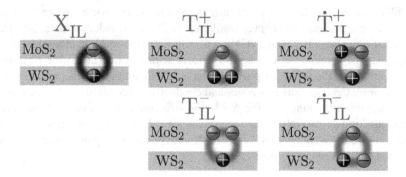

FIGURE 4.4 Possible spatial extent of interlayer trion states in a heterobilayer in comparison to the interlayer exciton X_{IL}. Reprinted with permission from Deilmann, T., & Thygesen, K. S. (2018). Interlayer trions in the MoS$_2$/WS$_2$ van der Waals heterostructure. *Nano Letters*, *18*(2), 1460–1465.

heterostructured regions (3 & 4) depicted the exciton resonance in both MoS$_2$ and WSe$_2$ 2D films, leading to an additional interlayer exciton peak, which was quenched at heterostructure due to the charge transfer process.

In addition to neutral excitons, the formation of charged trions at 2D heterostructures can influence or even dominate the photoluminescence spectra when free charges are present. Thus, the trions can be charged positively or negatively. The evidence of existence of trions at 2D hetero-interfaces is theoretically confirmed. Different kinds of combination of trions in MoS$_2$/WS$_2$ heterostructure are presented in Figure 4.4. The existence of bound interlayer trions below the neutral interlayer was predicted at the MoS$_2$/WS$_2$ heterostructure, which was governed by the neutral excitons and charged trions. It was found that the binding energies were 18 and 28 meV for the positive and negative interlayer trions, respectively, when both electron/hole pairs reside on the same layer (Figure 4.4). On the other hand, it was found that the electron/hole binding energy for interacting with interlayer exciton of the other layer is negligible [50].

4.3 THE ATOMIC STRUCTURES AT HETERO-INTERFACES

4.3.1 Homogenous Junctions

The hetero-interfaces between the same materials can be considered as a homogenous heterostructure. The properties of hetero-interfaces can be modulated by several strategies including but not limited to manipulating the geometry of hetero-interfaces, doping, controlling the strain at hetero-interfaces, passivation and controlling the dielectric constant at hetero-interfaces. The next sections give an overview of these factors.

4.3.1.1 Structural Properties at Hetero-Interfaces

Graphene is one of the most interesting 2D materials with different electronic properties. If the graphene is cut along the zigzag direction the produced nanoribbon (zGNR) behaves similar to a metal at nonmagnetic states, while an armchair graphene nanoribbon (aGNR) behaves similar to the 2D semiconductor materials. Thus, the hetero-interfaces between a zGNR and aGNR are the semimetal/semiconductor junctions. The presence of topological defects at hetero-interfaces (the ring structures) can affect the transport phenomenon at zGNR and aGNR hetero-interfaces. Furthermore, the transport properties and conductivity are highly influenced by the ring structure at the junction. The 5-7-5 ring structure is the most stable armchair-zigzag junction at the GNR hetero-interfaces [51, 52]. It was shown that a designed hetero-interfaces between zGNR and aGNR with pentagon-heptagon pairs have rectification characteristics (Figure 4.5a) [53].

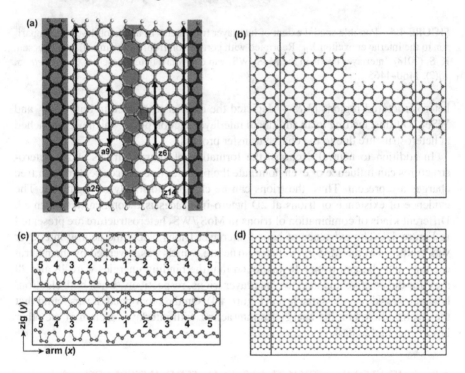

FIGURE 4.5 (a) The graphical scheme of aGNR (25) and zGNR (14) heterostructure. (b) The heterojunction between aGNR (20) with aGNR (9). (c) Top and side view of 5–5 black–blue phosphorene lateral heterostructure with octatomic-ring interface and a hexatomic-ring interface. (d) The atomic structure in GNM heterostructures. (a) Reprinted with permission from Li, X. F., Wang, L. L., Chen, Y., & Luo, K. Q. (2011). Design of graphene-nanoribbon heterojunctions from first principles. *Journal of Physical Chemistry C*, *115*(25), 12616–12624. (b) Reprinted with permission from Wang, J., Li, Z., Chen, H. C., Deng, G., & Niu, X. (2019). Recent advances in 2D lateral heterostructures. *Nano-Micro Letters*, *11*(1), 48. (c) Reprinted with permission from Li, Y., & Ma, F. (2017). Size and strain tunable band alignment of black-blue phosphorene lateral heterostructures. *Physical Chemistry Chemical Physics*, *19*(19), 12466–12472. (d) Reprinted with permission from Wang, J., Li, Z., Chen, H. C., Deng, G., & Niu, X. (2019). Recent advances in 2D lateral heterostructures. *Nano-Micro Letters*, *11*(1), 48.

Hetero-Interfaces in 2D-Based Devices

By altering the width of GNR hetero-interfaces the electronic properties can be tuned. The metallic properties were found at the GNR junction when the width is 6p+5 (p=1, 2...), while the semiconducting hetero-interfaces were created when the width is 6p+1 or 6p+3 (p=1, 2...). The variable conductance can be observed by altering the width of hetero-interfaces. For instance, the rectification is the main property of destructive interfaces and even the level of rectification can be tuned by modulating the width of GNR at two sides of hetero-interfaces, so that the interface can even behave similarly to the Schottky barrier. The rectification can be also controlled by alteration of the width of semiconductor part of GNR. The shorter semiconducting width has resulted in the stronger rectification properties. In this concept, it was demonstrated that the transport characteristics at the cross-bar T shaped heterojunction between armchair graphene nanoribbons can be tuned by adjusting the width of semiconducting hetero-interfaces (Figure 4.5b) [55].

The same dependence of electronic properties to the structural characteristics of hetero-interfaces was found at the black–blue phosphorene lateral heterojunction. The small lattice mismatch along the zigzag direction of the black and blue phosphorene contributed to the creation of two types of interfaces between black–blue phosphorene: octatomicring interface and hexatomicring interface [55]. The metallic features were found at the octatomicring interface, while the metal-to-semiconductor transition features were discovered at the hexatomicring interface after hydrogen passivation (Figure 4.5c) [55]. Similar to the graphene, the energy band at black–blue phosphorene interfaces can be tuned by changing the width of heterostructure component. Type-I band alignment is mostly found when the width is small. In this condition the CBM and VBM are mainly from the blue phosphorene, while when the width is tangibly large, the CBM and VBM are located at black phosphorene component of 2D heterostructure. The type-II band alignment at black–blue phosphorene hetero-interfaces is mostly observed when the width ranges from 2.0 ~ 3.1 nm or from 3.7 ~ 4.2 nm. In this condition the CBM is mainly located at the blue phosphorene component while the VBM is mainly located at the black phosphorene part.

Another strategy to modulate the 2D nano-materials bandgap is a creation of periodic nano holes on the single monolayer materials, which also helps to create heterojunctions in the 2D films. In this regard the novel graphene nanostructure is synthesized by using block polymer lithography method, which is called graphene nanomesh (GNM) [56]. This approach can open up the bandgap characteristics in graphene monolayer contributing to the semiconducting characteristics of graphene (Figure 4.5d) [54]. The GNM *p–n* junctions have also observed with the negative differential conductance (NDC) properties. This is a progressing stage in fabrication of electronic 2D based units without using electrostatic tuning. It was also found that the shape of holes (circular, square, and triangular holes) inside of 2D graphene is another structural factor, which can modulate the properties of 2D graphene films. However, the mechanical strength and ductility of 2D monolayer films are destructively affected by the presence of nano holes in 2D monolayer films. Thus, the heterojunctions based on GNM may have limitation in the practical applications.

4.3.1.2 Doping and Passivation

Doping is another strategy to manipulate the properties of 2D materials, and consequently, affects the properties of their lateral heterostructures. N-doped and P-doped graphene nanoribbons have been tremendously investigated. The hetero-interfaces of N-doped and P-doped graphene have been developed and tailored to make a Z-type heterojunction [55–57]. The contact resistance at aGNR and zGNR is small, and the barrier at hetero-interfaces is mostly attributed to donors and acceptors. A rectifying interface can also be developed on the semiconducting aGNR by doping the hetero-interfaces. The width of hetero-interfaces at aGNR $p–n$ junctions also determines the level of rectification. A positive rectification was observed when the length of the $p–n$ heterojunction was 3n and 3n + 2, while the negative rectification was realized at hetero-interfaces when the length of heterojunction was 3n + 1 [58]. Generally, this observation has confirmed that both level and direction of rectification can be modulated by proper adjustment of the hetero-interface length and doping conditions.

The edge modification of monolayer 2D materials is another strategy to modulate the electronic and magnetic properties and also to influence the edge magnetism and to reduce the edge states of 2D materials. The hydrogen atoms are mostly employed to passivate the graphene-based nanostructures. Oxygen is another element used to passivate the free side of the graphene. A heterojunction based on the H atom-terminated zGNR and O atom-terminated zGNR was fabricated to mimic the transport mechanisms in hetero-interfaces. The rectification effect was evidently observed at O/zGNR–H/zGNR heterojunction [59]. It has also affected the spin transport properties where the dual spinning behavior was characterized at O–zGNR–H/H_2–zGNR–H heterojunctions [60]. In addition, it was observed that the asymmetric edge hydrogenation can effectively change the electronic properties of the several different types of 2D monolayer materials. The manipulated 2D materials then were employed to develop novel heterojunctions. With the same strategy, the edge of zigzag MoS_2 was passivated by hydrogen atoms and then a heterojunction was developed between (zMoS_2NR–H) and zMoS_2NR. A perfect spin filter effect was observed at the developed hetero-interfaces. The spin filter effect escalated up to 95% spin polarization and the considerable negative differential resistance (NDR) was observed at hetero-interfaces [61]. Generally, the passivation of 2D monolayer films was found as the applicable approach for tuning the electronic characteristics of 2D films, thus it is also a useful technique to control and adjust the properties of 2D hetero-interfaces.

4.3.1.3 Strain and Dielectric Modulation

Applied strain at hetero-interfaces is effectively impact the electronic structure of 2D materials, and it has been realized as the effective strategy to tune the superlattices as the fundamental component of lateral heterojunctions. It was observed that when the applied strain is high enough (e.g. 20% for graphene), the in-plane strong covalent interactions of 2D materials preserve the 2D structure from bond-breaking. At large strain, due to difference between the elastic modulus of 2D monolayer film and its substrate, the formation of periodic rippled structures is expected. Interestingly, these 2D films with rippled structures behave similar to a superlattice containing two different materials with strain-modulated electronic structure. $ReSe_2$

Hetero-Interfaces in 2D-Based Devices

[62] and black phosphorous [63] with rippled structures are two examples of these 2D materials. By controlling the magnitude of strain, the direction of strain (tensile or compressive), and also by applying spatially controllable strain it is possible to have tunable bandgap semiconductors based on 2D materials. One of the main practical examples of using the strain-modulated electronic properties in 2D heterostructures is the device which is made based on the deposition of 2D monolayer films over dielectric nanopillar structures. In this method, nanopillars are covered and separated by 2D layered films. These nanopillars are called corrugated substrates. By controlling the size and separation distance of substrate pillars the strength and spatial pattern of the strain are controllable. Therefore, the electronic properties of 2D heterojunction are tunable. This approach is considered as a strong technique for fabrication of the microscale strained superlattices. By inducing strain distributions on 2D films, the electronic and optical properties of 2D materials improved considerably. By the same strategy a MoS_2 superlattice was made successfully over SiO_2-P/Si substrate (Figure 4.6) [64]. Schematic interpretations of the strained MoS_2

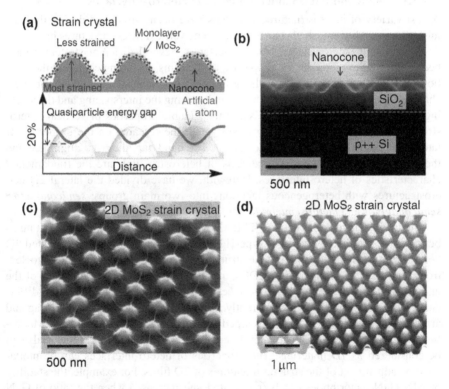

FIGURE 4.6 (a) The graphical scheme of strained MOS_2 film on SiO_2 nanocones and (b) corresponding cross-sectional SEM image of the same 2D heterostructure. (c) The overview SEM image of stained 2D MoS_2 over SiO_2 substrate and corresponding (d) AFM image of the same view. Reprinted with permission from Li, H., Contryman, A. W., Qian, X., Ardakani, S. M., Gong, Y., Wang, X., ... Zheng, X. (2015). Optoelectronic crystal of artificial atoms in strain-textured molybdenum disulphide. *Nature Communications*, 6, 8080.

film over SiO_2 nanocones are depicted in Figure 4.6a, where the regions on the tips of nanocones exhibit highest tensile strain while the areas between nanocones are less strained. By altering the distribution of strain values, the optical properties of 2D heterostructured films are modulated. The broadband light absorption from 677 nm (unstrained MoS_2) to 905 nm (most strained MoS_2) was characterized. The MoS_2 energy band gap inversely changed by the strain profile and became spatially modulated. It results in the formation of artificial atoms at the point of the peak strain. At this point, the strain-induced potential mimics the Coulomb potential around ions of the crystals. The cross-sectional SEM image of hetero-interfaces, the surface SEM image and the AFM image of the nanoarrays are respectively demonstrated in Figure 4.6b, c and d [64]. As a summary, the strain engineering of 2D material heterojunctions is a well-recognized strategy to control the energy gap variation and modulate the optical properties of this type of 2D heterostructures.

4.3.2 HETEROGENEOUS JUNCTIONS

4.3.2.1 Semiconductor/Semiconductor 2D Hetero-Interfaces

A vast variety of heterostructured materials have been introduced during the past few years with distinguished characteristics. The vertical heterostructures have been developed with heterogeneous hetero-interfaces. Several different strategies have been employed to modulate the properties of 2D films, including changing the rotation of different stacking layers, changing the component of 2D heterostructures, changing the layer thickness of 2D films, and altering the interspacing and the stacking mode of 2D heterostructures. While the tunable properties of vertical heterostructures have been tremendously investigated, the research on the properties of lateral heterostructures is mostly neglected. The band gap, the band alignment and the electronic properties of 2D materials of heterostructures are the fundamental characteristics of hetero-interfaces. Herewith, we have divided the lateral 2D heterostructures with heterogeneous junctions into two main groups: *semiconductor/semiconductor* and *metal/semiconductor* heterojunctions.

The lateral 2D heterostructured TMDCs have been the focus of quite high number of the research activities. The type-II band alignment is widely investigated 2D heterostructure since it has a unique structure for various electronic and optoelectronic applications. Graphene, TMDCs, and h-BN are different components of the lateral 2D heterostructures. The interfaces of heterogeneous junctions have different geometrical characteristics. Currently, various hetero-interfaces with zigzag and armchair geometries have been developed form monolayer TMDCs such as $MoSe_2$-WSe_2 and MoS_2-WS_2 [65]. The lateral hetero-interfaces between 2D materials can be considered as 1D materials. The properties of hetero-interfaces can be modulated by adjustment of the structural features of 2D films. For example, the stability of AlN-GaN nanoribbons can be modulated and increased when the ratio of GaN increases [66]. The band gap also is changed by alteration of the characteristics and composition of GaN at AlN-GaN heterostructure. Another example is the hetero-interfaces between arsenene and blue phosphorous or As_2P_2 heterojunction. This interface is composed of a junction between a zigzag arsenene monolayer and a zigzag blue phosphorene monolayer [67]. It was found the bandgap of heterostructured

2D films can be tuned by altering the components ratio of the structure. In other words the composition of hetero-interfaces affects the band alignment. The lateral hetero-interfaced As_2P_2 behaves similar to a quasi-type-II indirect semiconductor. The electronic properties of lateral As_mP_n heterostructure can also change from indirect to direct bandgap by increasing the width of heterostructure or by changing the component of 2D films. For example, the As_mP_n (m=n=4, 6, 8, 10) LHS behaves as type-II direct semiconductor, with the VBM localized around arsenene and the CBM localized around blue phosphorene. The reconfiguration of band alignment was also observed at the lateral hetero-interfaces of As_mP_{20-m} by changing the chemical composition of the materials. It was shown that when the m is an even number less than 10, CBM shifts downward at the lateral As_mP_{20-m} hetero-interfaces and VBM shifts upward. When the m is an even number higher than 10, the CBM in As_mP_{20-m} hetero-interfaces shifts upward and the VBM shifts downward.

Furthermore, the tensile strength along the hetero-interfaces has affected the bandgap of 2D heterostructure. The increase of hetero-interfaces width directly affects the bandgap values, when the bandgap first increases and then decreases with increasing the hetero-interfaces width. While the strain has tangible effects on VBM and CBM of 2D lateral hetero-interfaces, it is claimed that the tensile strain at the lateral 2D hetero-interfaces will not affect the direct bandgap of 2D heterostrcuture. The band alignment transition from type-II to type-I was reported when the strain at hetero-interfaces exceeds more than 6%. As it was mentioned, the desired band alignment is type-II where the facile charge transfer between 2D films of heterostructures contributes the high carrier mobility at the lateral hetero-interfaces. Thus, these types of hetero-interfaces have tremendous application in photovoltaic, optoelectronic, and photocatalytic devices.

4.3.2.2 Metal/Semiconductor (MS) Heterogeneous Junctions

The properties of 2D-based electronic and optoelectronic instruments by high extent are depended on electrical contacts between the 2D semiconductor component and metal connector part. The MS contact between graphene and other type of 2D materials such as phosphorene, BN and TMDCs can be easily developed by employment of the gapless graphene. This is one of the most well-known examples of MS junction in electronic devices based on 2D materials. The hetero-interfaces between graphene and BN can be developed either by armchair or zigzag atomic arrangement [68]. The intrinsic strength at hetero-interfaces is highly depended on misorientation angle between two components of heterostructure. As it was mentioned, the strain at hetero-interfaces can alter the bandgap of heterostructure. It was found that the uniaxial strain did not affect the bandgap of 2D heterostructures with zigzag interfaces. However, the bandgap of graphene/h-BN with armchair interface is changed by imposition of the uniaxial strain. These electronic and physical properties should be taken into account before the design of other novel lateral hybrid heterostructures. The study on the hetero-interfaces between graphene and h-BN has provided valuable information about the effect of atomic scale geometry on the hetero-interfaces properties. It was found that the left-right type lateral hetero-interfaces between graphene (left) and h-BN (right) contributes the rectification phenomenon during the charge transfer (Figure 4.7a). The NDR effect was observed when the

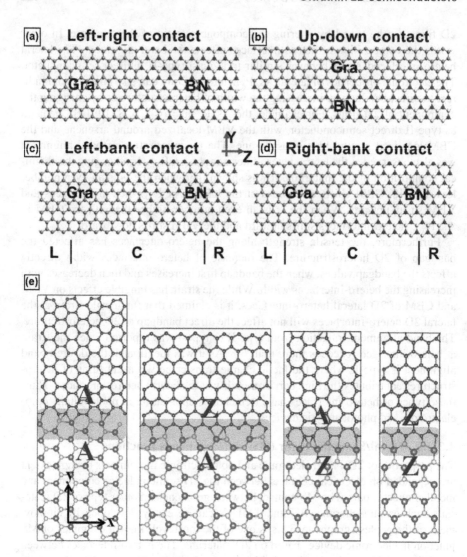

FIGURE 4.7 (a) The heterojunction of aGNR and zGNR with pentagon-heptagon pairs. (b) The heterojunction between aGNR-aGNR with passivated edge with hydrogen. (c) The top and side view of black and blue phosphorene lateral hetero-interfaces. (d) The graphical scheme of GNM heterostructures. (e) Four edge-contact geometries of graphene-MoS$_2$ heterostructures, including the armchair-armchair, zigzag-armchair, armchair-zigzag, and zigzag-zigzag, from right to left. (a, b, c, d) Reprinted with permission from An, Y. P., Zhang, M., Wu, D., Wang, T., Jiao, Z., Xia, C., ... Wang, K. (2016). The rectifying and negative differential resistance effects in graphene/h-BN nanoribbon heterojunctions. *Physical Chemistry Chemical Physics*, 18(40), 27976–27980. (e) Reprinted with permission from Sun, J., Lin, N., Tang, C., Wang, H., Ren, H., & Zhao, X. (2017). First principles studies on electronic and transport properties of edge contact graphene-MoS$_2$ heterostructure. *Comput. Mater.*, 133, 137–144.

Hetero-Interfaces in 2D-Based Devices

up-down hetero-interfaces were arranged between graphene (Top) and *h*-BN (down) (Figure 4.7b) [69]. The left back or right bank contacts facilitate both NDR effect and large rectification ratio (Figure 4.7 c, d) [69] and the level of NDR and rectification can be tuned by altering the contact structures. The small mismatch of phosphorene and graphene along the armchair direction can contribute to the development of in-plane phosphorene/graphene heterostructures (PNR/GNR). A typical PNR/GNR hetero-interfaces are demonstrated in Figure 4.7d [69]. By controlling the width of PNR/GNR hetero-interfaces it was possible to tune the band gap of the heterostructure. The hydrogen doping in hetero-interfaces can reduce the bandgap and induce the semiconductor to metal transition. These characteristics facilitate the development of two-probe graphene/phosphorene/graphene devices with tunneling transport characteristics.

The hetero-interfaces between graphene and MoS_2 can have different geometrical edge contact (armchair-armchair, armchair-zigzag, zigzag-zigzag, zigzag-armchair) (Figure 4.7e) [70]. Due to the gap state at hetero-interfaces, the electronic properties between graphene and MoS_2 are more metallic. The measurements have confirmed that four contact geometries have different potential values thus the interface plays different role in carrier transportation at the graphene/MoS_2 heterojunction. It was generally accepted that the growth preference on hetero-interfaces (C–S or C–Mo), the charge transfer mechanism, and mid-gap state at hetero-interfaces directly affect the electronic performance of graphene-MoS_2 junction.

4.3.3 ELECTRICAL CONTACT AT 2D SEMICONDUCTORS

A fine control is required over the flow of charge carriers through the semiconductor electronic instruments. The electric charges are injected into the semiconductor material through electrical contacts. The quality of electrical contact at the semiconductor-metal hetero-interfaces directly affects the performance of 2D-based electronic devices. In the case of units based on 2D materials the charge transfers at the hetero-interfaces between 2D films and metal connection fundamentally affect the properties of 2D devices. In some devices like field effect transistors (FET) and optical sensors the low contact resistance in 2D semiconductor-based instruments is critically important to achieve high "*on*" current, appreciable photo-responsivity and high-frequency operations [71, 72]. The chemical interaction and type of hetero-interfaces between the 2D film and metal contact are also highly determining factor in performance of devices. Due to the lack of dangling bonds at 2D materials, the formation of strong interfacial bonds between 2D films and metal contacts are highly difficult. It results in an increase of contact resistance at metal/2D film hetero-interfaces. To ultimately understand the charge-injection mechanisms between 2D films and metal electric contacts, we have to analyze several parameters including: the geometry and nature of the interfaces between 2D materials and metal contacts, the formation of Schottky barrier height and the contact resistance at the 2D hetero-interfaces between metal contact and semiconductors.

4.3.3.1 The Geometry of Interfaces

The metal contacts can be developed either as the top contacts configuration over 2D films or as the edge-contact configuration. In the top contact the metals are deposited over 2D films, while in the edge-contact configuration, the metal electrodes are in contact with edge of the 2D films [73, 74]. From practical point of view, most of the contacts on 2D materials have a combined structure of both top and edge configuration. Contrary to the bulk semiconductors with 3D structures (Figure 4.8a, b) [75], the 2D films with pristine surfaces are not capable of development of covalent bonds with metallic electrodes. Therefore, most of the developed top contacts are formed by vdW gap (Figure 4.8c) [75]. At the hetero-interfaces between bulk semiconductor and metal contacts, the inherent Schottky barrier is formed at hetero-interfaces (Figure 4.8b), while at the hetero-interfaces between 2D semiconductor and metal electric contact, an additional "tunnel barrier" is formed in front of the charge carrier barrier accompanied by Schottky barrier (Figure 4.8d) [75, 76]. Because of the developed tunnel barrier (Figure 4.8c, d), the contact resistance at the hetero-interfaces between

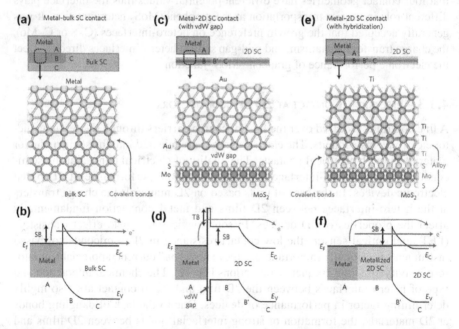

FIGURE 4.8 Different types of metal/semiconductor junction and their corresponding energy band diagrams. (a) and (b), respectively, show a typical metal/bulk semiconductor interface and its corresponding band alignment. (c) and (d) Metal/2D SC interface with vdW gap (Au–MoS$_2$ contact). (e) and (f) Metal/2D SC interface with hybridization (for example, Ti–MoS$_2$ contact, where MoS$_2$ under the contact is metallized by Ti). (b), (d) and (f), respectively, demonstrate the charge-injection mechanisms. From left to right: thermionic emission, thermionic field emission and field emission (tunneling). In (d), only thermionic emission is available. Reprinted with permission from Allain, A., Kang, J., Banerjee, K., & Kis, A. (2015). Electrical contacts to two-dimensional semiconductors. *Nature Materials*, *14*(12), 1195–1205.

Hetero-Interfaces in 2D-Based Devices

2D films and bulk metal electrodes is higher than that of 3D heterostructures (Figure 4.8a, b). To decrease the contact resistance at 2D semiconductor/metal electrode hetero-interfaces, the edge contact is recommended shorter bonding distance with stronger hybridization (orbital overlap) than top contacts. Furthermore, it was suggested that the incorporation of additional interface species can help to improve the bonding strength, and consequently, contributes the enhanced transmission. However, most of the metal contacts on 2D semiconductors are combined version of top and edge contact type. Furthermore, due to the large surface-area-to-edge-area ratio of the top contact, this type of the contact has a greater contribution to charge transfer. Thus, the strategies for the optimization of contact resistance at 2D semiconductor/metal hetero-interfaces should focus on reducing the tunnel barrier at the top contact configuration. For this matter, the hybridization is one of the main strategies to decrease the resistance at top metal contacts on 2D films. It was demonstrated that some specific metals form the covalent bonds to 2D semiconductors (Figure 4.8e) and then eliminate the vdW gap at the hetero-interfaces between 2D films and electric metal contacts. There are some examples showing the credibility of this strategy. For instance, Ni contact for graphene [77], Ti for 2D MoS_2 films [78], Pd for WSe_2 [79], Mo/W for MoS_2/WSe_2 heterostructure [80], and 2D Ti_2C for 2D MoS_2 semiconductor film [81]. The mechanism of reducing the contact resistance at 2D hetero-interfaces can be specific for each individual case, which can be demonstrated by Ti and Mo form nonlocalized overlap states in the original bandgap of 2D MoS_2 semiconductor and it effectively turns the MoS_2 film into a metallic material at the heterojunction [82].

In practical condition to develop close-perfect hetero-interfaces, the surface impurities should be removed from heterojunction. Furthermore, the additional annealing process is required to form strong covalent bonds between the metal contact and 2D semiconductor. For example, the annealing of 2D graphene devices, where the carbon atoms dissolves into the metallic top contacts and form strong covalent bonds by reduced contact resistance. As a drawback, the employment of annealing process contributes the formation of Fermi level pining at hetero-interfaces. It happens since the work function of the metal layer at interface is changed to the work function of metal-MoS_2 alloy [83]. Moreover, after the formation of metal-MoS_2 alloy a gap state is developed, which is the result of weakened bonding between Mo-S interlayer. These phenomena can significantly affect the Schottky barrier height at 2D semiconductor/metal hetero-interfaces. One of the introduced strategies is the employment of 2D material as the electrical contacts at 2D heterostructures. One of the practical cases is the employment of graphene family as the electrical contact for other 2D materials [84, 85]. Carbon nanotubes, graphene, and wide graphene ribbons have metallic characteristics, while some of the other allotropes and structures show semiconducting characteristics. In this technique, a semiconducting GNR structure is fabricated, while the other edge of GNR will be manipulated to show the metallic characteristics. The theoretical studies have demonstrated that the interface at the CNT/graphene, graphene/GNR and the monolayer/multilayer graphene has native sp^2 carbon-carbon bonds. This typical carbon-based metal contacts can facilitate the development of 2D based circuits with seamless carbon material electrical contacts. The same concept can be adapted for the other type of 2D semiconductor-based devices. As an example, the metallic 1T-MoS_2 has employed as the electrical contact for 2D semiconducting

2H-MoS$_2$ with resulted contact resistance of 0.2 kΩ μm. The development of sequential layered 2D TMDC films and metallic films by CVD method is the other strategy for development of seamless electrical contacts on 2D materials [86–88].

4.3.3.2 The Charge-Injection Mechanism

The mechanisms for charge injection to semiconductor materials can be divided to two main groups, including the thermionic emission over the Schottky barrier and the field emission (tunneling) across the Schottky barrier (Figure 4.9a) [75]. At the hetero-interfaces between metallic contact and TMDCs we mostly deal with the thermionic emission mechanisms at the low doping concentration. In thermionic emission-diffusion mechanism the *I-V* characteristics of metal/2D semiconductor junction is described as the function of Schottky junction. By the increase of doping phenomenon, the thermionic field emission starts to contribute to the charge-injection mechanism (Figure 4.9a) [75]. The similar charge injection is observed at

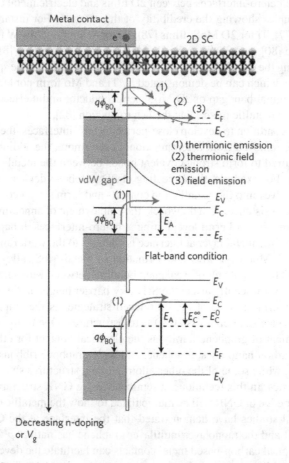

FIGURE 4.9 Charge-injection mechanisms by considering the level of hetero-interfaces doping. Reprinted with permission from Allain, A., Kang, J., Banerjee, K., & Kis, A. (2015). Electrical contacts to two-dimensional semiconductors. *Nature Materials*, *14*(12), 1195–1205.

Hetero-Interfaces in 2D-Based Devices

the hetero-interfaces between silicide contacts in advanced complementary metal–oxide–semiconductor (CMOS) devices [89]. The carrier recombination can also occur at the hetero-interfaces between the low-bandgap semiconductors (like 2D black phosphorous 0.3 eV) and metal electric connections. In the bulk materials, the diffusion region is expanded both vertically and laterally into the semiconductor film. However, in the metal/2D semiconductor junction with no hybridization states, the alteration of the band energy position only occurs on the lateral directions (Figure 4.8c, d, and Figure 4.9a) [75]. In this case, the injected charge carriers first encounter the flat-band region of the B at hetero-interfaces at the diffusion region (Figure 4.8c, d). In this condition, prediction of relative contributions of the different charge-injection mechanisms is not possible. By increasing doping factors, the energy band system is pushed from the flat-band condition into another band structure where the tunneling mechanism also contributes to the charge-injection process. By increasing the doping level, a flat-band condition is met. When the channel is doped beyond the flat-band condition, charge injection is mediated by a combination of thermionic emission and tunneling (Figure 4.9 up to down).

The experimental results have confirmed that the contact resistance at 2D semiconductor/metal hetero-interfaces are depended on three main parameters: the contact metal, the resistivity of the conduction channel between 2D semiconductor film and metal and finally on the number of layers of 2D films. Considering the combined effect of all these factors, this is difficult to compare contact resistance values from the different literatures. A more viable comparison may include the contact resistance measured at similar sheet resistance. Figure 4.10 demonstrates the contact resistance for top-contacted 2D semiconductors. The curves in Figure 4.10 correspond to the backgate voltages sweep, wherein the backgate dopes both the channel and the contact areas. From Figure 4.10 [75], it is understood that the contact resistance decreases with increasing the thickness of MoS_2 film at the constant sheet resistivity values. The observations showed that the contact resistance at 2D semiconductor-metal hetero-interfaces can be modulated by molecular and electrolyte doping. It is evaluated that the decreased contact resistance is attributed to the decreased channel resistivity. Furthermore, it was realized that the decrease of Schottky barrier width contributes the enhanced tunneling effect at the contact edges of hetero-interfaces. Generally, it is highly challenging target to decorate a low contact resistance at 2D semiconductor/metal hetero-interfaces. Several methods were employed toward the development of high-quality electrical contacts, where the most promising one is attributed to the seamless electrical contact. In this type of electrical contacts, the formation of native chemical bonding allows the facile charge transfer at the hetero-interfaces and thus the contact resistance is much lower than vdW hetero-interfaces.

4.4 DEVICES BASED ON HETEROSTRUCTURED 2D FILMS

4.4.1 ELECTRONIC DEVICES BASED ON HETEROSTRUCTURED 2D FILMS

2D films have been widely employed as the channel material in FET, since their ultrathin atomic scale nature offers ideal electrostatic control over the channel characteristics. A typical FET based on 2D materials employs h-BN film as the dielectric

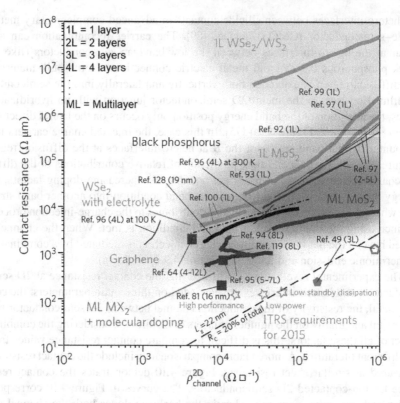

FIGURE 4.10 The calculated and measured contact resistance for different values of channel sheet resistivity for TMDCs, black phosphorous, and graphene. Reprinted with permission from Allain, A., Kang, J., Banerjee, K., & Kis, A. (2015). Electrical contacts to two-dimensional semiconductors. *Nature Materials*, 14(12), 1195–1205.

layer and then uses TMDCs film with large bandgap as the channel of FET device. These channel materials are usually traditional metal oxide films. The replacement of metallic oxide films with 2D materials leads to the occurrence of the technical problems such as the pinhole formation, the oxygen doping and interlayer defects [90]. A tunnel field effect transistor (TFET) based on SnSe$_2$/WSe$_2$ vertical heterostructure was fabricated (Figure 4.11a). The subthreshold swing of instrument was measured to be 80 mV dec^{-1} with a high I_{on}/I_{off} ratio of 10^6 (Figure 4.11b) [36]. The high performance of TFET device was attributed to the tuning of the backgate voltage facilitated by the band-to-band alignment (BTBT) switching effects.

Recently, the floating-gate memories were introduced with improved characteristics via employment of 2D heterostructured materials [91]. In this 2D heterostructure, the correct selection of floating gate guarantees the successful performance of memory unit. A programmable memory device has been recently introduced based on vertically stacked MoS$_2$/*h*-BN/graphene heterostructure where the graphene plays the role of floating gate (Figure 4.11c) [45]. By choosing an appropriate thickness for *h*-BN insulator, the *on/off* ratio over 10^9 has been observed. The 2D nature of device

Hetero-Interfaces in 2D-Based Devices

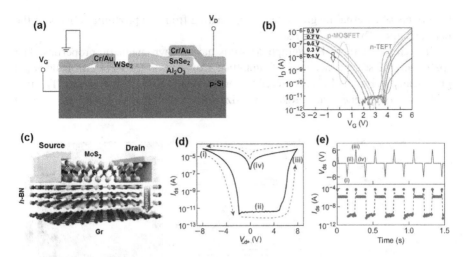

FIGURE 4.11 The heterostructured 2D electronic devices (a) the schematic of WSe$_2$/SnSe$_2$ heterostructured TFET device working based on BTBT mechanism and its (b) channel current. (c) The schematic of programmable floating-gate memory device based on h-BN/graphene heterostructure and (d) its hysteresis programming loop with its (e) applied cyclic voltage and current characteristics. (a, b) Reprinted with permission from Yan, X., Liu, C. S., Li, C., Bao, W. Z., Ding, S., Zhang, D. W., & Zhou, P. (2017). Tunable SnSe$_2$/WSe$_2$ heterostructure tunneling field effect transistor. *Small, 13*(34), 1701478. (c, d, e) Reprinted with permission from Vu, Q. A., Shin, Y. S., Kim, Y. R., Nguyen, V. L., Kang, W. T., Kim, H., ... Yu, W. J. (2016). Two-terminal floating-gate memory with van der Waals heterostructures for ultrahigh on/off ratio. *Nature Communications, 7*, 12725.

has also contributed to its stretchability. The programming (i), reading (ii), erasing (iii), and reading (iv) operations of the instrument were clearly observed by employment of repeated voltage pulses (Figure 4.11d). The hysteresis behavior of I_{ds}–V_{ds} plot (Figure 4.11e) was measured and related to the tunneling effect through h-BN film and also attributed to the asymmetric potential drop-in unit due to the resistance effect of MoS$_2$ film.

4.4.2 Magnetic Devices Based on 2D Heterostructured Materials

The magnetic priorities in 2D materials, including the coercivity, saturation magnetization, and Curie temperature (T$_C$) are dependent to the number of the layers of 2D structures and the size of magnetic domains. Thus, the magnetic properties of 2D materials are different from their bulks. The heterostructured 2D materials have tremendous applications and they are of great importance for study of spintronic properties [92]. These heterostructured 2D films can contribute to control the propagation of spin and valley (polarized) currents at the room temperature. The basic idea behind the employment of the heterostructured 2D films with ferromagnetic properties returns to this fact that a huge magnetic exchange field can be generated at the interface and therefore the regulation of the spin and valley pseudospin in 2D materials can be realized [93, 94]. The magnetic exchange field can also amplify the effects of external magnetic fields originating from the proximity effects of the heterojunctions between 2D materials.

2D heterostructured EuO/graphene films have shown magnetic properties. The spatial atomic scale conception of sublattice graphene on EuO 2D film is depicted graphically in Figure 4.12a. Different colors belong to the different atoms. The two sublattices of graphene are broken into size folders as a result of stacking on EuO

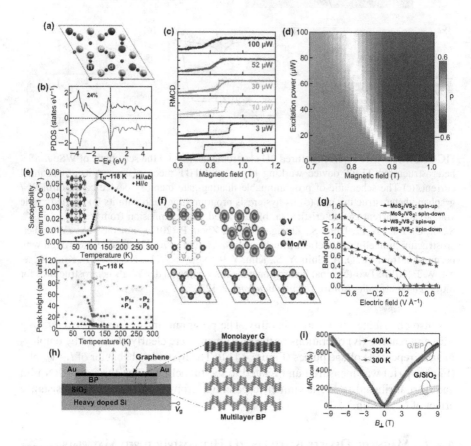

FIGURE 4.12 The magnetic properties of 2D heterostructured devices. (a) The broken lattice with six sublattice of graphene on EuO substrate. (b) The calculated values of spin polarization in 2D graphene over EuO film. (c) The RMCD value of WSe_2/CrI_3 heterostructure and (d) polarization (ρ) of WSe_2/CrI_3 heterostructure measured under sweeping external magnetic field and different excitation powers. (e) The effect of temperature on transition of several PL peaks of $FePS_3$ and the magnetic susceptibility of the same 2D material. (f) The spin density and possible magnetic ground states for XS_2/VS_2 structure and (g) the electric field dependency of the bandgap for MoS_2/VS_2 and WS_2/VS_2 heterostructures. (h) The graphical scheme of Gr/BP on Si/SiO_2 substrate and the (i) MR characteristics of the Gr on Si/SiO_2 and Gr/BP on Si/SiO_2 substrate. Reprinted with permission from Liu, Y., Zhang, S., He, J., Wang, Z. M., & Liu, Z. (2019). Recent progress in the fabrication, properties, and devices of heterostructures based on 2D materials. *Nano-Micro Letters*, *11*(1), 13.

Hetero-Interfaces in 2D-Based Devices

substrate as the calculated spin polarization is depicted these effects on in Figure 4.12b [95]. The approximate effects cause structural changes and contribute to enhance the magnetic moment of surface Eu atoms, thus causing the variable spin polarization on the graphene sublattices. This calculated spin polarization for the mentioned effects is about 24% in average. This spin polarization also affects the band structures of graphene films. The tunable magnetic exchange field (MEF) was characterized in WSe_2/multilayer CrI_3 heterostructure. It was fairly shown that the MEF can be tuned over arrange of 20 T by small changes in the laser excitation power [96]. The effect of excitation light power on the magnetization of CrI_3 in the WSe_2/CrI_3 heterostructure was evaluated via the reflection magnetic circular dichroism (RMCD) as the function of external magnetic field. The power dependent hysteresis loop behavior was evidently observed during the study of RMCD characteristics of heterostructured WSe_2/CrI_3 films. The power switchable valley properties are found in the study of PL spectra of heterostructured 2D films caused by the opto-magnetic effects of 2D WSe_2/CrI_3 films (Figure 4.12d) [96].

Finding suitable 2D materials with magnetic properties is always a challenging step toward the development of suitable heterostructured 2D materials [97, 98]. For example, CrI_3 has a layered structure with the thickness dependent magnetic characteristics. Another recently introduced 2D magnetic materials are called transition metal phosphorus trichalcogenides (TMPS) where the TM elements can be one of these metals: V, Mn, Fe, Co, Ni, or Zn. TMPS could also be feasibly exfoliated and their magnetic ground state is strongly depended on their TM elements. By the increase of layer thickness from bulk to the monolayer, an Ising-type antiferromagnetic ordering was observed in $FePS_3$ films by characterizing the transition at the Neel temperature (TN, 118 K) in the Raman peaks of this layered material (Figure 4.12e). Table 4.1 has summarized some of the 2D materials and their heterostructures with magnetic properties. Most of the well-known 2D materials with magnetic properties are in two various forms. The MX materials (M = Cr, Co, V, Mo, Mn, etc., and X = C, O, S, Se or N) and MAX (M = Cr or Fe; A = Ge, Si, Al, Sn and X = Te). Most of the well-known 2D materials show semiconductor characteristics.

The main challenge about the heterostructured 2D magnetic materials returns to their low T_c temperature, which is much lower than the room temperatures. To this aim it is highly fundamental to find heterostructured 2D materials with high T_c. The calculation has shown an ultrahigh T_c for ferromagnetic XS_2/VS_2 heterostructures [99]. It is critically important to determine the atomic arrangement for the different types of magnetic properties. The investigation of optimized configuration, magnetic arrangement, magnetic moments and energy relative to that of the ferromagnetic configuration for nonmagnetic (NM), ferromagnetic (FM), and antiferromagnetic (AFM) states, are respectively demonstrated in Figure 4.12f. It is expected that ferromagnetic state would be the most prevalent properties for the heterostructured XS_2/VS_2. The calculations have shown the T_c of 485 K for the MoS_2/VS_2 and the T_c of 487K for WS_2/VS_2 2D heterostructure. The semiconductor to metal transition is another capability of some of the 2D heterostructured magnetic films. It was observed that the electric field modulation of the magnetic properties of MoS_2/

TABLE 4.1
Calculated and Experimental Characteristics of 2D Magnetic Materials

Material	T_c (K)	Saturation magnetizations (μB/unit)	Bandgap (eV)	Exchange parameters
VSe$_2$ ML	> 330 K	15	55 m (15 K)	–
Fe$_3$GeTe$_2$	207 K (bulk) 130 K (ML)	–	–	–
TL Fe$_3$GeTe$_2$	~ 300 K (Vg = 1.75 V)	1.8	–	J_{ij} = 10 mV (Heisenberg model)
CrC ML	> 330 K	8	2.85	J_1 = 7.4 meV, J_2 = 14.7 meV (Heisenberg model)
MnO$_2$ ML	140 K 210 K (strained)	3 (<75 K)	3.41	J = 1.72 meV (Ising model)
Co$_2$S$_2$ ML	> 404 K	–	–	J_1 = 58.7 meV, J_2 = 15.8 meV (Ising model)
GdAg$_2$	85 K	5 (5 K)	–	–
MnS$_2$ ML	225 K, 330 K (strained)	3	0.69	–
MnSe$_2$ ML	250 K, 375 K (strained)	3	0.01	–
CrI$_3$	61 K (bulk) 45 K (ML) AFM (BL)	3	–	–
MoS$_2$/VS$_2$ HS	485 K	14.6	Tunable	–
WS$_2$/VS$_2$ HS	487 K	14.7	Tunable	–

Reprinted with permission from [100].

VS$_2$ and WS$_2$/VS$_2$ 2D heterostructures, was accompanied by semiconductor to metal transition [99]. These are highly important characteristics contributing to the generation of pure spin polarized currents (Figure 4.12g).

Magnetoresistance (MR) effect is another interesting characteristic of some of the 2D graphene-based heterostructured materials created by the hetero-interfaced effects. In this area, a highly stable graphene/black phosphorus (Gra/BP) 2D heterostructured device was fabricated where a monolayer graphene was located over multilayer BP (Figure 4.12h) [100]. The ultrahigh MR values of 775% were reported for this unit (Figure 4.12i) [100]. These calculated values became more interesting when it was found that the nonlocal MR value of the same heterostructure is 10,000% at the room temperature [100]. These huge values of magnetic properties were attributed to the enhanced flavor Hall effect induced by the BP channel. While the investigation on 2D heterostructured magnetic devices are in their early stage, it undoubtedly will open up a research field that bring up valuable scientific knowledge on the magnetization dynamics of 2D materials. This knowledge is critically important for development of magnetic-based electronic instruments such as magnetoresistive random-access memories.

4.4.3 Spintronic and Valleytronic 2D Heterostructured Devices

The excellent spin valley characteristics of 2D materials have contributed the spin injection, manipulation and detection capabilities in an integrated device to fabricate ultrafast nonvolatile logic circuits with ultra-low energy dissipation. Graphene is one of the 2D materials with outstanding long spin diffusion length. More recently the spin valve effect was also observed in heterostructured FM/2D materials (FM=Ni, Co) [101–104]. The most famous example of these materials are FM/Gr/FM sandwiched structures (Gr=graphene). The spin filtering effect is theoretically anticipated to occur at FM/Gr/FM sandwiched structures when the lattice mismatch at the hetero-interfaces of FM and graphene is small. The calculation has shown that only spin states at Dirac point in FM can be injected into graphene (Figure 4.13a) [105]. However, only minority of spins exist at the Dirac point. The negative tunneling magnetoresistance (TMR) of −0.85% was measured in the magnetic junction of NiFe/G–hBN/Co under the room temperature (Figure 4.13b) [106]. It confirms that the outputs of spin filters can be modulated by the different spin polarizations at interfaces. Noteworthy, the quality of hetero-interfaces has also direct impact on the performance of devices.

Monolayer TMDCs can show valleytronic characteristics in specific conditions. The external control of valleytronic devices is critically challenging since the lifting

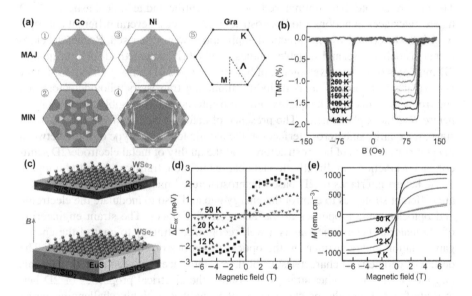

FIGURE 4.13 The spintronic and valleytronic characteristics of 2D heterostructured device. (a) The majority spin state of Co (1&2) and the minority spin states on Ni (3&4) and the Fermi states of the graphene. The color bar depicts the number of Fermi surface states. (b) The effect of temperature on TMR characteristics of NiFe/Gra–hBN/Co magnetic junction. (c) The graphical scheme of WSe$_2$ and WSe$_2$/EuS 2D heterostructured devices and their (d) valley splitting and (e) magnetization vs. the employed external magnetic field. Reprinted with permission from Liu, Y., Zhang, S., He, J., Wang, Z. M., & Liu, Z. (2019). Recent progress in the fabrication, properties, and devices of heterostructures based on 2D materials. *Nano-Micro Letters*, *11*(1), 13.

of the valley degeneracy in monolayer materials is by Zeeman splitting requires a strong external magnetic exchange field (MEF) and it only happens at the ultra-low temperatures [107]. It was recently demonstrated that deposition of 2D TMDCs on a ferromagnetic substrate contribute to utilization of the interfacial magnetic exchange field of ferromagnetic substrate to enhance the valley splitting in monolayer TMDCs. In this unit a monolayer WSe_2 was deposited on Si/SiO_2 and EuS substrates (Figure 4.13c) [108]. When the magnetic field (B) was perpendicular to 2D hetero-interfaces, the valley splitting was 1.5 meV for the WSe_2 film deposited on Si/SiO_2 substrate, while the same property increased to 3.9 meV when the EuS was employed as the substrate (Figure 4.13d, e). The dependence of ΔE_{ex} of 2D heterostructured WSe_2/EuS to magnetic field showed that the ΔE_{ex} tend to saturate at high magnetic field. Furthermore, it was found that at higher temperatures the ΔE_{ex} of 2D heterostructured WSe_2/EuS decreased continuously [108]. The behavior of 2D EuS was similar to the heterostructured one, where the field- and the temperature-dependent magnetization of EuS follow the same pattern. It confirms that the spintronic properties of 2D device are highly affected by the magnetic properties of EuS substrate.

CONCLUSION

The present chapter has summarized various scientific and technical aspects of 2D hetero-interfaces, junctions and heterostructured devices. Several different concepts were introduced and various practical applications were discussed. Recent developments in controllable, and scalable synthesis and fabrication methods of high-quality 2D materials offer the development of heterostructured 2D units with extraordinary performances in various areas including tunneling transistors, photodetectors and spintronic instruments. The quality of hetero-interfaces has tangible impacts on the hetero-interfaces phenomena. The presence of defects remained from the synthesis stage, the quality of hetero-interfaces at the atomic levels, the type of force between 2D component films of heterostructures and the quality of metal electrode/2D semiconductor junction can directly impact charge injection, transportation and trapping at the hetero-interfaces of 2D based heterostructured materials and devices. There are different strategies to control the energy gap and also to modulate the electronic and optoelectronic properties at the 2D hetero-interfaces. The strain engineering of 2D material heterojunctions is a well-recognized strategy to control the energy gap variation and adjustment of the optical properties of 2D heterostructures. The design of geometrical characteristics of hetero-interfaces between the 2D materials and monolayers is another strategy to control the electrical properties of 2D hetero-interfaces. From the practical point of view, it is a highly challenging target to decorate a low contact resistance at 2D semiconductor/metal hetero-interfaces. Several methods were employed toward the development of high-quality electrical contacts, where the most promising one is attributed to the seamless electrical contact. Generally, while the 2D heterostructured materials and devices are now well-recognized by the scientific communities, a considerable research and development activities should be allocated to further understand the hidden scientific parameters of physical and chemical phenomena at the 2D hetero-interfaces.

REFERENCES

1. Novoselov, K. S., Jiang, F., Schedin, D., Booth, T. J., Khotkevich, V. V., Morozov, S. V., & Geim, A. K. (2005). Two-dimensional atomic crystals. *Proceedings of the National Academy of Sciences of the United States of America, 102*(30), 10451–10453.
2. Tan, C. L., Cao, X. H., Wu, X. J., He, Q., Yang, J., Zhang, X., ... Zhang, H. (2017). Recent advances in ultrathin two-dimensional nanomaterials. *Chemical Reviews, 117*(9), 6225–6231.
3. Mak, K. F., Lee, C., Hone, J., Shan, J., & Heinz, T. F. (2010). Atomically thin MoS_2: A new direct-gap semiconductor. *Physical Review Letters, 105*(13), 136805.
4. Xie, L., Liao, M. Z., Wang, S. P., Yu, H., Du, L., Tang, J., ... Zhang, G. (2017). Graphene-contacted ultrashort channel monolayer MoS_2 transistors. *Advanced Materials, 29*(37), 1702522.
5. Tan, H. J., Xu, W. S., Sheng, Y. W., Lau, C. S., Fan, Y., Chen, Q., ... Warner, J. H. (2017). Lateral graphene-contacted vertically stacked WS_2/MoS_2 hybrid photodetectors with large gain. *Advanced Materials, 29*(46), 1702917.
6. Xiang, D., Liu, T., Xu, J. L., Tan, J. Y., Hu, Z., Lei, B., ... Chen, W. (2018). Two dimensional multibit optoelectronic memory with broadband spectrum distinction. *Nature Communications, 9*(1), 2966.
7. Hunt, B., Sanchez-Yamagishi, J. D., Young, A. F., Yankowitz, M., LeRoy, B. J., Watanabe, K., ... Ashoori, R. C. (2013). Massive Dirac fermions and Hofstadter butterfly in a van der Waals heterostructure. *Science, 340*(6139), 1427–1430.
8. Geim, A. K., & Grigorieva, I. V. (2013). Van der Waals heterostructures. *Nature, 499*(7459), 419–425.
9. Lin, C. Y., Zhu, X., Tsai, S. H., Tsai, S. P., Lei, S., Shi, Y., ... Lan, Y. W. (2017). Atomic-monolayer two-dimensional lateral quasi-heterojunction bipolar transistors with resonant tunneling phenomenon. *ACS Nano, 11*(11), 11015–11023.
10. Chen, X., Qiu, Y., Yang, H., Liu, G., Zheng, W., Feng, W., ... Hu, P. (2017). In-plane mosaic potential growth of large-area 2D layered semiconductors MoS_2-$MoSe_2$ lateral heterostructures and photodetector application. *ACS Applied Materials and Interfaces, 9*(2), 1684–1691.
11. Li, M. Y., Pu, J., Huang, J. K., Miyauchi, Y., Matsuda, K., Takenobu, T., & Li, L. (2018). Self-aligned and scalable growth of monolayer WSe_2–MoS_2 lateral heterojunctions. *Advanced Functional Materials, 28*(17), 1706860.
12. Chen, D. R., Hofmann, M., Yao, H. M., Chiu, S. K., Chen, S. H., Luo, Y. R., ... Hsieh, Y. P. (2019). Lateral two-dimensional material heterojunction photodetectors with ultrahigh speed and detectivity. *ACS Applied Materials and Interfaces, 11*(6), 6384–6388.
13. Tosun, M., Fu, D. Y., Desai, S. B., Ko, C., Kang, J. S., Lien, D. H., ... Javey, A. (2015). MoS_2 heterojunctions by thickness modulation. *Scientific Reports, 5*, 10990.
14. Yang, S. X., Wang, C., Sahin, H., Chen, H., Li, Y., Li, S. S., ... Tongay, S. (2015). Tuning the optical, magnetic, and electrical properties of $ReSe_2$ by nanoscale strain engineering. *Nano Letters, 15*(3), 1660–1666.
15. Frisenda, R., Navarro-Moratalla, E., Gant, P., Pérez De Lara, D., Jarillo-Herrero, P., Gorbachev, R. V., & Castellanos-Gomez, A. (2018). Recent progress in the assembly of nanodevices and van der Waals heterostructures by deterministic placement of 2D materials. *Chemical Society Reviews, 47*(1), 53–68.
16. Wang, H., Liu, F. C., Fu, W., Fang, Z., Zhou, W., & Liu, Z. (2014). Two-dimensional heterostructures: Fabrication, characterization, and application. *Nanoscale, 6*(21), 12250–12272.
17. Novoselov, K. S., Mishchenko, A., Carvalho, A., & Castro Neto, A. H. (2016). 2D materials and van der Waals heterostructures. *Science, 353*(6298), 9439.

18. Gong, C., Zhang, H. J., Wang, W., Colombo, L., Wallace, R. M., & Cho, K. (2015). Band alignment of two-dimensional transition metal dichalcogenides: Application in tunnel field effect transistors. *Applied Physics Letters, 107,* 053513.

19. Cai, Z., Liu, B., Zou, X., & Chen, H.-M. (2018). Chemical vapor deposition growth and applications of two dimensional materials and their heterostructures. *Chemical Reviews, 118*(13), 6091–6133.

20. Withers, F., Del Pozo-Zamudio, O., Mishchenko, A., Rooney, A. P., Gholinia, A., Watanabe, K., ... Novoselov, K. S. (2015). Light-emitting diodes by band structure engineering in van der Waals heterostructures. *Nature Materials, 14*(3), 301–306.

21. Xu, W. G., Liu, W. W., Schmidt, J. F., Zhao, W., Lu, X., Raab, T., ... Xiong, Q. (2017). Correlated fluorescence blinking in two-dimensional semiconductor heterostructures. *Nature, 541*(7635), 62–67.

22. Woods, C. R., Britnell, L., Eckmann, A., Ma, R. S., Lu, J. C., Guo, H. M., ... Novoselov, K. (2014). Commensurate–incommensurate transition in graphene on hexagonal boron nitride. *Nature Physics, 10*(6), 451–456.

23. Bampoulis, P., van Bremen, R., Yao, Q. R., Poelsema, B., Zandvliet, H. J. W., & Sotthewes, K. (2017). Defect dominated charge transport and Fermi level pinning in MoS_2/metal contacts. *ACS Applied Materials and Interfaces, 9*(22), 19278–19286.

24. Hattori, Y., Taniguchi, T., Watanabe, K., & Nagashio, K. (2018). Determination of carrier polarity in Fowler-Nordheim tunneling and evidence of Fermi level pinning at the hexagonal boron nitride/metal interface. *ACS Applied Materials and Interfaces, 10*(14), 11732–11738.

25. Le Quang, T., Cherkez, V., Nogajewski, K., Potemski, M., Dau, M. T., Jamet, M., ... Veuillen, J. (2017). Scanning tunneling spectroscopy of van der Waals graphene/semiconductor interfaces: Absence of Fermi level pinning. *2D Materials, 4*(3), 035019.

26. Purdie, D. G., Pugno, N. M., Taniguchi, T., Watanabe, K., Ferrari, A. C., & Lombardo, A. (2018). Cleaning interfaces in layered materials heterostructures. *Nature Communications, 9,* 5387.

27. Li, B., Huang, L., Zhong, M. Z., Li, Y., Wang, Y., Li, J., & Wei, Z. (2016). Direct vapor phase growth and optoelectronic application of large band offset SnS_2/MoS_2 vertical bilayer heterostructures with high lattice mismatch. *Advanced Electronic Materials, 2*(11), 1600298.

28. Liu, K. H., Zhang, L. M., Cao, T., Jin, C., Qiu, D., Zhou, Q., ... Wang, F. (2014). Evolution of interlayer coupling in twisted molybdenum disulfide bilayers. *Nature Communications, 5,* 4966.

29. Tan, H. J., Xu, W. S., Sheng, Y. W., Lau, C. S., Fan, Y., Chen, Q., ... Warner, J. H. (2017). Lateral graphene-contacted vertically stacked WS_2/MoS_2 hybrid photodetectors with large gain. *Advanced Materials, 29*(46), 1702917.

30. Bellus, M. Z., Ceballos, F., Chiu, H. Y., & Zhao, H. (2015). Tightly bound trions in transition metal dichalcogenide heterostructures. *ACS Nano, 9*(6), 6459–6464.

31. Drissi, L. B., Ramadan, F. Z., & Kanga, N. B. J. (2018). Optoelectronic properties in 2D GeC and SiC hybrids: DFT and many body effect calculations. *Materials Research Express, 5*(1), 015061.

32. Hong, X. P., Kim, J., Shi, S. F., Zhang, Y., Jin, C., Sun, Y., ... Wang, F. (2014). Ultrafast charge transfer in atomically thin MoS_2/WS_2 heterostructures. *Nature Nanotechnology, 9*(9), 682–686.

33. Hung, H. H., Wu, J. S., Sun, K., & Chiu, C. K. (2017). Engineering of many-body Majorana states in a topological insulator/swave superconductor heterostructure. *Scientific Reports, 7*(1), 3499.

34. Rivera, P., Schaibley, A. M., Jones, J. S., Ross, S. F., Wu, Aivazian, G., ... Xu, X. (2015). Observation of long-lived interlayer excitons in monolayer $MoSe_2$-WSe_2 heterostructures. *Nature Communications, 6,* 6242.

Hetero-Interfaces in 2D-Based Devices 141

35. Li, M., Esseni, D., Snider, G., Jena, D., & Xing, H. G. (2014). Single particle transport in two-dimensional heterojunction interlayer tunneling field effect transistor. *Journal of Applied Physics, 115*(7), 074508.

36. Yan, X., Liu, C. S., Li, C., Bao, W. Z., Ding, S., Zhang, D. W., & Zhou, P. (2017). Tunable $SnSe_2/WSe_2$ heterostructure tunneling field effect transistor. *Small, 13*(34), 1701478.

37. Yamaoka, T., Lim, H. E., Koirala, S., Wang, X., Shinokita, K., Maruyama, M., ... Matsuda, K. (2018). Efficient photocarrier transfer and effective photoluminescence enhancement in type I monolayer $MoTe_2/WSe_2$ heterostructure. *Advanced Functional Materials, 28*(35), 1801021.

38. Cai, Z., Cao, M., Jin, Z. P., Yi, K., Chen, X., & Wei, D. (2018). Large photoelectric-gating effect of two-dimensional van-der-Waals organic/tungsten diselenide heterointerface. *NPJ 2D Materials and Applications, 2*(1), 21.

39. Bertolazzi, S., Krasnozhon, D., & Kis, A. (2013). Nonvolatile memory cells based on MoS2/graphene heterostructures. *ACS Nano, 7*(4), 3246–3252.

40. Choi, M. S., Lee, G. H., Yu, Y. J., Lee, D. Y., Lee, S. H., Kim, P., ... Yoo, W. J. (2013). Controlled charge trapping by molybdenum disulphide and graphene in ultrathin heterostructured memory devices. *Nature Communications, 4*, 1624.

41. Zhang, Y., Yuan, Y. B., & Huang, J. S. (2017). Detecting 100 fW cm^{-1} light with trapped electron gated organic phototransistors. *Advanced Materials, 29*, 1603969.

42. Wang, J. L., Zou, X. M., Xiao, X. H., Xu, L., Wang, C., Jiang, C., ... Liao, L. (2015). Floating gate memory-based monolayer MoS_2 transistor with metal nanocrystals embedded in the gate dielectrics. *Small, 11*(2), 208–213.

43. Liu, Y. P., Lew, W. S., & Liu, Z. W. (2017). Observation of anomalous resistance behavior in bilayer graphene. *Nanoscale Research Letters, 12*(1), 48.

44. Liu, Y. P., Liu, Z. W., Lew, W. S., & Wang, Q. J. (2013). Temperature dependence of the electrical transport properties in few layer graphene interconnects. *Nanoscale Research Letters, 8*(1), 335.

45. Vu, Q. A., Shin, Y. S., Kim, Y. R., Nguyen, V. L., Kang, W. T., Kim, H., ... Yu, W. J. (2016). Two-terminal floating-gate memory with van der Waals heterostructures for ultrahigh on/off ratio. *Nature Communications, 7*, 12725.

46. Hong, X. P., Kim, J., Shi, S. F., Zhang, Y., Jin, C., Sun, Y., ... Wang, F. (2014). Ultrafast charge transfer in atomically thin MoS_2/WS_2 heterostructures. *Nature Nanotechnology, 9*(9), 682–686.

47. Chen, H. L., Wen, X. W., Zhang, J., Wu, T., Gong, Y., Zhang, X., ... Zheng, J. (2016). Ultrafast formation of interlayer hot excitons in atomically thin MoS_2/WS_2 heterostructures. *Nature Communications, 7*, 12512.

48. Hong, X., Kim, J., Shi, S. F., Zhang, Y., Jin, C., Sun, Y., ... Wang, F. (2014). Ultrafast charge transfer in atomically thin MoS2/WS2 heterostructures. *Nature Nanotechnology, 9*(9), 682–686.

49. Gong, Y. J., Lin, J. H., Wang, X. L., Shi, G., Lei, S., Lin, Z., ... Ajayan, P. M. (2014). Vertical and in-plane heterostructures from WS_2/MoS_2 monolayers. *Nature Materials, 13*(12), 1135–1142.

50. Deilmann, T., & Thygesen, K. S. (2018). Interlayer trions in the MoS_2/WS_2 van der Waals heterostructure. *Nano Letters, 18*(2), 1460–1465.

51. Wu, G. X., Meng, Q. Y., & Jing, Y. H. (2012). Computational design for interconnection of graphene nanoribbons. *Chemical Physics Letters, 531*, 119–125.

52. Wu, G. X., Li, C. L., Jing, Y. H., Wang, C., Yang, Y., & Wang, Z. (2014). Electronic transport properties of graphene nanoribbon heterojunctions with 5-7-5 ring defect. *Computational Materials Science, 95*, 84–88.

53. Li, X. F., Wang, L. L., Chen, Y., & Luo, K. Q. (2011). Design of graphene-nanoribbon heterojunctions from first principles. *Journal of Physical Chemistry C, 115*(25), 12616–12624.

54. Wang, J., Li, Z., Chen, H. C., Deng, G., & Niu, X. (2019). Recent advances in 2D lateral heterostructures. *Nano-Micro Letters, 11*(1), 48.
55. Li, Y., & Ma, F. (2017). Size and strain tunable band alignment of black-blue phosphorene lateral heterostructures. *Physical Chemistry Chemical Physics, 19*(19), 12466–12472.
56. Oswald, W., & Wu, Z. (2012). Energy gaps in graphene nanomeshes. *Physical Review. Part B, 85*(11), 115431.
57. Wang, X. R., Li, X. L., Zhang, L., Yoon, Y., Weber, P. K., Wang, H., ... Dai, H. (2009). N-doping of graphene through electrothermal reactions with ammonia. *Science, 324*(5928), 768–771.
58. Zheng, J. X., Yan, X., Yu, L. L., Li, H., Qin, R., Luo, G., ... Lu, J. (2011). Family dependent rectification characteristics in ultra-short graphene nanoribbon $p–n$ junctions. *Journal of Physical Chemistry C, 115*(17), 8547–8554.
59. Nazirfakhr, M., & Shahhoseini, A. (2018). Negative differential resistance and rectification effects in zigzag graphene nanoribbon heterojunctions: Induced by edge oxidation and symmetry concept. *Physics Letters. A, 382*(10), 704–709.
60. Cui, L. L., Long, M. Q., Zhang, X. J., Li, X., Zhang, D., & Yang, B. (2016). Spin-dependent transport properties of heterojunction based on zigzag graphene nanoribbons with edge hydrogenation and oxidation. *Physics Letters. A, 380*(5–6), 730–738.
61. Peng, L., Yao, K., Zhu, S., Ni, Y., Zu, F., Wang, S., ... Tian, Y. (2014). Spin transport properties of partially edge-hydrogenated MoS_2 nanoribbon heterostructure. *Journal of Applied Physics, 115*(22), 223705.
62. Yang, S. X., Wang, C., Sahin, H., Chen, H., Li, Y., Li, S. S., ... Tongay, S. (2015). Tuning the optical, magnetic, and electrical properties of $ReSe_2$ by nanoscale strain engineering. *Nano Letters, 15*(3), 1660–1666.
63. Quereda, J., San-Jose, P., Parente, V., Vaquero-Garzon, L., Molina-Mendoza, A. J., Agraït, N., ... Castellanos-Gomez, A. (2016). Strong modulation of optical properties in black phosphorus through strain-engineered rippling. *Nano Letters, 16*(5), 2931–2937.
64. Li, H., Contryman, A. W., Qian, X., Ardakani, S. M., Gong, Y., Wang, X., ... Zheng, X. (2015). Optoelectronic crystal of artificial atoms in strain-textured molybdenum disulphide. *Nature Communications, 6*, 8080.
65. Lee, J., Huang, J. S., Sumpter, B. G., & Yoon, M. (2017). Strain-engineered optoelectronic properties of 2D transition metal dichalcogenide lateral heterostructures. *2D Materials, 4*(2), 021016.
66. Zhang, Z., & Xu, Y. (2013). First-principles study on the structural stability and electronic properties of AlN/GaN heterostructure nanoribbons. *Superlattices and Microstructures, 57*, 37–43.
67. Li, Q. F., Ma, X. F., Zhang, L., Wan, X. G., & Rao, W. (2018). Theoretical design of blue phosphorene/arsenene lateral heterostructures with superior electronic properties. *Journal of Physics. Part D, 51*(25), 255304.
68. Ge, M., & Si, C. (2018). Mechanical and electronic properties of lateral graphene and hexagonal boron nitride heterostructures. *Carbon, 136*, 286–291.
69. An, Y. P., Zhang, M., Wu, D., Wang, T., Jiao, Z., Xia, C., ... Wang, K. (2016). The rectifying and negative differential resistance effects in graphene/h-BN nanoribbon heterojunctions. *Physical Chemistry Chemical Physics, 18*(40), 27976–27980.
70. Sun, J., Lin, N., Tang, C., Wang, H., Ren, H., & Zhao, X. (2017). First principles studies on electronic and transport properties of edge contact graphene-MoS_2 heterostructure. *Comput. Mater., 133*, 137–144.
71. Lopez-Sanchez, O., Lembke, D., Kayci, M., Radenovic, A., & Kis, A. (2013). Ultrasensitive photodetectors based on monolayer MoS_2. *Nature Nanotechnology, 8*(7), 497–501.

72. Krasnozhon, D., Lembke, D., Nyffeler, C., Leblebici, Y., & Kis, A. (2014). MoS$_2$ transistors operating at gigahertz frequencies. *Nano Letters, 14*(10), 5905–5911.

73. Song, S. M., Park, J. K., Sul, O. J., & Cho, B. J. (2012). Determination of work function of graphene under a metal electrode and its role in contact resistance. *Nano Letters, 12*(8), 3887–3892.

74. Wang, L., Meric, I., Huang, P. Y., Gao, Q., Gao, Y., Tran, H., ... Dean, C. R. (2013). One-dimensional electrical contact to a two-dimensional material. *Science, 342*(6158), 614–617.

75. Allain, A., Kang, J., Banerjee, K., & Kis, A. (2015). Electrical contacts to two-dimensional semiconductors. *Nature Materials, 14*(12), 1195–1205.

76. Kang, J., Liu, W., Sarkar, D., Jena, D., & Banerjee, K. (2014). Computational study of metal contacts to monolayer transition-metal dichalcogenide semiconductors. *Physical Review X, 4*(3), 031005.

77. Stokbro, K., Engelund, M., & Blom, A. (2012). Atomic-scale model for the contact resistance of the nickel-graphene interface. *Physical Review. Part B, 85*(16), 165442.

78. Popov, I., Seifert, G., & Tománek, D. (2012). Designing electrical contacts to MoS$_2$ monolayers: A computational study. *Physical Review Letters, 108*(15), 156802.

79. Kang, J., Sarkar, D., Liu, W., Jena, D., & Banerjee, K. (2012). A computational study of metal-contacts to beyond-graphene 2D semiconductor materials. *IEEE Int. Electron Dev. Meet.*, 407–410.

80. Kang, J., Liu, W., & Banerjee, K. (2014). High-performance MoS$_2$ transistors with low-resistance molybdenum contacts. *Applied Physics Letters, 104*(9), 093106.

81. Gan, L. Y., Zhao, Y. J., Huang, D., & Schwingenschlögl, U. (2013). First-principles analysis of MoS$_2$/Ti$_2$C and MoS$_2$/Ti$_2$CY$_2$ (Y=F and OH) all-2D semiconductor/metal contacts. *Physical Review. Part B, 87*(24), 245307.

82. Kang, J., Liu, W., Sarkar, D., Jena, D., & Banerjee, K. (2014). Computational study of metal contacts to monolayer transition-metal dichalcogenide semiconductors. *Physical Review X, 4*(3), 031005.

83. Leong, W. S., Nai, C. T., & Thong, J. T. L. (2014). What does annealing do to metal–graphene contacts? *Nano Letters, 14*(7), 3840–3847.

84. Ouyang, Y., & Guo, J. (2010). Carbon-based nanomaterials as contacts to graphene nanoribbons. *Applied Physics Letters, 97*(26), 263115.

85. Kang, J., Sarkar, D., Khatami, Y., & Banerjee, K. (2013). Proposal for all-graphene monolithic logic circuits. *Applied Physics Letters, 103*(8), 083113.

86. Kappera, R., Voiry, D., Yalcin, S. E., Branch, B., Gupta, G., Mohite, A. D., & Chhowalla, M. (2014). Phase-engineered low-resistance contacts for ultrathin MoS$_2$ transistors. *Nature Materials, 13*(12), 1128–1134.

87. Kappera, R., Voiry, D., Yalcin, S. E., Jen, W., Acerce, M., Torrel, S., ... Chhowalla, M. (2014). Metallic 1T phase source/drain electrodes for field effect transistors from chemical vapor deposited MoS$_2$. *APL Materials, 2*(9), 092516.

88. Lin, Y. C., Dumcenco, D. O., Huang, Y. S., & Suenaga, K. (2014). Atomic mechanism of the semiconducting to metallic phase transition in single-layered MoS$_2$. *Nature Nanotechnology, 9*(5), 391–396.

89. Banerjee, K., Amerasekera, A., Dixit, G., & Hu, C. (1997). Temperature and current effects on smallgeometrycontact resistance. *IEEE Int. Electron Dev. Meet.*, 115–118.

90. Li, M. Y., Chen, C. H., Shi, Y., & Li, L. J. (2016). Heterostructures based on two-dimensional layered materials and their potential applications. *Materials Today, 19*(6), 322–335.

91. Wang, S., He, C., Tang, J., Lu, X., Shen, C., Yu, H., ... Zhang, G. (2019). New floating-gate memory with excellent retention characteristics. *Advanced Electronic Materials, 5*(4), 1800726.

92. Liu, Y. P., Lew, W. S., & Sun, L. (2011). Enhanced weak localization effect in few-layer graphene. *Physical Chemistry Chemical Physics*, *13*(45), 20208–20214.

93. Qi, J. S., Li, X., Niu, Q., & Feng, J. (2015). Giant and tunable valley degeneracy splitting in $MoTe_2$. *Physical Review. Part B*, *92*(12), 121403.

94. Wei, P., Lee, S., Lemaitre, F., Pinel, L., Cutaia, D., Cha, W., ... Chen, C. T. (2016). Strong interfacial exchange field in the graphene/EuS heterostructure. *Nature Materials*, *15*(7), 711–716.

95. Yang, H. X., Hallal, A., Terrade, D., Waintal, X., Roche, S., & Chshiev, M. (2013). Proximity effects induced in graphene by magnetic insulators: First-principles calculations on spin filtering and exchange-splitting gaps. *Physical Review Letters*, *110*(4), 046603.

96. Seyler, K. L., Zhong, D., Huang, B., Linpeng, X., Wilson, N. P., Taniguchi, T., ... Xu, X. (2018). Valley manipulation by optically tuning the magnetic proximity effect in WSe_2/CrI_3 heterostructures. *Nano Letters*, *18*(6), 3823–3828.

97. Lee, J. U., Lee, S., Ryoo, J. H., Kang, S., Kim, T. Y., Kim, P., ... Cheong, H. (2016). Isingtype magnetic ordering in atomically thin $FePS_3$. *Nano Letters*, *16*(12), 7433–7438.

98. Du, J., Xia, C. X., Xiong, W. Q., Wang, T., Jia, Y., & Li, J. (2017). Two-dimensional transition-metal dichalcogenides-based ferromagnetic van der Waals heterostructures. *Nanoscale*, *9*(44), 17585–17592.

99. Liu, Y. P., Yudhistira, I., Yang, M., Laksono, E., Luo, Y. Z., Chen, J., ... Loh, K. P. (2018). Phonon-mediated colossal magnetoresistance in graphene/black phosphorus heterostructures. *Nano Letters*, *18*(6), 3377–3383.

100. Liu, Y., Zhang, S., He, J., Wang, Z. M., & Liu, Z. (2019). Recent progress in the fabrication, properties, and devices of heterostructures based on 2D materials. *Nano-Micro Letters*, *11*(1), 13.

101. Dlubak, M. B., Martin, R. S., Weatherup, H., Yang, H., Deranlot, C., Blume, R., ... Robertson, J. (2012). Graphene-passivated nickel as an oxidation-resistant electrode for spintronics. *ACS Nano*, *6*(12), 10930–10934.

102. Iqbal, M. Z., Iqbal, M. W., Siddique, S., Khan, M. F., & Ramay, S. M. (2016). Ramay, room temperature spin valve effect in $NiFe$/WS_2/Co junctions. *Scientific Reports*, *6*, 21038.

103. Wang, Z., Sapkota, D., Taniguchi, T., Watanabe, K., Mandrus, D., & Morpurgo, A. F. (2018). Tunneling spin valves based on Fe_3GeTe_2/hBN/Fe_3GeTe_2 van der Waals heterostructures. *Nano Letters*, *18*(7), 4303–4308.

104. Zhang, H., Ye, M., Wang, Y., Quhe, R., Pan, Y., Guo, Y., ... Lu, J. (2016). Magnetoresistance in Co/2D MoS_2/Co and Ni/2D MoS_2/Ni junctions. *Physical Chemistry Chemical Physics*, *18*(24), 16367–16376.

105. Karpan, V. M., Giovannetti, G., Khomyakov, P. A., Talanana, M., Starikov, A. A., Zwierzycki, M., ... Kelly, P. J. (2007). Graphite and graphene as perfect spin filters. *Physical Review Letters*, *99*(17), 176602.

106. Zhao, C., Norden, T., Zhang, P. Y., Zhao, P., Cheng, Y., Sun, F., ... Zeng, H. (2017). Enhanced valley splitting in monolayer WSe_2 due to magnetic exchange field. *Nature Nanotechnology*, *12*(8), 757–762.

107. Stier, A. V., McCreary, K. M., Jonker, B. T., Kono, J., & Crooker, S. A. (2016). Exciton diamagnetic shifts and valley Zeeman effects in monolayer WS_2 and MoS_2 to 65 Tesla. *Nature Communications*, *7*, 10643.

108. Luo, Y. K., Xu, J. S., Zhu, T. C., Wu, G., McCormick, E. J., Zhan, W., ... Kawakami, R. K. (2017). Opto-valleytronic spin injection in monolayer MoS_2/few-layer graphene hybrid spin valves. *Nano Letters*, *17*(6), 3877–3883.

5 Photonic and Plasmonic Devices Based on Two-Dimensional Semiconductors

5.1 INTRODUCTION

Properties of nanostructured materials can be fundamentally changed upon reduction of their dimensionality down to the two-dimensional (2D) scale. The low-dimensionality has brought distinct features to 2D materials, where the atoms, electrons, and holes are tightly restricted to the 2D limits [1, 2]. Strong light mater interaction and ultra-high charge carrier mobility are the most well-known electronic phenomenon which are ultimately affected by the 2D electronic characteristics of ultrathin 2D materials [3]. The last decade has witnessed the advent and introduction of various types of 2D materials with outstanding and unexpected photonic and optoelectronic applications. Graphene as the first member of the 2D materials with its semi-metallic characteristics interacts with a wide range of electromagnetic wavelengths [4]. The graphene's ultra-high carrier mobility properties and high transparency are some of the main characteristics of optoelectronic devices based on graphene 2D materials [5]. The structure of graphene with its zero bandgap and its linear dispersion near the Dirac point is the host of several nano-photonic properties [6]. The highly sensitive reaction to optical signals and various types of light-matter interactions in graphene is originated from the capability of engineering of electronic structure of graphene. Moreover, graphene can support the localized and propagating plasmons at the terahertz and mid-infrared range, since the Fermi energy level of graphene can be accurately controlled [7, 8]. While the plasmonic characteristics of the metal-based plasmonic structures are not tunable, the plasmonic response of graphene-based devices could be controlled by use of the external electrostatic biasing gating [9]. Graphene is also capable of employing in modulator, photonic and photosensors originated from the broadband absorption of this ultrathin carbon structures [10–12]. However, the metallic nature of graphene has restricted the light generation functionalities of 2D graphene, thus the other 2D materials are considered to be replaced by the graphene.

Transition metal dichalcogenides (TMDCs) are the other type of layered 2D semiconductors with distinguished electronic and optoelectronic characteristics. 2D layered TMDCs have demonstrated much higher carrier mobility and several improved electronic characteristics compared with those of the similar bulk semiconductors [13]. While the carrier mobility of TMDCs is lower than that of graphene, their bandgap tunability (1.0 ~ 2.5 eV) makes them much suitable options for the light emitters and absorbents in a wide range of illumination spectrum compared with graphene [14]. However, the light harvesting capability of 2D TMDCs is still limited due to the dimensional limitations. Since the efficient utilization of 2D materials is still far from the practical schemes, another strategy should be taken into account for solving the challenges of introduction of 2D materials to everyday applications and into electronic instruments. Tailoring and integration of 2D materials into plasmonic nanostructures can be the main employed strategy to tackle these challenges. The hybrid 2D plasmonic structures have shown considerable improvement in the light detection and energy harvesting characteristics [15]. In this view, high number of plasmonic structures can be utilized to successfully design and fabricate photonic devices with the capability of controlling the light-matter interactions in 2D based optoelectronic units. The other main privilege of integration of 2D materials into plasmonic structures returns to the capability of rapid charge carriers and hot electron transfer from the plasmonic structures into 2D materials [16, 17]. This optical characteristic of 2D integrated plasmonic device has facilitated numerous applications in photovoltaic and photocatalysis [18]. Furthermore, presence of the different types of quasi-particles in the plasmonic units as well as coupling phenomenon between excitons and plasmons in 2D structures have introduced novel optoelectronic applications [19, 20]. Considering the scientific and practical observations, it inspires more research and development studies for future applications of 2D plasmonic-based devices. Thus the present chapter has targeted both theoretical and practical aspects of tailoring 2D materials into plasmonic structured instruments. The chapter will focus on the properties and plasmonic characteristics of 2D materials and the fabricated 2D integrated plasmonic utensils. Then, the privilege of coupling plasmonic characteristics of 2D materials with their excitonic irradiations will be explained and novel photonic devices with distinguished optoelectronic characteristics will be reviewed.

5.2 PLASMONIC 2D NANOSTRUCTURES

5.2.1 PRINCIPLES

The light interaction with free electrons in the nanostructured metal materials can collectively excite the electron cloud as a plasmon [21]. The diffraction limit of the light can be broken by the plasmon and light interactions. As a result, it can be localized in subwavelength dimensions, creating strong enhanced electron magnetic field (EM) and optical near-field enhancement [22, 23]. Under some occasions and conditions the collective oscillation of electron cloud becomes in phase with the outer optical field, helping to trap the light waves at the interfaces of metals. This phenomenon is known as the surface plasmon polariton (SPP) [24]. The energy localization on the surface of the metal will be accompanied by the induced strong electromagnetic waves. Based on the propagation condition, the plasmonic effects can be divided into localized surface plasmons resonance (LSPR) and the surface plasmon polariton [25].

The LSPR refers to the light localization phenomenon within the nanostructures. In LSPR the collective oscillation of electrons in resonance with incident light generates plasmons. When the metal nanoparticles are illuminated with light sources, the electric field starts oscillation, which consequently induces the coherent oscillation of conducting electrons. Due to this phenomenon, the charges are polarized on the surface of nanoparticles [26]. The size of nanoparticles has ultimate impact on the resonance type. Specifically, when the particle size is less than 15 nm, the spectral response is dominating by the absorption peaks [27]. However, when the size of nanoparticles is greater than 15 nm the spectral response is caused by the scattering phenomenon [27].

SPP is the occurrence of light localization in the form of propagating waves at the metal-dielectric interface [28]. In fact, SPPs are the propagated electromagnetic waves along the metal-semiconductor interfaces [28]. The generated waves are in visible or infrared spectrum range. The intensity of generated waves decreases proportional to the distance from the metal-semiconductor heterointerfaces.

The geometrical features of plasmonic structures determine the type of generated plasmons. Plasmon is typically excited on two different types of substrates. (i) nanoparticles and their aggregates are one of the main sources of excitement of plasmonic generated carriers. (ii) The plasmons can also be generated on 2D and 3D plasmonic nanoarrays. These patterned nanostructures are highly interesting for practical optoelectronic instruments in the real-world applications.

There are several following privileges for 2D or 3D nano-array based devices:

i. Unlike the colloidal nanoparticles, the nanoarrays with their periodic geometry over the solid-state substrates facilitate the patterned design of 2D nanostructures.

ii. Coupling properties of 2D nanostructures and plasmonic partners can be precisely modulated by the control of geometry, size and the distance of hybrid 2D plasmonic structures. Coupling of the plasmonic modes is therefore restricted to the 2D heterointerfaces between 2D materials and plasmonic nanostructures.

iii. Compared with plasmonic nanoparticles, nanoarrays are highly stable and reliable design, thus facilitate the modulation of light spectrum and management of the enhanced electromagnetic fields. It makes possible to design the plasmonic structures with tunable capabilities for performance at the specific and desired wavelength of light.

iv. The synthesis and control of colloidal nanostructures such as silver and gold nanoparticles and nanorods faced several challenges. The synthesis methods are not efficient enough to produce the same nanoparticles with similar geometry, thus the light-interaction phenomena with plasmonic nanoparticles are not precise and the final results may not be reproducible.

The geometrical feature of 2D materials is the host of several fundamental light-matter interaction phenomena. The quantum confinement effect is one of the main outcomes of decreasing the size of materials to 2D scales [29]. This effect is observed when the size of the nanostructured material is comparable to the wavelength of the electron. The two dimensionality considerably affects the electronic structures and

148 Ultrathin 2D Semiconductors

optical properties of 2D materials, which is ultimately distinct from their 3D materials [30]. Therefore, the photonic and plasmonic properties of 2D materials are highly interesting for novel electronic applications. The surface plasmon (SP) mode of graphene is well recognized in the infrared regions of light spectrum. Furthermore, the fabricated instruments based on excitonic low-dimensional TMDCs/plasmonic structured systems have been the host of coupling effect phenomenon [31]. Thus, the 2D/plasmonic hybrid structured devices have introduced tremendous plasmonic/photonic properties. The plasmonic properties are not only restricted to the metallic structures. Graphene plasmons are quite different from the plasmons in metal nanostructures. The next section will specifically focuse on photonic and plasmonic characteristics of 2D nanostructures.

5.3 PLASMONIC PHENOMENA IN GRAPHENE

Metallic nanostructures including silver and gold are the most predominant group of materials for application in plasmonic instruments. However, the metallic based plasmonic devices have faced several technical constraints. The first challenge returns to the wavelength dependency of plasmonic properties of nanostructured metallic materials. It is difficult to find the optimum of the wavelength for the best plasmonic performance. It is due to non-uniform particle size distribution and geometrical features of plasmonic nanoparticles [32]. Furthermore, the metal plasmonic-based devices suffer from the large Ohmic losses during operation [33]. The main reason for Ohmic losses returns to the limitation of carrier mobility, surface roughness of metal plasmonic structures, grain surface, geometrical features of metallic nanostructures and the presence of impurities [33].

Graphene is considered as a promising 2D material for application in terahertz to mid-infrared plasmonic photonic instruments because of its high carrier mobility and conductivity [34]. Plasmonic science is interested in study of the excitation, propagation, and utilization of the collective oscillation of electron or other carriers, which have several types of polaritons. In the case of graphene, surface plasmon polaritons (SPPs) investigate the plasmonic phenomenon at the insulator/graphene interfaces. The privilege of plasmonic graphene over conventional plasmonic metal nanostructures are as follows:

i. **Tunability:** The plasmon mass in graphene increases proportionally with the level of Fermi energy [35]. It is originated from the nature of charge carriers in graphene. The level of doping of graphene determines the optical response of graphene-based devices. Thus, by controlling the doping level, we can modulate the generated plasmon mass in graphene.

ii. **Strong field confinement:** The generated plasmons in graphene propagate at the speed comparable to the Fermi velocity. The plasmon's velocity is much smaller than the light speed, and consequently, the wavelength of the graphene's plasmon is typically shorter than the light wavelength by 1 to 3 orders of magnitude [36].

iii. **Long lifetime and low losses:** The high conductivity nature of graphene has resulted in a long optical relaxation time. Compared with gold (10^{-14}

Plasmonic Devices Based on 2D semiconductors 149

s), the optical relaxation time of graphene is almost 10 times longer (10^{-13} s), thus the plasmon dissipation is much less and the plasmons lifetime is longer [36].

iv. **Perfect structure:** The structure of graphene is ideally perfect, which means that the strong chemical bonds between carbons make the graphene a perfect platform for generation of plasmons with different wavelengths. However, the structural imperfection can put a restriction on the perfect plasmonic performance of graphene.

5.3.1 Structural Design of 2D Graphene for Plasmonic Tuning

The experimental excitation of plasmons in graphene faces difficulties originated from the light subwavelength confinement. Due to the lack of sufficient momentum, the photons of lights always fail to excite plasmons directly [37]. The wavelength of free-space photons is always too long to be able to have higher momentum than plasmons. One of the main strategies to excite localized plasmon modes is to confine the plasmon to the metal surface. To this aim the graphene is patterned to create discontinuities in the electric field permittivity. The graphene ribbons are one of the desired designs for photonic and plasmonic applications of graphene. In this structure, plasmons with specific half-wavelengths can couple with the light and generate the effective charge dipoles to create the required force for collective oscillation of electrons or other carriers [37].

Recent research has shown that it is possible to control plasmon resonance wavelength to a large degree by controlling the excitation wavelength [38]. To this aim the plasmon resonance of graphene nanoribbons on the substrate of SiC was excited with different light wavelengths. Scattering-type SNOM was used to record the backscattered radiation from the tip of the illuminated atomic force microscope tip. The backscattered radiation was monitored simultaneously to record the infrared near-field images of the surface with nanoscale details. Figure 5.1a shows the results of a near-field image, which depicts the dependence of fringe space on the dielectric constant of the SiC substrate [38]. The results confirmed that the fringe spacing is increased when the excitation wavelength of light increased. The general understanding will reinforce this idea that the carrier density in narrow graphene ribbons is greater than that graphene ribbons with longer width.

Furthermore, it was observed that the excitation wavelength could also be controlled by creating structural defects or by the design of graphene with specific structural features. The plasmonic characteristics of a graphene ribbons with the structural defects were experimentally investigated [39]. The observations showed that the produced IR pattern by the line defects was different with the patterns of the long-gated defect-free graphene. The IR image showed the characteristic fringes on both sides of the boundaries of created defect in graphene structure. It vividly confirms that the plasmon resonance pattern of graphene can be altered when the structural defects are introduced into the graphene structure (Figure 5.1b) [39].

The alteration of structural design of graphene can tangibly affect the plasmonic characteristics of 2D graphene. The graphene ribbons are the most prevalent design of 2D graphene materials employed for photonic applications of the

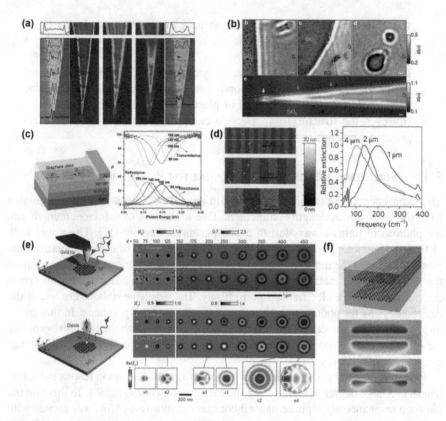

FIGURE 5.1 (a) Near-field resonance image of localized mode of graphene ribbon produced with different imaging wavelengths. (b) Photograph of graphene under IR spectrum region. The defects and boundaries are recognized by patterns. (c) Schematic of device with graphene discs and corresponding graph of transition and reflection spectra versus the diameter of the disk. (d) AFM image of graphene ribbons with different widths and the corresponding plasmon excitation vs the frequency of light. (e) Measurement and simulation of the localized plasmon resonance of graphene disks on the SiO$_2$ substrate with different diameters of graphene disks. (f) Schematic of ribbon pairs of graphene and corresponding plasmonic resonance modes. (a) Reproduced with permission from Chen, J., Badioli, M., Alonso-Gonzalez, P., Thongrattanasiri, S., Huth, F., Osmond, J., ... Koppens, F. H. (2012). Optical nano-imaging of gate-tunable graphene plasmons. *Nature*, 487(7405), 77–81. (b) Reproduced with permission from Fei, Z., Rodin, A. S., Andreev, G. O., Bao, W., McLeod, A. S., Wagner, M., ... Basov, D. N. (2012). Gate-tuning of graphene plasmons revealed by infrared nano-imaging. *Nature*, 487(7405), 82–85. (c) Reproduced with permission from Fang, Z., Wang, Y., Schlather, A. E., Liu, Z., Ajayan, P. M., de Abajo, F. J., ... Halas, N. J. (2014). Active tunable absorption enhancement with graphene nanodisk arrays. *Nano Letters*, 14(1), 299–304. (d) Reproduced with permission from Ju, L., Geng, B., Horng, J., Girit, C., Martin, M., Hao, Z., ... Wang, F. (2011). Graphene plasmonics for tunable terahertz metamaterials. *Nature Nanotechnology*, 6(10), 630–634. (e) Reproduced with permission from Nikitin, A. Y., Alonsogonzález, P., Vélez, S., Mastel, S., Centeno, A., Pesquera, A., ... Hillenbrand, R. (2016). Real-space mapping of tailored sheet and edge plasmons in graphene nanoresonators. *Nature Photonics*, 10(4), 239–243. (f) Reproduced with permission from Christensen, J., Manjavacas, A., Thongrattanasiri, S., Koppens, F. H., & de Abajo, F. J. (2011). Graphene plasmon wave guiding and hybridization in individual and paired nanoribbons. *ACS Nano*, 6(1), 431–440.

Plasmonic Devices Based on 2D semiconductors

graphene-based devices. The alteration of graphene ribbon's width is one of the most important strategies to tune the plasmonic and photonic properties of graphene [40]. Graphene nanoribbons and disk arrays are two main structures, which brought novel photonic properties to graphene. To promote the optical absorption of graphene to the visible and infrared regions of the light, beam lithography technique was employed to design an array of graphene nanodisk on the In-In_2O_3/ BaF_2 substrate [40]. The ion gel was employed to alter and tune the graphene resonance over a wide range of terahertz frequency by using the electrostatic doping. In a practical case, the ion gel was employed in the plasmonic device as the gate electrolyte. The FTIR measurements (transmittance, reflectance) confirmed the prominent absorbance caused by the plasmonic excitation of graphene nanodisk array [40]. It was observed experimentally that the changes of Fermi energy were accompanied by the increase of absorbance and reflectance peaks. The absorption peak of plasmons can be moved to higher level of energies. In this structure the peaks in the spectra of nanodisk arrays can match the calculated plasmon energy modes of the single nanodisks. By using this strategy, the absorption efficiency of graphene nanodisks was considerably enhanced in the infrared regions of the light spectrum. Different designs of nanodisks have increased the light absorption efficiency from 3 to 30 % [40]. Thus it was found as a promising approach to tune the light absorbance by graphene.

The effect of nanoribbon's width on the plasmon resonance and light-plasmon coupling of graphene-based plasmonic devices was investigated by Fourier transform infrared spectroscopy [41]. The designed unit contains several versions of graphene ribbon arrays with different widths. The analyzing of transmission spectra of graphene ribbons revealed that the decrease of nanoribbon's width was accompanied by the red shift of transmission spectra (Figure 5.1d). As a general rule, there is a relation among the geometrical features of graphene ribbon, the level of doping and the plasmon frequency. The relation between the plasmon frequency and the width of graphene ribbon can be expressed by $W^{-1/2}$ where the W is the width of graphene nanoribbon [41]. The plasmon frequency has another relation with the carrier density of graphene nanoribbon, which was expressed by $n^{1/4}$ [41].

Another research has investigated the effect of size of graphene nanodisks on the plasmon mode of graphene-based disk nano-resonators, which were fabricated over the SiO_2 substrate [42]. The unit consisted of two sets of the graphene disks with variable diameter from 50 nm to 450 nm. Two different illumination sources were employed to excite the plasmonic resonance of graphene nanostructure. The AFM images depicted the near-field features of the graphene disks in the form of the bright and dark circular objects. The plasmonic resonance of small nanodisks of graphene was recognized by the dark central spots with the bright fringes. By the increase of the size of the disk, the dimensions of bright regions were increased, as for the largest disk the stripped schemes of bright/dark rings were patterned. The increase of wavelength of illumination source slightly but not fundamentally altered the size of the dark and bright regions, confirming the size-induced effect of graphene nanodisks on the plasmonic characteristics of graphene resonator (Figure 5.1e). The numerical electromagnetic analysis and calculations fairly matched with experimental results [42].

Strong interlamination coupling is another photonic phenomenon affecting the plasmonic performance of graphene-based photonic instruments. The practical measurements have confirmed that the plasmonic modes of the graphene nanoribbon pairs are distinguished from the single graphene sheet [43]. The results of imaging the electric near-field (Figure 5.1f) reaffirmed that the plasmonic interactions in coplanar ribbon pairs resulted in the increase of hybridized states. The binding and anti-binding combination modes of monopoles were observed during the study of electric near-field characteristics of graphene ribbon pairs [43]. It was also found that the doping of the graphene nanoribbon resulted in the extending of life time and further confinement of the plasmons compared with those in the metallic plasmonic materials [43]. The paired plasmonic characteristics of graphene nanoribbons and the interaction between plasmons of paired graphene nanoribbons were found as distinguished properties for efficient infrared sensing and signal processing applications.

5.3.2 PLASMONIC TUNING OF HYBRID 2D GRAPHENE-BASED DEVICES

The plasmonic tuning of photonic 2D graphene in optical frequency remains a serious challenge, since the doping methods can only alter the modulation rate from terahertz to MIR region. The investigation on plasmonic behavior of hybrid 2D/nanostructured material revealed that the plasmonic resonance of metallic nanostructures and 2D materials interact with each other, just in case that 2D materials and plasmonic metal nanoparticles are brought together and form a heterostructured junction. One of the main technical strategies is the employment of specific metallic nanostructured patterns for joining to the 2D materials. This typical design facilitates excellent tuning abilities over a wide spectral range [44].

In a practical case, active plasmonic tuning for both resonance frequency and quality factors in a near-infrared region were achieved by hybridization of 2D graphene nanosheets with gold nanorods (Figure 5.2a) [45]. The quality factor of plasmonic tuning effect of 28 % was gained by the appropriate electrical gating of hybrid unit [45]. The in-plane electric field near to the end of the golden rod was enhanced tangibly. Thus, the end of the golden rod was recognized as the plasmonic hot points in hybrid 2D plasmonic structure. It is estimated that 14 % of integrated field intensity of graphene was related to these two active plasmonic hot points, while the number of induced electrons in this area is only tens of active electrons [45]. The interaction between graphene and metallic plasmons can efficiently help to control the plasmon resonances at various optical frequencies.

Technologically, the electrons are mostly concentrated at the sharp corners, and consequently, these regions have high intensity of electrons and electric field. The bowtie is therefore a preferable structure to support strong electrical hot spots [46]. Considering the effect of structural design, a field effect transistor (FET) device was designed with highly tunable carrier concentration. The interaction of incident light with graphene was intensified on hybrid 2D structure [46]. It was found that the heterostructured 2D graphene plasmonic unit showed the best performance at mid-infrared wavelengths [46]. The graphical scheme of voltage-controlled optical transistor based on the plasmonic hybrid 2D graphene is shown in Figure 5.2b. The effect

Plasmonic Devices Based on 2D semiconductors

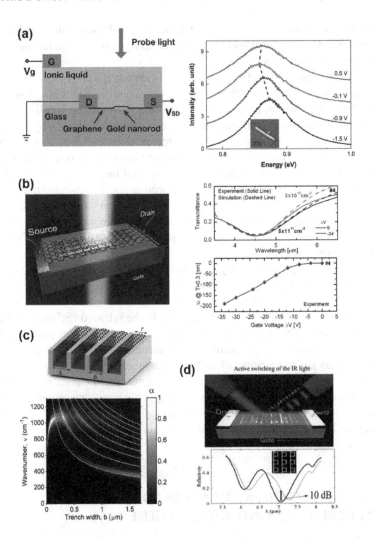

FIGURE 5.2 (a) Schematic of graphene-gold nanorod hybrid structures accompanied by corresponding Rayleigh scattering spectra of device at different gate voltages. (b) Schematic of illustration of electrically modulated plasmonic antenna with hybrid structure. The effect of gate voltage on plasmonic resonance width. (c) Schematic illustration of graphene nanoribbons on the metal grating structure, and the absorption map of hybrid graphene-based device versus the trench width. (d) Schematic of hybrid gated graphene Fano-resonant nanostructure accompanied by the reflection spectra of device under electrical switching. (a) Reproduced with permission from Kim, J., Son, H., Cho, D. J., Geng, B., Regan, W., Shi, S., ... Wang, F. (2012). Electrical control of optical plasmon resonance with graphene. *Nano Letters*, *12*(11), 5598–5602. (b) Reproduced with permission from Emani, N. K., Chung, T. F., Ni, X., Kildishev, A. V., Chen, Y. P., & Boltasseva, A. (2012). Electrically tunable damping of plasmonic resonances with graphene. *Nano Letters*, *12*(10), 5202–5206. (c) Reproduced with permission from Zhao, B., & Zhang, Z. M. (2015). Strong plasmonic coupling between graphene ribbon array and metal gratings. *ACS Photonics*, *2*(11), 1611–1618. (d) Reproduced with permission from Dabidian, N., Kholmanov, I., Khanikaev, A. B., Tatar, K., Trendafilov, S., Mousavi, S. H., ... Shvets, G. (2015). Electrical switching of infrared light using graphene integration with plasmonic Fano resonant metasurfaces. *ACS Photonics*, *2*(2), 216–227.

of gate voltage on plasmonic resonance of device is also demonstrated in Figure 5.2b. The role of graphene in 2D hybrid unit is the damping of plasmonic resonance in the mid-infrared spectral range.

The efficient damping and the tuning of plasmonic resonance can be achieved by alteration of design of hybrid plasmonic instrument. In this view, a new structure was fabricated which facilitated the coupling between the localized plasmon resonance in metals and plasmonic resonance in graphene ribbons [47]. Thus, the structural absorption increased to a high degree (Figure 5.2c). The graphics in Figure 5.2c depict the variation of the absorption versus the trench width at heterostructured plasmonic graphene device at the constant incident angles. The appearance of plasmonic-assisted bright band at absorption spectra revealed the capability of graphene-based plasmonic unit for heterostructured optoelectronic applications [47]. The waveguide modulators based on graphene materials are one of the promising applications of graphene-based plasmonic utensils for ultrafast optical communications.

Active tuning of plasmons in graphene-based plasmonic units is the practical application of these types of instruments in the infrared light region. In fact, since the optical response of graphene is weak, most of the research has focused on the enhancement of optical absorption of graphene in the infrared regions for optoelectronic applications. The optical Fano-resonance unit composed of graphene hybrid structures was tuned through electric gating technique. The reflectivity spectrum of device versus the wavelength of light is presented in Figure 5.d [48]. The rapid injection of charge carriers into graphene facilitated the ultrafast response of the graphene-based instrument. Furthermore, the specific metasurface design of hybrid photonic device facilitated the large modulation depth in reflection. Therefore such design of the hybrid graphene-based plasmonic unit is quite promising approach for the light modulation.

5.4 PLASMONIC PHENOMENON IN 2D MATERIALS BEYOND THE GRAPHENE

5.4.1 PLASMON TUNING BY DOPING OF 2D MATERIALS BEYOND THE GRAPHENE

5.4.1.1 Doping of 2D MoS$_2$

Doping is a highly capable technique to excite or alter the plasmonic properties of 2D materials. In this method, ionic species are introduced into structure of 2D materials or are used as the doping elements. To alter the plasmonic properties of 2D MoS$_2$ nanoflakes, Li$^+$ ions were successfully introduced into MoS$_2$ 2D flakes with electrochemical method [49]. The plasmonic properties of Li$^+$ ion-doped MoS$_2$ nanoflakes were investigated via observation of optical properties and absorption spectra of ion-doped nanostructures in ultraviolet to visible (UV-Vis) regions of light spectrum. It was found that the Li$^+$ ion-doped MoS$_2$ nanoflakes showed plasmon resonance in the visible and near ultraviolent region [49]. The study on PL spectra of MoS$_2$ nanoflakes showed that the intercalation of Li$^+$ ions into 2D MoS$_2$ has altered the PL spectra of nanoflakes. Owing to the plasmon resonance of highly doped MoS$_2$

nanoflakes, the PL spectra of MoS$_2$ nanoflakes changed by altering the intercalating voltage. By decreasing the intercalation voltage from 0 to −10 V, the light emission of MoS$_2$ nanoflakes decreased and finally became negligible at −4.0 V (Figure 5.3a). The different emission characteristics of ion-doped MoS$_2$ nanoflakes was attributed to the phase conversion from 2H to 1T phase [49]. It was predicted that higher numbers of free electrons were generated in MoS$_2$ nanoflakes with mixed crystalline structure. It was found that under the low intercalation voltage the MoS$_2$ phase is 1T.

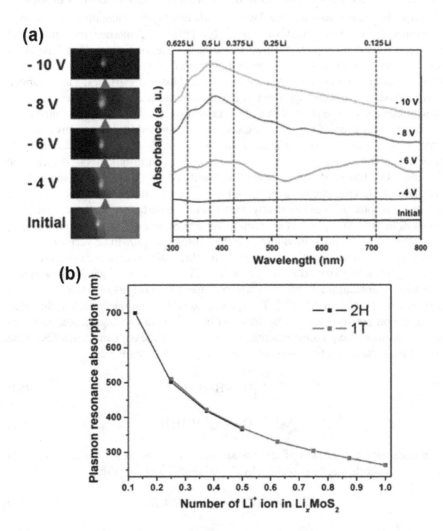

FIGURE 5.3   (a) Photoluminescence (PL) images of 2D monolayer MoS$_2$ film accompanied by the absorption spectrum of 2D MoS$_2$ nanoflakes. (b) Graph of plasmon resonance peak positions versus the number of Li$^+$ ions in 2D MoS$_2$ flakes. Reproduced with permission from Wang, Y., Ou, J. Z., Chrimes, A. F., Carey, B. J., Daeneke, T., Alsaif, M. M., ... Kalantar-Zadeh, K. (2015). Plasmon resonances of highly doped two-dimensional MoS$_2$. *Nano Letters*, *15*(2), 883–890.

On the contrary, a significant phase conversion from 2H phase to 1T phase occurred when the applied intercalation voltage decreased to − 6.0 V. This phase transformation enabled the efficient plasmonic modulation over a wide range of spectral range (Figure 5.3e). The plasmonic modulation can therefore be facilitated by the several other techniques including gas physisorption, chemical doping and surface functionalization methods.

5.4.1.2 Surface Functionalization of 2D Surface Oxide of Galinstan Alloy

A surface functionalization method was recently employed to manipulate the absorption properties of surface oxide of a liquid alloy [50]. In this method, the liquid metal galinstan (GaInSn) droplet was used as the base material for reactions. The ultrasonic cavitation on the surface of galinstan droplet led to the separation of surface oxide of galinstan (Ga_2O_3) from the liquid metal host nanoparticles. The separated surface oxides then were employed as template for nucleation and growth of nanostructured Se and Ag on the surface of Ga_2O_3 nanosheets [50]. By detail, sonochemical functionalization of Ga_2O_3 nanosheets was performed in the aqueous solutions containing Ag and Se ions. The ultrathin surface oxide film of galinstan is fairly unique and can be separated from the host alloy by ultrasonic-assisted technique (Figure 5.4a). The SEM studies of the functionalized Ga_2O_3 nanosheets showed distribution of alloying elements on the surface of Ga_2O_3 nanosheets. The silver and selenium nanoparticles are evidently observed on the surface of the nanosheets synthesized in aqueous solutions containing Ag and Se ions. Thus, the sonochemical reactions cause the nucleation and help the subsequent growth of various types of nanostructures. The acoustic cavitation in the ultrasonic process can drive chemical reactions including oxidation and reduction. The explosion of gas bubbles generates high temperature and pressure which causes the pyrolysis of water into H and OH radicals (Equation 5.1) [50]. The quenching rate is extremely high at the interfacial region between the gas bubbles and bulk solution causing the nucleation of nanostructures. The possible reaction for nucleation of the Ag nanostructures on the nanosheet surface can be expressed as shown [50] in Equation 5.2:

$$H_2O \rightarrow H^0 + OH \tag{5.1}$$

$$Ag^+ + H_2O \rightarrow Ag^0 + OH + H^+ \tag{5.2}$$

The nucleation and growth of amorphous and crystalline Se on the surface of 2D Ga_2O_3 nanosheets can be explained by Equations 5.3 and 5.4 [50]:

$$SeCl_4 + H_2O \rightarrow H_2SeO_3 + 4HCl \tag{5.3}$$

$$4H^0 + H_2SeO_3 \rightarrow Se + 3H_2O \tag{5.4}$$

Characterization of functionalized nanosheets by Raman spectroscopy confirmed the presence of Ag and Se nanostructures on the surface of Ga_2O_3 films. Figure 5.4b

Plasmonic Devices Based on 2D semiconductors 157

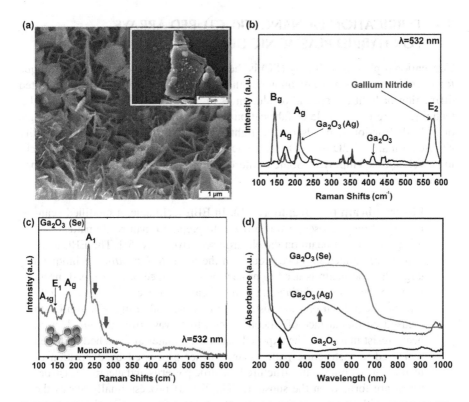

FIGURE 5.4 (a) Sonochemical functionalized and extracted Ga_2O_3 nanosheets. (b) The Raman spectra of Ga_2O_3 and Ga_2O_3 (Ag) nanosheets. (c) Raman spectra of Ga_2O_3(Se) nanosheets. (d) UV-Vis spectra of Ga_2O_3, Ga_2O_3 (Ag) and Ga_2O_3(Se) nanosheets. (a) Reproduced with permission from Karbalaei Akbari, M., Hai, Z., Wei, Z., Ramachandran, R. K., Detavernier, C., Patel, M., ... Zhuiykov, S. (2019). Sonochemical functionalization of the low dimensional surface oxide of galinstan for heterostructured optoelectronic applications. *Journal of Materials Chemistry C*, 7(19), 5584–5595.

depicts the silver Raman peaks at 172 cm^{-1} and 215 cm^{-1} that are the characteristics of the surface enhanced Raman scattering (SERS) of silver on Ga_2O_3 nanosheets [50]. It was observed that the SERS properties of Ag could intensify the Raman characteristic peaks of the Ga_2O_3 nanosheets. The Raman spectra of Se containing nanosheets showed the characteristic peaks of crystalline and amorphous Se on the surface of Ga_2O_3 nanosheets (Figure 5.4c). The UV-Vis absorbance spectra of functionalized Ag containing nanosheets shows an absorbance peak at the wavelengths of $\lambda = 300$ nm (Black arrow). This band is the absorbance peak of Ga_2O_3. The characteristic surface plasmon resonance of metallic Ag is monitored by detection of the broad peak around $\lambda = 450$ nm in the Ga_2O_3 (Ag) nanosheets (arrow on Ag line in Figure 5.4d). The absorption spectrum of the Ga_2O_3 (Se) nanosheets demonstrates a broad peak all over the visible light (Figure 5.4d).

5.5 FABRICATION OF NANOSTRUCTURED ARRAYS FOR HYBRID PLASMONIC DEVICES

Conventional photolithography (PL) is the most available and common technique for microfabrication of semiconductor instruments. However, due to the limited diffraction of light the special resolution of PL techniques are restricted. The PL techniques are not capable of fabrication of ultra-precise devices since the angle of light diffraction depends on the light wavelength. To fabricate high-quality nanoscale featured 2D and 3D electronic and optoelectronic devices novel techniques should be employed. Some of these techniques are introduced below. They are as follows:

i. **Electron-beam lithography (EBL).** In EBL technique, a modified scanning electron microscope was utilized to prepare a nanoscale patterns by using an electron beam on the electron sensitive layer [51]. The EBL techniques are divided into two groups. In the *projection printing* technique a large electron beam is used to project on the surface of a designed mask, while in the *direct printing* an electron beam spot is employed to directly scheme the pattern on the substrate. In this method, a fine electron beam focused on the surface of silicon wafer which was covered with electron beam resist material. The exposed part of the material is etched and then remained nanostructure patterns are further processed by deposition of metals. In so doing, nanostructured metallic patterns and electrodes (structures) are formed on the substrate. The lift-off process finally leaves the surface with patterns on the surface of substrate. Considering the angstrom scale dimension of an electron beam, the EBL is capable of nanofabrication with ~10 nm precision. Thus, the EBL has high spatial resolution and capable of fabrication of instruments with complex geometrical features. The main challenge for EBL technique returns to patterning speed, since the electron beam requires focusing on one point and thus it takes a long time to pattern the surface. This is the main limitation of EBL for the large-scale production and wafer-scale fabrication of electronic devices and it is also recognized as high-cost operating techniques.

ii. **Focused ion beam (FIB) lithography.** In FIB technique the ions are used for surface patterning. In this method, the focused ion beams draw the desired pattern on the surface of substrate. To prepare a stable and fine spot of radiative beams, the heavier ions similar Be^+, Ga^+ and He^+ are used for direct writing [52]. In FIB technique the penetration depth of ions is less than electrons, thus the larger energy and power is employed in the range of 100 kV to 200 kV to prepare more stronger and focused beams for patterning of surface. The FIB resolution is higher than the resolution of EBL technique. Despite the capability of FIB for fabrication of nanostructures with higher resolution, the large-scale fabrication of nano devices is still a challenging target for the commercial usage of FIB technique.

iii. **Dip-pen lithography (DPN).** In DPN technique, the tip of an AFM is used to directly design the patterns on the surface of substrate by delivering the chemical reagents to different kinds of substrates including metal, insulator, and conductors. Moreover, the biological and soft organic substrates can be used in this method [53]. The work environment in the DPN technique is inert atmosphere, thus organic soft substrates can be patterned by this method without destroying their natural structures. Several different types of the inks can also be employed in the DPN process, some of the famous examples are organic molecules, metal ions, and biological polymer. The technique is capable of development of pattern on the substrates with ~15 nm resolution. The advanced DPN techniques are more precise to address the complexity of complicated nanostructures, like cantilever arrays. However, the main challenge of the DPN is its single pen configuration, which requires an inherent serial fabrication. While the advent of multiple-pen DPN methods can increase the efficiency of this nanofabrication technique, still the expensive system, low general throughput and limited accessibility to the ink materials are the main challenges for further development of this nanofabrication method.

iv. **Laser interface lithography (LIL).** LIL as a similar method to the maskless photolithography is used for large-scale fabrication of nanostructures [54]. Contrary to a conventional photolithography, which uses monochromatic light for patterning over a masked photoresist film, LIL employs multiple laser beams for patterning on the photoresist substrate. The interference of laser beams and then intensified field intensity are two main factors which make it possible to pattern the surface of the photoresist film by the superposition of multiple laser beams [54]. The lack of photomask during LIL process enabled the creation of the different patterns with various sizes based on interference of laser beams. The resolution of technique is not limited by the diffraction of light, but it is restricted by the light wavelength [54]. The limitation of LIL technique returns to the wavelength of the light, where the minimum number of periodic pattern of nanostructures is restricted by the half wavelength of the light [54]. Therefore, the most precise and efficient LIL technique needs to use expensive deep ultraviolet (UV) lasers to pattern nanostructures. Furthermore, for the patterning of a large exposure area, researchers need to employ long coherence laser spectra with higher resolution. Thus, the usage of shorter wavelength would be able to help in production of more precise and repeatable pattern. The light specifications and properties are considered as the main challenge for development of large-scale nano-patterning by LIL system.

v. **Nanosphere lithography (NSL).** The NSL method is based on the deposition of spherical collides on the surface of the substrate. The spherical collides will later be dried and then form hexagonal closed pack (hcp) monolayer structures on the surface [55]. The monolayer is later transferred to the surface by deposition techniques similar spin coating and dip coating to be used as the mask for fabrication of different patterned structures. The materials are deposited at the empty spaces between hcp patterned nanostructures. Then the sonication is employed to remove the mask. Finally the surface of the patterned film was cleaned by ultra-sonication leaving the ordered nanostructures over the substrate for further processing. Since the NSL is a combination of *top-down* and *bottom-up* nanofabrication approaches, it is more flexible for inexpensive and high-throughput nano-patterning of the surface [55]. Compared with the other nanofabrication techniques, the NSL is cheaper and capable of high-throughput fabrication of nanostructures. However, the main restriction of technique returns to the limited capability for generation of versatile nanostructures with precise geometries and features.

vi. **Nanoimprint lithography (NIL).** The NSL method is based on the mechanical deformation of the imprint fluid and following hardening process [56]. In this technique a patterned stamp mechanically press the surface of the liquid fluid which is deposited over the substrate. After the hardening process and removing the stamp, the patterned nanostructures will appear on the surface. Contrary to the lithography process, the limitation of NIL method returns to the patterns on the mold and not to the diffraction patterns of the illumination source in the photolithography based techniques. The curing source in NIL can be supported by thermal heating or ultraviolet lights [56]. In thermal NIL, a thermoplastic polymer is employed as the imprint fluid material. The UV-Nil technique uses a UV-sensitive polymer that is employed as imprint fluid to pattern the substrate. Generally, the technique is low-cost with the acceptable resolution and high-throughput for fabrication of nanostructures with sub-5.0 nm pattern fabrication capabilities. Considering the fabrication characteristics of NIL system, the method is considered as the promising approach for semiconductor fabrication industry. NIL was already employed for fabrication of optical storage device, hard disk media, light emitting diodes, microfluidic devices, and biosensors.

vii. **Anodic aluminum oxide template (AAO).** The AAO technique is based on the nano-patterning of substrate with metal oxide nanopores. In AAO, the template is made based on metal anodization in anodic solution [57]. The final results would be in the form of periodically developed nonporous with pore size in the range of 7 nm to 300 nm [57]. Thus, the method is recognized as the capable technique for fabrication of nanorings, nanopillars, nanonodes, and nanotubes. While aluminum was the first employed metal in AAO, zirconium, and titanium are the other types of the metals employed for the anodic etching of nanostructured patterns on the surface [58]. However, there are some restrictions for AAO. The main challenge returns to the difficulty of generation and development of long-range ordered nanopores and nanoarrays on the large-area of substrate. Another restriction returns to the natural geometry of holes of metal oxide pores. For example, the template nanopores have rounded shape with hexagonal developed pattern.

All of these nanofabrication techniques are employed to pattern the surface of substrates for further fabrication of nanopatterned plasmonic instruments. To choose an ideal fabrication technique for nanopatterning of plasmonic devices several parameters should be taken into account, including fabrication cost, the throughput of the technique, the substrate features, the dimension and precision of nano-patterns, and finally, the area and scale of substrate [30].

5.6 APPLICATION OF PLASMONIC DEVICES BASED ON ULTRATHIN 2D MATERIALS

5.6.1 PLASMONIC TWO-DIMENSIONAL AU-WO$_3$-TIO$_2$ HETEROJUNCTION

Electrically responsive plasmonic instrument benefiting from the privilege of surface plasmon excited hot carries have supported fascinating applications in the visible light-assisted technologies. The properties of plasmonic devices can be appropriately tuned by controlling the charge transfer. It can be attained by intentional architecture of the metal-semiconductor (MS) interfaces [59]. Semiconductor materials with high-density states in their conduction bands, such as TiO_2, are considered as one of the main options for development of MS junction. TiO_2 with its fast electron injection capability can be joined into plasmonic Au nanostructures. In most of the cases, Au nanoparticles are employed as the metal component in MS junction. However, there are several main difficulties. The visible light absorption of Au nanoparticles is restricted to limited wavelengths, which is originated from the localized surface plasmon resonance. To address this limitation, it is highly desired to design the plasmonic MS planar heterojunction films with broadband visible light sensing. However, the excitation of SP cannot be attained on the ultra-smooth surface of Au films. To tackle this problem the idea of deposition of 2D ultrathin TiO_2 semiconductor films over the granular Au substrate was introduced [59]. Despite several undeniable advantages, one of the main drawbacks of the visible light responsivity of the MS-based plasmonic devices is the elevated undesirable dark current of device. The dark current is originated from the thermionic emission. To control the heterointerfaces, the proper design of metal-insulator-semiconductor (MIS) is required. The main mechanism is the control of the surface states of the semiconductor component. To alter the surface state, approximately ~ 0.7 nm thick 2D WO_3 film was fabricated as an intermediate component between Au substrate and 2D TiO_2 semiconductor to develop MIS structure. The mechanisms of plasmonic assisted charge generation at heterostructured Au-TiO_2 and Au-WO_3-TiO_2 films are depicted in Figure 5.5b, c, and d, respectively [59]. The plasmonic hot carries can be transferred directly into the conduction band (CB) of TiO_2, or individually create an electric field which is called a hot spot. The hotspot can generate an electron in CB of TiO_2 and leaves a hole in metal component of plasmonic heterojunction. By using an intermediate insulator ultrathin film of WO_3, the heterointerfaces between Au and TiO_2 and the Schottky barrier were modified. As the evident consequence, the dark current of plasmonic device was decreased tangibly and therefore the photoresponsivity of plasmonic devices increased (Figure 5.5e, f, and g) [59]. Improvement over 13.4 % in EQE for the 3.5 nm thick TiO_2 samples was observed when the 0.7 nm thick WO_3 film was employed as an intermediate layer between Au and 2D TiO_2 films [59]. The enhanced photoresponsivity and EQE for the Au–WO_3–TiO_2 heterostructured devices can be attributed to the manipulated Schottky junction of samples.

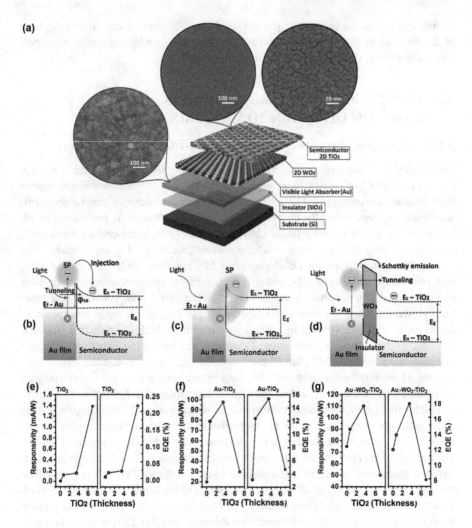

FIGURE 5.5 (a) Graphical scheme of heterostructured plasmonic Au-WO$_3$-TiO$_2$ device, accompanied by the SEM images of granular Au film, and ultrathin TiO$_2$ film as the semiconductor component of device. (b) Schematic diagram shows the generation of hot carriers, followed by the injection or tunneling of electrons to the conduction band of the adjacent semiconductor. (c) The decay of SP and generation of electrons in the conduction band of the semiconductor and corresponding holes in the metal. (d) The charge control at MIS structure when the insulator WO$_3$ intermediates and controls the charge carriers by altering the Schottky barrier at heterointerfaces. The photoresponsivity of (e) TiO$_2$ (f) Au-TiO$_2$ and (g) Au-WO$_3$-TiO$_2$ devices. Reproduced with permission from Karbalaei Akbari, M., Hai, Z., Wei, Z., Detavernier, C., Solano, E., Verpoort, F., & Zhuiykov, S. (2018). ALD-developed plasmonic two-dimensional Au-WO$_3$-TiO$_2$ heterojunction architectonics for design of photovoltaic devices. *ACS Applied Materials and Interfaces*, 10(12), 10304–10314.

5.6.2 Plasmonic-Assisted 2D Photodetector Devices

Ultrathin thickness, distinctive charge carrier mechanisms, and transparency are some of the fundamental characteristics of 2D materials, which made them appealing candidates for various applications. Monolayer of graphene and 2D MoS_2 are two well-recognized 2D materials capitalizing on the privilege of plasmonic-assisted charge generation. The gapless graphene is capable of absorption of light spectra over a wide range of wavelengths from UV and Vis light to the infrared and terahertz regions [60]. One of the main characteristics of 2D materials is their ultrafast transfer of charge carriers and ultrasensitive detection of received photons. Ultrathin TMDCs are also a big group of 2D units with unique optoelectronic characteristics. The direct bandgap of 2D TMDC semiconductors facilitated the efficient light harvesting [60]. Considering the band gap tunability of TMDCs, there are particular advantages for 2D TMDCs films for strong light absorption in the visible range of the light. It is highly interesting approach to recombine the plasmonic nanomaterials with 2D films and crystals to develop heterogeneous plasmonic instruments. These 2D hybrid devices use the benefits of plasmonic phenomenon at ultrathin films accompanied by the privileges of strong light-matter interaction and sensitive photon detection in 2D films [60].

Specifically, photodetector was recently reported, which work based on the gate tunable intrinsic photoresponse mechanism of graphene ribbons (Figure 5.6a) [61]. The unit represents a field effect transistor in which the graphene ribbon acts as the channel of the FET structure. The measurement of temperature-dependent current in a gate-biased device showed that the photocurrent of FET device can be modulated by electrostatic doping of graphene. The test was carried out to probe the photoresponse of the unit induced by generation of hot carriers and the heat dissipation of photons [61]. Asymmetric metallization was found as efficient approach to break the mirror symmetry of built-in potential in graphene [61]. Based on the measured optical properties of the developed instrument, the maximum photoresponsivity of 6.1 mAW^{-1} at the wavelength of 1.55 μm was measured for the FET device [61].

The direct band gap of TMDCs offers visible light detection due to the quantum mechanical confinement. There are several devices which use the privileges of 2D TMDCs for photo sensing. To this aim, a variety of 2D semiconductors are deposited or grown over different type of substrates. For example, Figure 5.6b demonstrates one of the photodetectors fabricated based on the monolayer MoS_2 films [62]. The MoS_2 based photodetector showed the photoresponsivity of 880 AW^{-1} under the illumination of 561 nm light [62]. The dark current measurements showed that the ratio of photocurrent intensity at *on-off* state was more than 4 when the bias voltage of 8.0 V was imposed [62]. The outstanding light sensing in the wavelength range of 400 to 800 nm was attributed to the improved electron mobility in MoS_2 monolayer films.

The intrinsic light absorption and photoresponsivity are determined by the electronic band structure of 2D materials. Considering the direct bandgap of 2D materials, the photoresponsivity of 2D material-based devices are highly limited to a specific spectrum of light. The developed metal/2D semiconductor heterostructures can tangibly widen the optical absorption limits of 2D materials, owing to the plasmonic-assisted phenomena. The plasmonic photodetectors work based on two

164 Ultrathin 2D Semiconductors

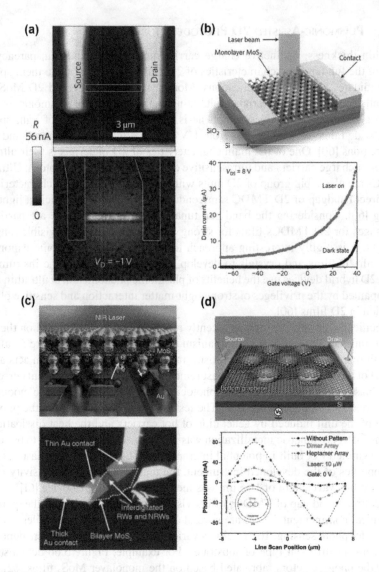

FIGURE 5.6 (a) The image of graphene-based photodetector modulated by gate bias voltage. (b) Schematic of monolayer MoS$_2$ photodetector and the gate response of the device. (c) Schematic of plasmonic bilayer coupled MoS$_2$ photodetector with integrated nano wires. (d) Graphical scheme of sandwiched plasmonic nanoantenna between monolayer graphene films accompanied by photocurrent characteristics of devices. (a) Reproduced with permission from Freitag, M., Low, T., Xia, F., & Avouris, P. (2012). Photoconductivity of biased graphene. *Nature Photonics*, *7*(1), 53–59. (b) Reproduced with permission from Lopezsanchez, O., Lembke, D., Kayci, M., Radenovic, A., & Kis, A. (2013). Ultrasensitive photodetectors based on monolayer MoS$_2$. *Nature Nanotechnology*, *8*(7), 497. (c) Reproduced with permission from Wang, W., Klots, A., Prasai, D., Yang, Y., Bolotin, K. I., & Valentine, J. (2015). Hot electron-based near-infrared photodetection using bilayer MoS$_2$. *Nano Letters*, *15*(11), 7440–7444. (d) Reproduced with permission from Fang, Z., Liu, Z., Wang, Y., Ajayan, P. M., Nordlander, P., & Halas, N. J. (2012). Graphene-antenna sandwich photodetector. *Nano Letters*, *12*(7), 3808–3813.

Plasmonic Devices Based on 2D semiconductors

distinguished mechanisms. The plasmonic resonance of metal particles can facilitate the generation of hot electrons [63]. The following jumping or tunneling of generated charge carriers into 2D semiconductor films can tangibly affect the generation of photocurrents. The other mechanism is the enhancement of photon absorption due to the near-field enhancement effects induced by the plasmonic phenomena (Figure 5.6c) [63]. In this view, the MoS_2 plasmonic photodetector was developed on resonant and non-resonant wires, where the dipolar resonant absorption occurs at 1250 nm [63]. Here the nanostructured metallic wires are the plasmonic component of photodetector. To assure that the photocurrent is owing to the generated hot electrons by resonant wire coupling, the laser energy of illuminating source was designed to avoid the indirect photon absorption of bilayer MoS_2 film. Here, the Schottky barrier still plays the main role in the control of hot electron injection into 2D semiconductor films. The plasmonic device based on bilayer MoS_2 film has gained the high photoresponsivity of 5.2 AW^{-1} at the illumination wavelength of 1070 nm [63].

Plasmonic Fano resonance devices have been developed for 2D photodetection [64]. The instrument is based on the graphene photodetectors where the plasmonic heptamer antennas were sandwiched between two single layers of graphene films. The fabricated heptamer antennas were in fact the nanodisks with varying diameter from 80 to 180 nm corresponding to the Fano-resonant frequency of 650 to 950 nm [64]. The device with nanodisk antenna showed 800-fold improvement of photoresponsivity compared with the photoresponsivity of the same photodetector without nanodisk antenna. The internal quantum efficiency of device was also increased more than 20 % in the visible and near-infrared spectral range when nanodisk antenna was employed in photodetectors [64] (Figure 5.6d). Other instruments were also used the advantages of hybrid plasmonic/2D structures for broadband detection of the light. In these devices various type of plasmonic nanostructures with different geometries (nanodisks, dimers, and gratings) were fabricated by e-beam lithography techniques. The polarization dependent photoresponsivity can be selectively enhanced by usage of the plasmonic nanostructure with various geometries. In another application the plasmon resonance enhanced multicolor photodetector was developed, where the back gated graphene transistor on Si/SiO_2 substrate was decorated randomly by plasmonic nanoparticles [65]. Photoresponsivity of the unit was increased more than 400 % and photocurrent of the device reached to 2.2 mA W^{-1} [65]. The plasmonic resonant absorption facilitated the enhancement of quantum efficiency of plasmonic graphene-based photodetectors by up to 1500 % [65]. By altering the size and geometry of nanoparticles, the specific detection of multicolor lights was facilitated.

CONCLUSIONS

The present chapter has summarized some of the distinguished characteristics of plasmonic-based 2D nanomaterials by focusing on 2D graphene, TMDCs, and ultrathin metal oxide films. The plasmonic devices are divided into different categories based on the materials characteristics, coupling mechanisms, and plasmonic tuning methods.

First, the principles of plasmonic resonance were introduced and the main concepts were analyzed. Based on the different mechanisms of plasmonic excitement we have divided the plasmonic 2D materials into two individual groups including graphene-based plasmonic instruments and plasmonic 2D materials beyond the graphene. In the case of graphene, two main mechanisms for tuning of plasmonic properties were individually introduced. The control of geometrical features of 2D graphene was one of the main employed mechanisms for tuning the plasmonic properties. The relations between the width of graphene nanoribbons and the plasmon resonance were discussed and the practical cases were introduced. Another main mechanism for tuning the plasmonic characteristics of graphene is development of the hybrid 2D graphene-based units. In this view, several graphene hybrid nanostructures were fabricated where the interface between the 2D graphene and metallic antenna or developed patterns was a source for generation of plasmonic resonance under illumination of visible and infrared light spectra.

The second group of 2D plasmonic devices was fabricated based on the 2D materials beyond the graphene. In this group of 2D optical instruments, the plasmonic properties were tuned by doping 2D films with ionic agents or by the surface functionalization of ultrathin films with plasmonic particles. The fabrication of nanostructured hybrid plasmonic photonic instruments is another important strategy to facilitate the plasmonic charge generation from the metallic component to the 2D semiconductor films. Finally, several practical examples of plasmonic devices based on 2D materials were introduced and their working mechanisms were subsequently explained. It was observed that the performance of instruments based on 2D plasmonic nanostructures was tangibly improved compared with the properties of similar 2D materials in non-plasmonic modes. Generally, the 2D-based plasmonic devices are in their early stage and further explorations are required in both fabrication techniques and design of novel 2D plasmonic materials.

REFERENCES

1. Velick, M., & Toth, P. S. (2017). From two-dimensional materials to their heterostructures: An electrochemist's perspective. *Applied Materials Today, 8*, 68–103.
2. Frantzeskakis, E., Rödel, T. C., Fortuna, F., & Santander-Syro, A. F. (2017). 2D surprises at the surface of 3D materials: Confined electron systems in transition metal oxides. *Journal of Electron Spectroscopy, 219*, 16–28.
3. Tan, C. L., Cao, X. H., Wu, X. J., He, Q., Yang, J., Zhang, X., ... Zhang, H. (2017). Recent advances in ultrathin two-dimensional nanomaterials. *Chemical Reviews, 117*(9), 6225–6231.
4. Akinwandeh, D., Huyghebaert, C., Wang, C. H., Serna, M. I., Goossens, S., Li, L. J., ... Koppens, F. H. L. (2019). Graphene and two-dimensional materials for silicon technology. *Nature, 573*(7775), 507–518.
5. Romagnoli, M., Sorianello, V., Midrio, M., Koppens, F. H. L., Huyghebaert, C., Neumaier, D., ... Ferrari, A. C. (2018). Graphene-based integrated photonics for next-generation datacom and telecom. *Nature Reviews Materials, 3*(10), 392–414.
6. Grigorenko, A. N., Polini, M., & Novoselov, K. S. (2012). Graphene plasmonics. *Nature Photonics, 6*(11), 749–758.
7. Xiong, L., Forsythe, C., Jung, M., McLeod, A. S., Sunku, S. S., Shao, Y. M., ... Basov, D. N. (2019). Photonic crystal for graphene plasmons. *Nature Communications, 10*(1), 4780.
8. Bonaccorso, F., Sun, Z., Hasan, T., & Ferrari, A. C. (2010). Graphene photonics and optoelectronics. *Nature Photonics, 4*(9), 611–622.

Plasmonic Devices Based on 2D semiconductors

9. García de Abajo, F. J. (2014). Graphene plasmonics: Challenges and opportunities. *ACS Photonics*, *1*(3), 135–152.

10. Shu, H., Su, Z., Huang, L., Wu, Z., Wang, X., Zhang, Z., & Zhou, Z. (2018). Significantly high modulation efficiency of compact graphene modulator based on silicon wave-guide. *Scientific Reports*, *8*(1), 991.

11. Chen, X., Shehzad, K., Gao, L., Long, M., Guo, H., Qin, S., ... Wang, X. (2019). Graphene hybrid structures for integrated and flexible optoelectronics. *Advanced Materials*, 1902039. doi: 10.1002/adma.201902039.

12. Liu, C. H., Chang, Y. C., Norris, T. B., & Zhong, Z. (2014). Graphene photodetectors with ultra-broadband and high responsivity at room temperature. *Nature Nanotechnology*, *9*(4), 273–278.

13. Mak, K. F., & Shan, J. (2016). Photonics and optoelectronics of 2D semiconductor transition metal dichalcogenides. *Nature Photonics*, *10*(4), 216–226.

14. Liu, Y., Duan, X., Huang, Y., & Duan, X. (2018). Two-dimensional transistors beyond graphene and TMDCs. *Chemical Society Reviews*, *47*(16), 6388–6409.

15. Li, X., Zhu, J., & Wei, W. (2016). Hybrid nanostructures of metal/two-dimensional nanomaterials for plasmon-enhanced applications. *Chemical Society Reviews*, *45*(11), 3145–3187.

16. Furube, A., & Hashimoto, S. (2017). Insight into plasmonic hot-electron transfer and plasmon molecular drive: New dimensions in energy conversion and nanofabrication. *NPG Asia Materials*, *9*(12), e454.

17. Li, Y., Chi, C., Shan, H., Zheng, L., & Fang, Z. (2017). Plasmonic of 2D nanomaterials: Properties and applications. *Advancement of Science*, *4*, 1600430.

18. Zhou, N., López-Puente, V., Wang, Q., Polavarapu, L., Pastoriza-Santos, I., & Xu, Q. (2015). Plasmon-enhanced light harvesting: Applications in enhanced photocatalysis, photodynamic therapy and photovoltaics. *RSC Advances*, *5*(37), 29076–29097.

19. Nerl, H. C., Winther, K. T., Hage, F. S., Thygesen, K. S., Houben, L., Backes, C., ... Nicolosi, V. (2017). Probing the local nature of excitons and plasmons in few-layer MoS_2. *npj 2D Materials and Applications*, *1*(1), 2.

20. Thygesen, K. S. (2017). Calculating excitons, plasmons, and quasiparticles in 2D materials and van der Waals heterostructures. *2D Materials*, *4*(2), 2.

21. Siroki, G., Lee, D. K. K., Haynes, P. D., & Giannini, V. (2016). Single-electron induced surface plasmons on a topological nanoparticle. *Nature Communications*, *7*, 12375.

22. D'Acunto, M., Fuso, F., Micheletto, R., Naruse, M., Tantussi, F., & Allegrini, M. (2017). Near-field surface plasmon field enhancement induced by rippled surfaces. *Beilstein Journal of Nanotechnology*, *8*, 956–967.

23. Dutta, A., Kildishev, A. V., Shalaveh, V. M., Boltasseva, A., & Marinero, E. E. (2017). Surface-plasmon opto-magnetic field enhancement for all-optical magnetization switching. *Optical Materials Express*, *7*(12), 4316.

24. Zayatsalgor, A. V., Smolyaninov, I., & Maradudin, A. A. (2005). Nano-optics of sur-face plasmon polaritons. *Physics Reports*, *408*(3–4), 131–314.

25. Sui, M., Kunwar, S., Pande, P., & Lee, J. (2019). Strongly confined localized sur-face plasmon resonance (LSPR) bands of Pt, AgPt, AgAuPt nanoparticles. *Scientific Reports*, *9*(1), 16582.

26. Haes, A. J., & Van Duyne, R. P. (2004). A unified view of propagating and local-ized surface plasmon resonance biosensors. *Analytical and Bioanalytical Chemistry*, *379*(7–8), 920–930.

27. Li, M., Cushing, S. K., & Wu, N. (2015). Plasmon-enhanced optical sensors: A review. *Analyst*, *140*(2), 386–406.

28. Raether, H. (1988). Surface plasmons. *Springer Tracts in Modern Physics*. Berlin: Springer, *111*, 1.

29. Stanford, M. G., Rack, P. D., & Jariwala, D. (2018). Emerging nanofabrication and quantum confinement techniques for 2D materials beyond graphene. *npj 2D Materials and Applications*, 2(1), 20.

30. Kasani, S., Curtin, K., & Wu, N. (2019). A review of 2D and 3D plasmonic nanostructure array patterns: Fabrication, light management and sensing applications. *Journal of Nanophotonics*, 8(12), 2065–2089.

31. Mueller, T., & Malic, E. (2018). Exciton physics and device application of two-dimensional transition metal dichalcogenide semiconductors. *npj 2D Materials and Applications*, 2(1), 29.

32. Butler, S. Z., Hollen, S. M., Cao, L., Cui, Y., Gupta, J. A., Gutiérrez, H. R., ... Goldberger, J. E. (2013). Progress, challenges, and opportunities in two-dimensional materials beyond graphene. *ACS Nano*, 7(4), 2898–2826.

33. Boriskina, S. V., Cooper, T. A., Zeng, L., Ni, G., Tong, J. K., Tsurimaki, Y., ... Chen, G. (2017). Losses in plasmonics: From mitigating energy dissipation to embracing loss-enabled functionalities. *Advances in Optics and Photonics*, 9(4), 775–828.

34. Singh, A. N., Devnani, H., Jha, S., & Ingole, P. P. (2018). Fermi level equilibration of Ag and Au plasmonic metal nanoparticles supported on graphene oxide. *Physical Chemistry Chemical Physics*, 20(41), 26719–26733.

35. Iranzo, D. A., Nanot, S., Dias, E. J. C., Epstein, I., Peng, C., Efetov, D. K., ... Koppens, F. H. L. (2018). Probing the ultimate plasmon confinement limits with a van der Waals heterostructure. *Science*, 360(6386), 291–295.

36. Du, D., Luo, X., & Qiu, T. (2016). Graphene plasmonics. In V. K. Thakur & M. K. Thakur (Eds.), *Chemical functionalization of carbon nanomaterials chemistry and applications* (pp. 477–495). Boca Raton, FL: CRC Press Press.

37. Hwang, E. H., & Sarma, S. D. (2007). Dielectric function, screening, and plasmons in two-dimensional graphene. *Physical Review. Part B*, 75(20), 205418.

38. Chen, J., Badioli, M., Alonso-Gonzalez, P., Thongrattanasiri, S., Huth, F., Osmond, J., ... Koppens, F. H. (2012). Optical nano-imaging of gate-tunable graphene plasmons. *Nature*, 487(7405), 77–81.

39. Fei, Z., Rodin, A. S., Andreev, G. O., Bao, W., McLeod, A. S., Wagner, M., ... Basov, D. N. (2012). Gate-tuning of graphene plasmons revealed by infrared nano-imaging. *Nature*, 487(7405), 82–85.

40. Fang, Z., Wang, Y., Schlather, A. E., Liu, Z., Ajayan, P. M., de Abajo, F. J., ... Halas, N. J. (2014). Active tunable absorption enhancement with graphene nanodisk arrays. *Nano Letters*, 14(1), 299–304.

41. Ju, L., Geng, B., Horng, J., Girit, C., Martin, M., Hao, Z., ... Wang, F. (2011). Graphene plasmonics for tunable terahertz metamaterials. *Nature Nanotechnology*, 6(10), 630–634.

42. Nikitin, A. Y., Alonsogonzález, P., Vélez, S., Mastel, S., Centeno, A., Pesquera, A., ... Hillenbrand, R. (2016). Real-space mapping of tailored sheet and edge plasmons in graphene nanoresonators. *Nature Photonics*, 10(4), 239–243.

43. Christensen, J., Manjavacas, A., Thongrattanasiri, S., Koppens, F. H., & de Abajo, F. J. (2011). Graphene plasmon wave guiding and hybridization in individual and paired nanoribbons. *ACS Nano*, 6(1), 431–440.

44. Ansell, D., Radko, I. P., Han, Z., Rodriguez, F. J., Bozhevolnyi, S. I., & Grigorenko, A. N. (2015). Hybrid graphene plasmonic waveguide modulators. *Nature Communications*, 6, 8846.

45. Kim, J., Son, H., Cho, D. J., Geng, B., Regan, W., Shi, S., ... Wang, F. (2012). Electrical control of optical plasmon resonance with graphene. *Nano Letters*, 12(11), 5598–5602.

46. Emani, N. K., Chung, T. F., Ni, X., Kildishev, A. V., Chen, Y. P., & Boltasseva, A. (2012). Electrically tunable damping of plasmonic resonances with graphene. *Nano Letters*, 12(10), 5202–5206.

47. Zhao, B., & Zhang, Z. M. (2015). Strong plasmonic coupling between graphene ribbon array and metal gratings. *ACS Photonics*, *2*(11), 1611–1618.
48. Dabidian, N., Kholmanov, I., Khanikaev, A. B., Tatar, K., Trendafilov, S., Mousavi, S. H., ... Shvets, G. (2015). Electrical switching of infrared light using graphene integration with plasmonic Fano resonant metasurfaces. *ACS Photonics*, *2*(2), 216–227.
49. Wang, Y., Ou, J. Z., Chrimes, A. F., Carey, B. J., Daeneke, T., Alsaif, M. M., ... Kalantar-Zadeh, K. (2015). Plasmon resonances of highly doped two-dimensional MoS_2. *Nano Letters*, *15*(2), 883–890.
50. Karbalaei Akbari, M., Hai, Z., Wei, Z., Ramachandran, R. K., Detavernier, C., Patel, M., ... Zhuiykov, S. (2019). Sonochemical functionalization of the low dimensional surface oxide of galinstan for heterostructured optoelectronic applications. *Journal of Materials Chemistry C*, *7*(19), 5584–5595.
51. Chen, Y. (2015). Nanofabrication by electron beam lithography and its applications: A review. *Microelectronic Engineering*, *135*, 57–72.
52. Machalett, F., & Seidel, P. (2019). Focused ion beams and some selected applications. In Jagadish, C. (Ed.), *Encyclopedia of applied physics*. 2–38, Wiley-VCH Verlag GmbH & Co. KGaA, Weinheim.
53. Salaita, K., Wang, Y., & Mirkin, A. M. (2007). Applications of dip-pen nanolithography. *Nature Nanotechnology*, *2*(3), 145–155.
54. Fischer, J., & Wegener, M. (2013). Three-dimensional optical laser lithography beyond the diffraction limit. *Laser and Photonics Reviews*, *7*(1), 22–44.
55. Colson, P., Henrist, C., & Cloots, R. (2013). Nanosphere lithography: A powerful method for the controlled manufacturing of nanomaterials. *Journal of Nanomaterials*, *2013*, 948510.
56. Sreenivasan, S. V. (2017). Nanoimprint lithography steppers for volume fabrication of leading-edge semiconductor integrated circuits. *Microsystems and Nanoengineering*, *3*, 17075.
57. Wang, X., & Han, G. R. (2003). Fabrication and characterization of anodic aluminum oxide template. *Microelectronic Engineering*, *66*(1–4), 166–170.
58. Lee, W. J., & Smyrl, W. H. (2003). Oxide nanotube arrays fabricated by anodizing processes for advanced material application. *Current Applied Physics*, *8*(6), 818–821.
59. Karbalaei Akbari, M., Hai, Z., Wei, Z., Detavernier, C., Solano, E., Verpoort, F., & Zhuiykov, S. (2018). ALD-developed plasmonic two-dimensional $Au-WO_3-TiO_2$ heterojunction architectonics for design of photovoltaic devices. *ACS Applied Materials and Interfaces*, *10*(12), 10304–10314.
60. Ahmed, S., & Yi, J. (2017). Two-dimensional transition metal dichalcogenides and their charge carrier mobilities in field-effect transistors. *Nanomicro Letters*, *9*(4), 50.
61. Freitag, M., Low, T., Xia, F., & Avouris, P. (2012). Photoconductivity of biased graphene. *Nature Photonics*, *7*(1), 53–59.
62. Lopezsanchez, O., Lembke, D., Kayci, M., Radenovic, A., & Kis, A. (2013). Ultrasensitive photodetectors based on monolayer MoS_2. *Nature Nanotechnology*, *8*(7), 497.
63. Wang, W., Klots, A., Prasai, D., Yang, Y., Bolotin, K. I., & Valentine, J. (2015). Hot electron-based near-infrared photodetection using bilayer MoS_2. *Nano Letters*, *15*(11), 7440–7444.
64. Fang, Z., Liu, Z., Wang, Y., Ajayan, P. M., Nordlander, P., & Halas, N. J. (2012). Graphene-antenna sandwich photodetector. *Nano Letters*, *12*(7), 3808–3813.
65. Kim, U. J., Yoo, S., Park, Y., Shin, M., Kim, J., Jeong, H., ... Park, S. (2015). Plasmon-assisted designable multi-resonance photodetection by graphene via nanopatterning of block copolymer. *ACS Photonics*, *2*(4), 506–514.

6 Memristive Devices Based on Ultrathin 2D Materials

6.1 INTRODUCTION TO MEMRISTOR CHARACTERISTICS

The ultrafast growth of information technology is accompanied by generation of the large amount of informative data. The data processing and storage are highly dependent on the development of new materials, devices, and their related advanced technologies. Today silicon-based memories are facing their technological limits, furthermore, the complementary metal oxide semiconductor (CMOS) devices are also suffering from the physical dimensional retractions and technological problems [1, 2]. Neumann technology-based computers emulate the human brain working mechanisms to operate information technology and processing. The human brain consists of a complicated network of neurons interconnected through biological synapses [3]. Therefore memorizing, learning, reminding, processing, recognition, understanding and multi task operations are all performed with the lowest operation energy. The power consumption of human brain is in the range of 1 to 10 fj for each individual data transfer process by a synaptic event, which is at least five orders of magnitude lower than the energy consumption of the existing conventional computers [4]. Thus, from an energy point of view, conventional von Neumann technology architecture is not efficient since the memory and processing units are physically separated. Consequently, it is highly required to develop new nanostructured materials with improved properties and also to improve the new devices to circumvent the technological restrictions related to the storage capacity of memory devices. Nanoscaled nonvolatile memristors with upgrading performance, enhanced processing speed, low energy consumption, and long-term endurance are one of the main candidates to be replaced by current memory technology [5]. However, requirements demanded for size reduction have imposed more pressure to scale down memristive devices to meet requirements of novel electronic technology.

The resistive switching (RS) behavior of 2D materials has also been explored recently. It was found that the employment of 2D materials in memory devices could potentially tackle the current limitations in memory devices [6]. Recently, the nonvolatile characteristics of 2D materials have been extensively investigated and characterized. The 2D semiconductor layered materials and ultrathin oxide films possess a wide range of properties in various fields that also can be accompanied by the memristive characteristics. These multiple advantageous properties of 2D materials have provided unprecedented opportunities to use ultrathin 2D films for different memristive applications. For example, the voltaic driven memristors are

171

one of the most common resistive switching materials, in them the electrical signals are employed to trigger the resistive switching behavior of 2D semiconductor based devices [6]. Apart from electrical-pulse-triggered memristive devices, the outstanding optical properties of 2D semiconductor photodetectors have also created incredible opportunities for multifunctional integration of optical sensing, data storage, and processing into one single device [7]. Furthermore, in contrast to regular memristor devices, the optical stimulation of 2D materials enabled the light-assisted storage of information at considerable low programing voltages for secured transportation of information [7]. Moreover, the combination of both electrical readout voltage and optical stimulation of memristors has the capability of enlarging the ratio between the low resistance state (LRS) and high resistance state (HRS) current of devices, hence tangibly improving the data storage capacity and also enabling the multilevel data storage capability [8]. The advantages of optical assisted memristors are not merely attributed to the above-mentioned points. The optical assisted resistive switching brings the benefits of employment of wide bandwidth materials and also assists the reduction of electrical loss during sensory data transmission [9]. In addition, light-assisted resistive switching is also a promising technique for digital processing functionalities of memristors. In this case light sources can be used as the individual controlling parameter. The combination of electrical inputs and optical pulses has facilitated the capability of arithmetic operation, logic processing and the coincident detection operation of devices [7, 8]. Moreover, optically stimulated sensor-memristor devices created a substantial improvement in the light recognition and image sensory and processing technologies [10].

The present chapter focuses primarily on the RS mechanism in solid-state electronics of ultrathin films and 2D materials. The design and fabrication techniques of memristor devices are introduced and both advantages and disadvantages of each fabrication design are discussed. Several practical examples are reviewed to address the microstructural effects of resistive switching in ultrathin semiconductor films. In doing so, the resistive switching phenomena in various 2D memristor devices have been investigated and the memristive properties have been reviewed. The device performance of 2D based memristor devices and their RS mechanism are separately summarized and categorized.

6.2 SOLID STATES ELECTRONICS OF RESISTIVE SWITCHING

6.2.1 THE MISSING MEMRISTOR

The first observation of resistive switching characteristics was announced at 2008, when the HP laboratory found memristive switching in TiO_2 metal oxides systems [11]. This system was the stripped 50 nm TiO_2 film with 50 nm half-pitch. The Pt nanowire was fabricated via nanoimprint lithography to build up a cross-sectioned vertical Pt/50 nm TiO_2/Pt stacked structure (Figure 6.1a). Typical bipolar switching behaviors with 10^3 *on/off* ratio and the rapid switching (50 ns) were the main characteristics of the first metal oxide memristor devices (Figure 6.1b). The current-voltage *I-V* characteristic curve of Pt-TiO_2-Pt memristor was quite special. The initial *I-V* curve of the device in its pre-switching state exhibits a rectifying

Memristive Devices Based on 2D Materials

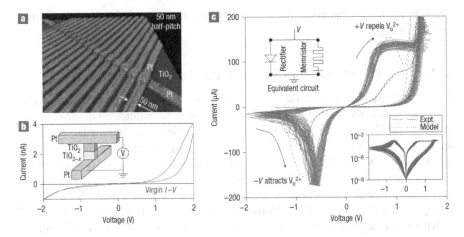

FIGURE 6.1 The bipolar, non-volatile switching of TiO_2 based memristor device. (a) The typical AFM image of the device with nano-cross print structure with 50 nm half-pitch. The Pt nanowire was printed by lithography technique to make a sandwiched Pt/TiO_2 (50 nm)/Pt device. (b) The typical I-V curves of device in pre-switching state with a rectifying characteristics. The graphical inset shows the sandwiched TiO_2 film over TiO_{2-x} film. The electrical voltage is applied on top electrode while the bottom Pt electrode was grounded. (c) The typical cyclic switching curve represented for 50 cyclic curve with high degree of repeatability. The inset shows the typical log scale switching *I-V* curves. The other upper inset is the equivalent circuit model. Consist of a rectifier in parallel with memristor device. Reprinted with permission from Tan, H., Liu, G., Zhu, X., Yang, H., Chen, B., Chen, X., ... Li, R. W. (2015). An optoelectronic resistive switching memory with integrated demodulating and arithmetic functions. *Advanced Materials, 27*(17), 2797–2703.

characteristic (Figure 6.1b). After a single irreversible forming step, the multiple switching curves, shown in Figure 6.1c, depicted a high degree of repeatability. In an *on state*, when the device switches to a low resistance state (LRS), the *I-V* curves show symmetric hyperbolic behavior (Sinh-like curve), while in an *off state*, when the device switches to high resistance state (HRS), the device exhibits asymmetric rectifying response. The switching mechanism was attributed to the interactions at Pt/TiO_2 interface [11]. It was experimentally shown that the electrical conduction in Pt/TiO_2/Pt thin-film devices was controlled by spatially heterogeneous metal/oxide electronic barrier. The main resistive mechanism in this device is based on migration and drift of positively charged oxygen vacancies which acting as native dopants to form locally conductive channels through the electronic barrier [11]. The concentration and distribution of oxygen vacancies in TiO_2 film controlled the resistive characteristics of devices, including its rectification and switching polarity. Thus, the memristive characteristics of Pt/TiO_2/Pt device were controlled by engineering the charge trapping/detrapping of oxygen vacancies in ultrathin, active material of memristor, so they were called valence change memory (VCM) devices [12]. It was the initiation step for the fabrication and characterization of one of the fundamental missing component of circuit systems, that is, memristors that later opened up one of the most fundamental research fields in materials

174 Ultrathin 2D Semiconductors

science, electronic, memory, synaptic, optoelectronics, and neuromorphic computing technologies.

6.2.2 MEMRISTIVE MATERIALS AND THEIR RS MECHANISMS

For identification of the precise mechanism of resistive switching in the memristor devices, caution should be taken into consideration to investigate all of the engaging parameters. There are several reactions and interactions, which may happen either individually or simultaneously in the memristor components due to the strong electric field and joule heating during resistive switching (RS). The solid-state electrochemical interaction of Ag and Cu electrodes with the resistor body and following formation and dissolve of filamentary conductive channels are the well-recognized electrochemical assisted RS mechanism [13]. This mechanism indicates that the RS behaviors are dependent not only on the main resistive body but also on the interactions of conductive electrodes and resistive materials [13]. In another RS mechanism, the formation of new material phases or phase transformation in the insulating body of memristor device during the initial electroforming process is fundamental characteristic of the phase changing memory devices [12, 13]. It is another evidence, which shows that the engaged parameters in RS phenomenon do not follow rudimentary principles or mechanisms. Hence, identifying the actual switching mechanism is a challenging topic but also fundamental process toward the realization of switching mechanism and engineering of memristor device. To understand the switching mechanisms, the type of mobile species should be recognized. Furthermore, it should be addressed where the locations of mobile species are in the memristor structures and how the driving force push forward the motion of mobile species and how the mobile species move under the effect of driving force [12, 13]. The next lines in this chapter will clarify and address the above-mentioned questions.

6.2.2.1 Anion-Based Memristors and Their RS Mechanisms

Most of the memristive devices can be classified according to the switching mechanisms or based on the type of switching materials. The anion-based devices are one of the main groups of switching material including oxide semiconductors and insulators, transition metal oxides, complex oxide materials, large band gap and non-oxide insulators such as chalcogenides and nitrides [12–15]. In the most of metal oxide materials the positively charged oxygen vacancies or equivalently "oxygen anions" are the mobile species which change the valence states of metal oxides and finally alter their resistance values and affects the resistive characteristics of them. However, the resistive switching is not merely restricted to the oxygen anions or vacancies, there are variety of defects which alter the transport mechanisms in insulating materials [16]. For example, while the oxygen deficiency (oxygen vacancy) acts as a native dopant in TiO_x insulator and thus facilitates the semiconducting characteristics of TiO_x [17], the oxygen excess in $Co_{1-x}O$ cause the *p-type* conductivity [18]. Mobility and concentration of the native dopants including oxygen vacancies or cation interstitials are adequately enough and sufficiently high to cause the considerable motion of charge carriers and switch the resistance of resistive materials

Memristive Devices Based on 2D Materials

at operating temperature of the most memristor devices [19]. These phenomena are confirmed by tracing of the charge carriers motion in the memristor microstructure or observed by microscopic observation, especially in the case of transition metal oxides [20]. The anion-based switching mechanism was observed in a large group of materials including, TiO_x, ZrO_x, HfO_x, VO_x, NbO_x, MoO_x, MnO_x, WO_x, CuO_x, NiO_x, AlO_x, SiO_x, SnO_x, BiO_x, FeO_x, GaO_x, as well as in variety of insulators, perovskites ($Ba_{0.7}Sr_{0.3}TiO_3$, $SrTiO_3$, $SrZrO_3$, $La_{0.33}Sr_{0.67}FeO_3$) and rare earth oxides [21]. The anion-based switching mechanism is not merely restricted to the oxide materials, it is also observed among the non-oxide insulator materials, nitride (AlN), selenides (ZnSe) and tellurides (ZnTe) [22].

By imposing the electrical bias on a switching memristor structure two main phenomena occur: a joule heating activated phenomenon and an induced electric field [23]. The generated current density by the joule heating source is typically efficient enough (about $> 10^6$ A cm^{-2}) to facilitated switching in anion-based memristor devices [24]. The joule heating mechanism has been observed in both bipolar and unipolar RS mechanisms. In practical condition, both joule heating and electric field–induced migration of charge carries coexist in all memristive switching process. However, the dominance of each mechanism is dependent on the device structure, resistive materials and a variety of technical parameters [25, 26]. Considering the relative importance and the share of each joule heating mechanism and electric field–induced mechanism, four different types of resistive switching have been classified for RS behavior of the oxide based memristor devices: bipolar linear, bipolar nonlinear, nonpolar bistable and nonpolar threshold switching [1].

In bipolar RS, the switching process shows directional behavior depending on the polarity of applied voltages. To be precise, the *on* (set) and *off* (reset) states occur at the different polarities of working voltages [27]. The switching mechanism in bipolar RS is usually regarded as electrochemical migration of ions and redox reactions. Thus, the polarity of applied voltage affects the switching phenomena in bipolar RS-based devices [27]. In bipolar switching the electric field plays the significant role, while the thermal effects are much more significant in unipolar switching [28]. The semiconductor oxides and perovskites are usually experienced the bipolar RS mechanisms. Figure 6.2a and b depicts two main bipolar RS mechanisms of anion-based devices. Figure 6.2a shows a typical bipolar switching where the growth and retraction of the conductive channel happen vertically under the effects of imposed electric field by either drift or migration of mobile carries [1]. The rectifying behavior of *I-V* curves in *off state* is caused by the creation of Schottky-like barriers at the metal/insulator interface, while the symmetric nonlinear curves in the *on state* is due to the formation of residual tunnel barrier (Figure 6.2a) [1]. In another type of bipolar switching mechanism (Figure 6.2b) conduction channel is formed and connects the upper and bottom electrodes during the switching process [29]. The combined effects of electric field and joule heating change the composition, volume, and geometry of conduction channel in memristor component and control the resistance of memristor device. The vertical drift of conductive channel is facilitated by the electric field–induced migration of mobile charges, while the joule heating phenomenon assists and enhance the effect of lateral diffusion of mobile carries [30].

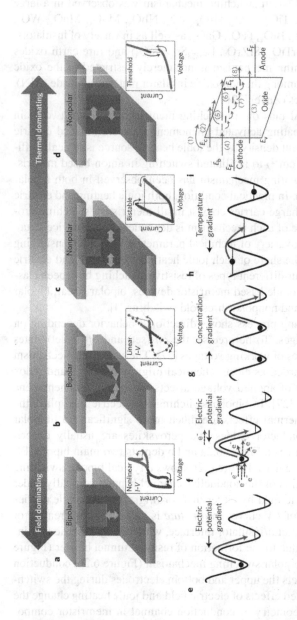

FIGURE 6.2 The graphical description of RS mechanisms with ion and electron transfer description, the driving forces for RS process, the electrical characteristics of memristors and finally the schematic representation of ion trapping and detrapping mechanisms in metal/oxide/metal sandwiched structure. (a–d) The simplified schematic of conduction channel (red) in active layer of memristor (blue) for typical switching memristors. The gray arrows show the ideal path for ionic motion. (a) *I-V* graph shows the typical bipolar nonlinear RS. (b) Bipolar, linear, nonvolatile RS. (c) The nonpolar bistable RS and (d) nonpolar threshold switching. (e–h) Schematically demonstrate the mechanisms for oxygen ion motion and drift for (e) electric potential gradient, (f) electro-migration, (g) Concentration gradient (Fick diffusion), (h) and temperature gradient (thermophoresis). The ion motion mechanisms may contribute each other to produce the (a–d) RS typical behavior. (i) Depicts the possible electron transport mechanisms based on the possible transport paths. (1) Schottky emission (2) Fowler–Nordheim tunneling, (3) Direct tunneling, (4) tunneling from cathode to traps, (5) emission from traps to the conduction band (Poole–Frenkel emission), (6) tunneling from trap to conduction band, (7) trap-to-trap hopping or tunneling, (8) tunneling from traps to anode. EF, Fermi energy level; Ev, valence band; Ec, conduction band; Eb, Schottky barrier height; Et, trap barrier height. Reprinted and reproduced with permission from Yang, J. J., Strukov, D. B., & Stewart, D. R. (2013). Memristive devices for computing. *Nature Nanotechnology*, 8(1), 13–24.

In the so-called unipolar RS, the switching direction depends on the amplitude of the applied voltage and is independent of the voltage polarity [30]. Therefore in unipolar switching, the HRS to LRS switching is achieved by applying a threshold voltage, which is larger than the voltage for reset process. To preserve the memristor device and to avoid the permanent damage from excess current, the compliance current is set during the RS measurements. The compliance current also determines the working power consumption of the memristor device. In unipolar RS the set process is controlled by compliance current, which is practically adjusted by the employment of a series resistor to the system [30]. The physical interpretation of unipolar RS includes the increased role of joule heating effects on migration of ionic components, thus the polarity of applied voltage does not play any role in the unipolar RS behavior of memristor device. Highly insulator metal oxides and binary metal oxides and their corresponding semiconductors are the main candidates for unipolar RS behavior [30].

The formation/rapture of filamentary conductive channel is one of the main mechanisms of RS in memristor devices. In filamentary conductive memories, the switching with a laterally uniform switching region across the entire device area is not included, but the switching happens by the formation of ultra-narrow conductive channels which connect the conductive electrodes of memristor device [31]. The solid-state formation of conductive filaments (CFs) can be categorized into three individual groups. These divisions are originated from the different migration mechanisms of mobile species. In the valence change memories (VCM) the RS originates form the migration of mobile species such as oxygen vacancy defects under applied bias voltage [32]. The electrochemical metallization (ECM) of metal cations species is another mechanism of CFs formation where the development of conductive channel of metallic cations are responsible for RS behavior of memristors [33]. The last mechanism is attributed to the formation and rapture of CFs under thermal joule heating, which is the fundamental basis of thermochemical memristor (TCM) devices [34]. Figure 6.2c shows a nonpolar, nonvolatile resistive switching. From the physical point of view, purely electric breakdown leads to the flowing of high current through the memristor body, which is called the electric field–assisted soft breakdown of dielectric component of memristor. Then, the joule heating effects cause a temperature gradient in resistor body, which ultimately leads to the attraction of oxygen vacancies to the conduction channel region to turn the device *on* (Figure 6.2c). The reset switching is estimated to be a fully thermally activated and assisted process [35]. The created concentration gradient causes a thermal-diffusion process assisting the thermal rapture of conduction filament and also reducing the free surface energy of the filament [36]. The thermally or electric field–assisted phase change process is also reported as the main mechanism of the nonpolar, nonvolatile resistive switching in filamentary RS materials [37] (Figure 6.2c).

The threshold switching in nonpolar RS devices is another thermally assisted RS mechanism, which in fact is a volatile switching process (Figure 6.2d). The formation of conductive filament is the result of sudden phase transformation from insulating to conducting state [38]. This insulator to metallic phase transformation happens at the certain current level accompanied by a steplike increase of current in *I-V* curve. The RS behavior of device is ultimately dependent on the current level, as the decrease of current level was accompanied by the phase transformation from metallic to insulating state. The threshold switching is observed in metal-insulator transition materials like VO_2, NbO_2, and TiO_2 [39–41]. The negative differential resistance phenomenon is frequently observed in this type of threshold switching where the increase of voltage is accompanied by the decrease of electric current [42]. Therefore, in these materials the insulator to metal transition is facilitated by the localized high-temperature joule heating phenomenon converting the heated region of insulator into a metallic phase [38]. The locally heated regions (metallic) can be cooled down by the decrease of device current returning the device again to insulating state. The example of the thermally assisted RS process is presented in Figure 6.2d.

The atomic motion and rearrangement during RS process is facilitated by either the individual or combined impacts of four different driving forces. The driving forces are (1) electric potential gradient due to field electric, (2) concentration gradient, (3) electron kinetic energy and the (4) temperature gradient. The charged dopants can move under the influence of a strong electric field [43]. Figure 6.2e shows how an oxygen anion drifts and jumps from trapping sites to another one, at the opposite direction of imposed electric potential gradient. The motion of atoms is also possible when the electron wind affects the microstructure [44]. The movement of high speed electrons under the impact of high electric field is likely to happen by momentum transfer mechanism called literally the electron wind mechanism [44] (Figure 6.2f). Recently, it was demonstrated that the resistance of Au nanowire was reversibly switched by the momentum transfer mechanism [44]. Both electron kinetic energy and electric field are polarity-dependent factors thus they can cause bipolar resistive switching. The joule heating could impose several combined effects on the motion of atoms in the structure. The drift and diffusion of atoms are tangibly affected by the temperature gradient (Figure 6.2 h). The effect of a high temperature gradient is therefore similar to the electric potential gradient and also similar to the element concentration gradient. Both of these parameters induce the mass transfer of atoms in structure. The same is applicable for the high temperature gradient causing the drift and transfer of atoms (Figure 6.2h). It is known that the electric field–assisted motion of dopants and electro-migration of atoms happens by the drift of atoms and dopants vertically along the conduction channels between the top and bottom electrodes [45]. Contrary to the electric field–assisted mass transfer, the generated temperature gradient by joule heating assists the lateral drift and diffusion of atoms within the conduction channel. At practical conditions, mass transfer and charge diffusion occurs in both mechanism since the vertical transfer of atoms within the conduction channel and the lateral drift of atoms toward the outward of conductive channel happen simultaneously. However, the dominance of each mechanism in RS process is determined by the dominance of electric field or joule heating assisted drift mechanisms (Figure 6.a, b, c, and d). The schematic of the recommended electron transport path in the metal/semiconductor/metal devices is shown in Figure 6.2i. In the Schottky emission dynamics (1), the thermally activated electrons jump over the Schottky barrier into the conduction band of adjacent semiconductor. Fowler–Nordheim (F–N) tunneling (2) is another electron transfer mechanism in which the electrons tunnel from the cathode electrode into the conduction band of adjacent semiconductor [46, 47]. A high electrode field is necessary to be employed on interfaces to facilitate the

Memristive Devices Based on 2D Materials

F–N tunneling. When the thickness of semiconductor oxide film is ultrathin, electrons can directly pass through the semiconductor layer and move from the cathode into anode electrode [48]. The direct tunneling of electrons is depicted in Figure 6.2i (3). Furthermore, when the semiconductor or insulator layer contains trapping states, additional conduction mechanisms are activated. The trapping sites may be created by disorder in materials, non-stoichiometric compounds, dopants, and impurities. The trap-assisted mechanisms can cover several stages including (4) tunneling of electrons from cathode to trap sites, (5) the Poole-Frenkel emission which is the emission of trapped electrons from trapping sites to the adjacent anode [49], and (6) tunneling from the trapped site of semiconductor to the conduction band of anode. The trap-to-trap tunneling (7) happens with various mechanisms ranging from the Mott hopping between localized states to metallic conduction via extended states [50]. The tunneling from trap sites to anode electrode (8) is another possible mechanism for transfer of the trapped charge carries. The detailed information of some selected anion-based memristor devices are summarized in Table 6.1.

6.2.2.2 Cation-Based Memristors and Their RS Mechanisms

The cation-based memristor devices are known with several different titles originating from their working mechanisms. This group of the memristor devices is often called electrochemical metallization memory, conducting bridging RAM, programmable metallization cell or atomic switches [51]. While the first version of cation-based memristors were introduced in the 1970s, further development was extensively performed during the 1990s. The metallic cations are well recognized as the mobile spices in cation-based memristor device. In this cation-based memristor devices, the active electrodes are made of electrochemical active materials. Cu and Ag are one of the main examples of electrode materials in cations based memristors. Furthermore, the insulator doping with electrochemical active dopants can facilitate the mobility of metallic cations in memristor device and then generates the RS behavior [51]. The counter electrodes are therefore selected from the electrochemically inert metallic materials. Co, Mo, Au, W, Pt, Cr, Ru, Ir, doped Si poly-crystals, and TiW are some of the electrochemical inert materials for counter electrodes in the cation-based memristors [52]. Table 6.2 gives detailed information about the cation-based memristor devices.

The insulating component in cation-based memristor devices belong to various type of materials, including sulfides, selenides, iodides, tellurides, ternary chalcogenides, vacuum, and even water. Great efforts have been devoted to replace the common insulators with the metal oxides. The main reason for choosing metal oxide semiconductors returns to the capability of these materials for applications in CMOS devices. By shifting from the traditional electroactive materials into inexpensive oxide materials, the switching voltage increases to the operating voltage of CMOS devices. The capability of CMOS devices for application in the large-scale integrated circuits for nonvolatile memories is highly interesting strategy for te development of cation-based memristor devices. There is quite strong resemblance between the switching mechanism of cation-based memristors and their anion-based counterparts. The formation of nanoscale channels, which host the electrochemically active metals, creates microstructural changes in the main resistor body of

TABLE 6.1

Characteristics of Anion-Based Memristor (VCM) Devices

Insulators	Bottom electrode	Top electrode	RS mechanism
MgO	Pt	Pt	Unipolar
TiO_x	Ru, Pt	Al, Pt	Non/Uni/Bipolar
ZrO_x	P$^+$-Si, n$^+$-Si	Pt, Cr	Uni/Bipolar
HfO_x	TiN	TiN	Bipolar
VO_x	N/A	N/A	Threshold
NbO_x	P$^+$-Si	Pt	Unipolar
TaO_x	Pt, Ta	Pt, Ta	Bipolar
CrO_x	TiN	Pt	Bipolar
MoO_x	Pt	Pt-Ir	Uni/Bipolar
WO_x	W, FTO	TiN, Au	Bipolar
MnO_x	Pt	Al, TiN	Bipolar
FeO_x	Pt	Pt	Non/Bipolar
CoO_x	Pt	Pt	Nonpolar
NiO_x	Pt	Pt	Nonpolar/Threshold
CuO_x	TiN, TaN, SRO, Pt	Pt	Bipolar
ZnO_x	Pt, Au	TiN, Ag	Bipolar
AlO_x	Ru, Pt	Pt, Ti	Unipolar/Bipolar
GaO_x	ITO	Pt, Ti	Bipolar
SiO_x	Poly-Si, TiW	Poly-Si, TiW	Unipolar
SiO_xN_y	W	Cu	Bipolar
GeO_x	ITO, TaN	Pt, Ni	Bipolar
SnO_2	Pt	Pt	Unipolar
BiO_x	Bi	Ag, Cu, W, Re	Bipolar
SbO_x	Pt	Sb	Unipolar/Bipolar
SmO_x	TiN	Pt	Bipolar
GdO_x	Pt	Pt	Unipolar
YO_x	Al	Al	Unipolar
CeO_x	Pt	Al	Bipolar
EuO_x	TaN	Ru	Uni/Bipolar
PrO_x	TaN	Ru	Bipolar
ErO_x	TaN	Ru	Unipolar
DyO_x	TaN	Ru	Unipolar
NdO_x	TaN	Ru	Unipolar
$Ba_{0.7}Sr_{0.3}TiO_3$	$SrRuO_3$	Pt, W	Bipolar
$SrTiO_3$	$SrRuO_3$, Au, Pt	Au, Pt	Bipolar
$SrZrO_3$	$SrRuO_3$	Au	Bipolar
$BiFeO_3$	$LaNiO_3$	Pt	Bipolar

Source: Reprinted with permission from [1].

device. The bipolar switching is the most common RS mechanism in cation-based devices confirming the dominating impact of electrical field–assisted switching in RS devices [53]. One of the iconic cation-based memristor devices is $Ag/H_2O/Pt$ unit [54]. An Ag electrode oxidation occurs during the set switching when a high voltage is imposed on the electrochemical active Ag electrode. The Ag atoms are

Memristive Devices Based on 2D Materials

TABLE 6.2
Characteristics of Selected Cation-Based Memristor (VCM) Devices

Insulators	Bottom electrode	Top electrode	RS mechanism
Sulfides			
Ge_xS_x	W	Ag	Bipolar
As_2S_3	Au	Ag	Bipolar
Cu_2S	Cu	Pt	Bipolar
$Zn_xCd_{1-x}S$	Pt	Ag	Bipolar
Iodides			
AgI	Pt	Ag	Bipolar
$RbAg_4I_5$	Pt	Ag	Bipolar
Ge_xSe_y	W	Ag, Cu	Bipolar
Tellurides			
Ge_xTe_y	TiW	Ag	Bipolar
Ternary chalcogenides:			
Ge-Sb-Te	Mo	Au, Ag	Bipolar
Oxides			
Ta_2O_5	Pt	Cu	Bipolar
SiO_2	W	Cu	Uni/Bipolar
HfO_2	Pt	Cu	Bipolar
WO_3	Pt	Cu	Bipolar
ZrO_2	Ag	Au	Bipolar
$SrTiO_3$	Pt	Ag	Bipolar
TiO_2	Pt	Ag	Bipolar
CuOx	Cu	Al	Unipolar
ZnO	Pt, Al- doped	Cu	Bipolar
Al_2O_3	Al	Cu	Bipolar
MoO_x	Cu	Pt	Bipolar
GdO_x	Pt	Cu doped MoOx	–
Others			
MSQ	Pt	Ag	Bipolar
Doped organic semiconductors	Pt	Cu	Bipolar
Nitrides	Pt	Cu	Bipolar
Amorphous Si	P^+-Si	Ag	Bipolar
Carbon	Pt	Cu	Bipolar
Vacuum gaps	$RbAg_4I_5$/Ag, Ag_2S/Ag	W, Pt	Bipolar

Source: Reprinted with permission from [1]

oxidized into Ag^+ cations and then dissolved into electrolytes and moves toward the Pt electrode to facilitate the switching process and turn the device *on*. The imposed opposite voltage on the electrochemically inert Pt electrode dissolves the conductive Ag filaments in the electrolyte and returns unit to the HRS and then tunes the device *off* [53]. The RS is not merely restricted to the electric field–assisted switching mechanisms, the unipolar RS behavior is also observed in the switching performance of the cation-based memristors. The $Cu/Ta_2O_5/Pt$ system is another example of the cation-based memristor devices with the temperature dependent switching behavior [43]. It is also shown that the reset process in this unit is likely thermal-diffusion assisted phenomenon [43]. It was observed that the switching phenomenon happens at the interface between the inert Pt electrode and the conduction channel of memristor. The reality is that the precise determination of switching mechanism is dependent on the several technical factors, including the materials selection, device fabrication, and operational cycles and conditions. Due to the variety of engaged parameters, the switching mechanism in each memristor device could be individual and specific.

6.3 FABRICATION OF MEMRISTOR DEVICES BASED ON 2D MATERIALS

6.3.1 Device Structure

6.3.1.1 Vertical Memristors

Most of the memristor devices already discussed, especially 2D-based memristors, are often made with a vertical sandwiched structure. This type of device is in fact a resistive martial sandwiched between two conductive electrodes as presented in Figure 6.3a. Because of vertical structure, the thickness of 2D-based memristor

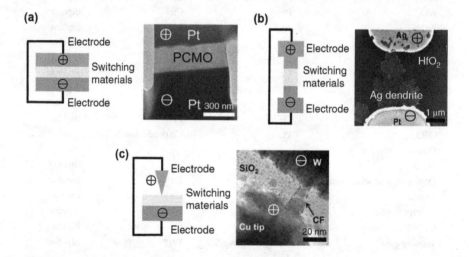

FIGURE 6.3 The schematic representation and actual structure of (a) vertical memristors structure, the SEM image shows the sandwiched PSMO active materials between upper and bottom Pt electrodes. (b) The lateral memristor structure. The SEM image depicts the formation of a dendrite like conductive Ag channel. The conductive Ag channel was developed in HfO_2 active resistive layer between electrochemical active Ag electrode and noble Pt electrode. (c) The tip-based memristor structures. The SEM image shows the SiO_2 active memristor layer between the positive Cu tip electrode and negative tungsten (W) electrode. Reprinted and reproduced with permission from Yang, J. J., Pickett, M. D., Li, X., Ohlberg, D. A., Stewart, D. R., & Williams, R. S. (2008). Memristive switching mechanism for metal/oxide/metal nanodevices. *Nature Nanotechnology*, 3(7), 429–433.

Memristive Devices Based on 2D Materials

material can be scaled down to a few nanometers to a subnanometer. It results in the tangible decrease of operating voltage and higher integration density of devices [55]. The considerably low switching voltage of vertical memristors has tangibly decreased the power consumption of devices [55]. The concept of vertical structure is not merely restricted to the ultrathin layered singular vertical structures. The 2D composite materials are also the main group of 2D vertical memristor materials [56]. In the composite 2D memristor materials, the combined 2D quantum dots (QDs) or interlayered 2D structures memristor device are employed as the charge trapping sites in memristor devices [56]. This strategy has taken to improve the switching speed, retention time, power consumption and endurance of memristor devices [56]. Both VCM and ECM restive switching mechanisms are observed in the memristive performance of 2D composite devices. The switching mechanisms of 2D based composite memristor devices are mostly attributed to two individual mechanisms. One of them is the development and rapture of conductive filaments assisted by the migration of different ionic components in 2D composite structure. Another RS mechanism of 2D composite memristor device is attributed to the charge trapping-detrapping due the interfaces phenomenon between 2D material components of memristor [57].

6.3.1.2 Lateral Memristors

In fact, the first report of application of 2D materials in memristor devices was related to the application of single-layer graphene or thin graphitic strips containing a transversal insulating nanogap [58]. The bridging and breaking the nanogap under the employed electric field was the RS mechanism in the first introduced lateral 2D memristor device [58]. Since the reset operation voltage in the lateral memristors is proportional to the length of channels, thus the power consumption is higher than that of vertical structure. The fabrication restrictions are also the other main challenges, where the control of nanogap length and the device variability opened the questions and concerns about the result stability of devices. By the development of advanced fabrication techniques, several new cases of the lateral memristor devices based on 2D materials are introduced recently [59, 60]. Lateral memristor-based on 2D MoS_2 film is turned *on* by the gate voltage in the newly developed design [59]. With this type of design, new functionalities and improved performance was observed. Furthermore, the control of properties of 2D materials and the control of charge carrier migration (vacancies) through the gain boundaries are the other strategies to improve the performance of lateral memristor devices which are attained by the development of advanced fabrication techniques of 2D materials [60]. Aiming to further improvement of 2D based memristors, different memristor designs, various 2D materials, new advanced fabrication techniques and novel fabrication and integration techniques of 2D materials into actual devices are developing [59, 60]. A typical lateral memristor device is shown in Figure 6.3b.

6.3.1.3 Tip-Based Memristors

The tip-based structured memristor device utilizes a conductive and mostly metallic electrode tips on the active layer of memristors as the opposite electrode. Figure 6.3c shows a typical tip-based memristor device in which the Cu metallic tip was inserted

184 Ultrathin 2D Semiconductors

into a SiO_2 ultrathin film with a tungsten (W) back electrode [61]. In this type of memristors, the ion migration occurs around the Cu tip, which is more convenient for device preparation, especially for laboratory and research targets. In this typical design a better control is provided for the resistive switching during the dynamic observation of RS process. The STM redox-based lithography techniques is of the recently developed techniques for dynamic observation of RS behavior in ultrathin memristor films [61].

6.3.2 ATOMIC-LAYERED 2D MATERIALS FOR MEMRISTOR DEVICES

A full description about the progress and development of conventional solid-state memristor devices and their active layers are discussed previously in Table 6.1 and Table 6.2. Compared with conventional solid-state memristors, the active layers in 2D-based memristor devices mostly focus on graphene, graphene derivatives, insulating materials such as h-BN, transition metal dichalcogenide (TMD) semiconductors, metal oxide semiconductors and 2D composite materials [62]. Each group of 2D memristor materials has different memristive characteristics and RS mechanism. Hence, each individual 2D based device should be investigated and discussed separately to understand the RS mechanism in 2D based memristor devices. Apart from the natural characteristics of materials using for classification of 2D based memristor device, authors also can categorize 2D devices based on their morphology, crystallinity and fabrication techniques [62]. In this view, 2D films can be selected in the different main groups. In the first group, single- or few-layered two flakes in which the grain boundaries and intrinsic defects like vacancies are responsible for RS characteristics of 2D materials. The second group belongs to 2D materials synthesized on the various chemical solution–based techniques. The third groups are 2D QDs, which act as the charge trapping sites in memristor materials and are usually mixed in polymer active layers. The fourth group of 2D memristive materials belong to deposited ultrathin 2D films of memristive materials. This type of 2D materials can be either crystalline or amorphous with respect to the deposition techniques. Chemical vapor deposition (CVD) [63], atomic layer deposition (ALD) [64], pulsed laser deposition [65] and other ultrathin film deposition techniques are the most versatile methods for fabrication of ultrathin 2D films for memristor applications.

6.3.3 FABRICATION TECHNIQUES OF 2D MATERIALS FOR MEMRISTORS

The synthesis and fabrication of 2D materials has been achieved mainly by several main methods including the mechanical cleavage of 2D materials from their host material [66], the CDV [63], the solution exfoliation and chemical synthesis methods [67], and recently by ALD of ultrathin 2D films [64]. The mechanical exfoliation has the capability of preparation of high crystalline ultra-pure 2D films with high quality. A large group of 2D based electronic devices has been made over last decade by using the mechanically exfoliated 2D materials. The weak van der

Waals (vdWs) force between the layers of these materials facilitated the mechanical exfoliation of their fundamental layers [66]. Although most of the 2D materials can be mechanically exfoliated and cleaved from their host materials with the highest level of crystallinity, the size controllability issue is accompanied by the low yield in mechanical exfoliation technique which has restricted this method as reliable technique for large-scale engineering targets. On the other hand, the solution-based synthesis has always been one of the main synthesis methods of 2D materials for memristor applications. Liquid phase exfoliation and lithium ion intercalation techniques are highly versatile methods for large-scale exfoliation of 2D materials with low cost and high scalability [67]. One of the main challenges is integration and employment of liquid phase synthesized 2D materials into the practical memristor devices. While spin coating and drop casting are widely used for deposition of 2D materials on desired substrates, the precise control of atomic thickness and conformal deposition of 2D films over substrates are not easily controllable so the precise control of memristor properties is not feasibly attained. As replacement methods, there are several cases, which report the large-scale synthesis with precise thickness controllability for high-quality 2D materials with CVD techniques [63]. However, this technique faces synthesis challenges for several 2D materials, thus the deficiencies of this method should be responded effectively. It is highly promising that other advanced techniques are more versatile for deposition of conformal 2D materials hence the outstanding breakthrough for the development of 2D based memristors is expected to be achieved. In this regard, ALD and molecular beam epitaxy (MBE) [64, 68] are among novel techniques for the large-scale deposition of high-quality 2D materials with boosting applications in electronic, memristive and optoelectronic industry.

6.3.4 ELECTRODES FOR 2D-BASED MEMRISTOR DEVICES

A detailed discussion has been made in previous section on the different types of electrodes and their effects on the characteristics of memristive devices. For 2D materials, some other requirements should also be taken into account. Considering the fact that the 2D films have ultrathin thickness and high reactivity with other materials, the electrodes should be selected wisely. Moreover, the interaction between electrodes and 2D films should be optimized for study of resistive characteristics of 2D materials. It is highly important when the electrochemically active electrodes or transparent flexible electrodes are used either as substrates or electrodes for 2D based memristors. It is worth nothing that in this case the attention must be devoted and concentrated on the physical aspects and interactions between 2D films and conductive electrodes. For example, the difference in work function of 2D materials and metal electrodes can lead to the creation of interface barriers between 2D films and memristor component and fundamentally change the resistive behavior of devices. Table 6.3 has summarized the properties of some of the memristor devices based on 2D materials beyond the graphene. Please note that Table 6.3 does not list all of the 2D memristor devices.

TABLE 6.3
Summary of Voltaic Memristor Devices Based on 2D Materials Beyond Graphene

Active layer materials	Electrodes	Fabrication method	RS mechanism	Switching voltage (V)	ON/OFF ratio
MoS_2-PVP	Al, rGO	Solution	SCLC	3.5	10^2
MoS_2 @ ZIF-8	rGO, rGO	Solution	Charge trapping	3.5	10^4
PtAg + MoS_2	Al, ITO	Solution	Charge trapping	−5	–
MoS_2-PMMA	Au, G	CVD	Charge trapping	5	10^3
MoS_2-PVA	Ag, Ag	Solution	SCLC	3	10^2
MoS_2-PCBM	ITO, Al	Solution	Charge trapping	2	10^2
MoS_2-GO	ITO, Al	Solution	SCLC	1.5	10^2
GO/MoS_2/GO	Au, Al	Solution	SCLC	3.5	10^4
MoS_2 nanofibers	rGO, rGO	Solution	SCLC	–	10^2
MoS_2	Ag, Au	-	ECM, CFs	–	–
MoO_x/MoS_2	Ag, Ag	LPE	Charge trapping	0.1–0.2	10^6
$MoS_{2-x} O_x$	G,G	Mech-exfoliation	Ion migration	1.5	10^2
Monolayer MoS_2	Au, Au	CVD	Grain boundary	3.5	10^3
1T-MoS_2	Ag, Ag	Mech-exfoliation	Lattice distortion	0.1	10^2
MoS_2 atomristor	Au, Au	CVD	CFs	1	10^4
MoS_2 double-layer	Cu, Au	MOCVD	ECM	0.1–0.2	–
h-BN	ITO, Ag	CVD	ECM	0.7	10^2
h-BN-PVOH	ITO, Ag	Solution	SCLC	0.78	10^2
h-BN	Ti, Cu	CVD	Grain boundary	0.4	10^6
h-BN atomristor	Au, Au/Ni	CVD	CFs	–	10^7
BP QDs-PVP	Ag, Au	Solution	SCLC	−1.2	10^4
PMMA/BP QDs/ PMMA	Al, Al	Solution	CFs	2.8	10^7
$MoSe_2$	Au, ITO	Solution	Charge trapping	3.8	10^3
WSe_2	Ag, graphene	Mech-exfoliation	Vacancies	0.2–0.5	10^2
$MoTe_2$	Ti/Ni, Ti/Au	Mech-exfoliation	Lattice distortion/ phase transition	2.3	10^2
WS_2	Pd, Pt	Solution	Vacancies	0.6	–

Source: Reprinted with permission from [21].

6.4 ELECTRICAL-PULSE-TRIGGERED MEMRISTOR DEVICES BASED ON ULTRATHIN 2D MATERIALS

Memristors based on ultrathin films and 2D materials are the promising candidates and alternatives for the next generation of memristors and synaptic devices. Specifically, the flexibility, transparency and their developed fabrication techniques are the main advantages of employment of 2D materials in memristor devices [69]. Because of the advantages of two-dimensionality, a host number of unprecedented properties is expected to be realized by using 2D materials in memristors. The new

Memristive Devices Based on 2D Materials

capabilities of advanced fabrication techniques let us to scale down the thickness of 2D materials to few nanometers and then build up heterostructured 2D films and nanostructures. So far the technological concerns of researchers and engineers are focused on several fundamental characteristics of memristor devices, including power consumption, reliable performance for reasonable cyclic applications, speed, production costs and stability [70]. The proposed new 2D materials should satisfy and improve the performance of memristors toward the smaller, faster, and cheaper units [70]. The next lines will focus on the recent advancement in fabrication of memristor devices based on ultrathin film and 2D materials beyond the graphene and its derivatives. In so doing, the fabricated devices are investigated based on their different memristive characteristics. Since the driving forces for triggering the memristive properties are not merely originated from the applied electrical pulses, authors initially focus on the memristor devices, which use the voltaic pulses to trigger the memristive characteristics. After that the optoelectronic memristors based on 2D materials will be introduced and discussed.

6.4.1 Complementary Resistive Switching (CRS) in TaO_x, HfO_x, and TiO_x Devices

Redox-based resistive random access memories (ReRAM) offer several capabilities for application in memories, logic devise and neuromorphic applications [71]. The main advantages of ReRAM return to their low power consumption, subnanosecond switching time and their scalability for size reduction to the atomic level [71]. Two fundamental RS mechanisms of ECM and VCM devices are observed in the oxide-based memristors. The transition metal oxides and refractory oxides such as HfO_2, TiO_2, and Ta_2O_5 are well known for their VCM resistive switching mechanisms [72]. The formation of conductive filaments and the switching phenomenon are widely accepted to be due to the redox reactions and the transport of oxygen ions or/and oxygen vacancies as mobile donors [72]. During the STM studies of RS behavior of refractory oxide-based memristors, it was observed that the CFs are partially re-oxidized after their formation and subsequently a disklike layer with the thickness of few nanometers is formed which seems to be the responsible for RS behavior of films [72]. Figure 6.4 demonstrates the switching behavior studied by scanning tunneling microscope (STM) on the 2 nm thick amorphous TaO_x film deposited on Ta bottom electrode. The STM tip was used as the electrode to apply cathodic and anodic bias at a fixed tip position over the TaO_x film to trigger the RS events (Figure 6.4a) and then the time dependence of tunneling current was evaluated. By applying −3 V, strong increase of current accompanied by the LRS indicated the occurrence of RS phenomenon. The measured high conductance of TaO_x film cannot be merely attributed to the formation of oxygen-deficient conductive filaments. The establishment of a metallic quantum point contact between STM tip and the surface of TaO_x film was introduced as the main reason for considerable increase of conductance values [72]. It was shown that the formation of metallic Ta bridge and breakdown of vacuum provides the direct galvanic contact via Ta atoms (Figure 6.4 A, B, C, and D). Applying sufficient voltage and facilitating the mobile transfer of the metal cations inside TaO_x film were the main principal requirement criteria to facilitate the development of atomic point

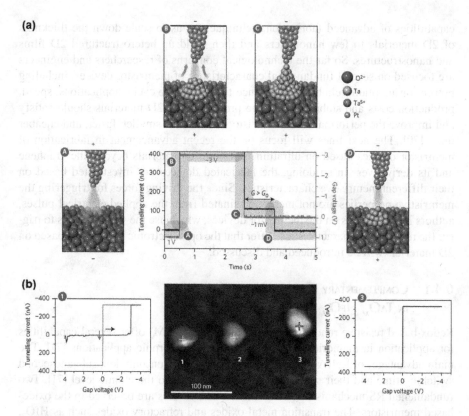

FIGURE 6.4 The schematic of scanning tunneling switching when the voltage of conductive STM tip is negative under ultrahigh vacuum (UHV) conditions. (a) The dependence of tunneling current to the time of measurement for 2 nm thick TaO_x film. The voltage is changing stepwise from +1 V to −3 V to −1 mV. The calculated conductance (G) is from the current at the tip of −1 mV which exceeds the quantum conductance (G_0). It confirms the temporarily formation of metallic phases at contact point due to diffusion of Ta cations to negatively charged STM tip. The graphics in Figure 6.4a shows the corresponding sequential stages for approaching, formation of the Ta bridge, complete "Ta" bridge connection, and detachment and deterioration of the Ta bridge when the voltage is reversed. (b) The STM image of the surface of TaO_x layer for time-resolved STM measurements when negative −1 V is employed to the tip with a tunneling current set point to 1 nA. The highly conductive regions are the location where the local RS was performed. The characteristics of *I-V* curves show that the regions are either purely metallic or a mixed metallic-semiconducting behavior. Reprinted with permission from Wedig, A., Luebben, M., Cho, D. Y., Moors, M., Skaja, K., Rana, V., … Valov, I. (2006). Nanoscale cation motion in TaO_x, HfO_x and TiO_x memristive systems. *Nature Nanotechnology*, *11*(1), 67–74.

contact. In fact, the required voltages for RS are used to increase the chemical potential of the electrons (Fermi level) to above the required energy level for reduction of Ta^{5+} to Ta atoms [72]. It was found that the Ta interstitials in the reduced amorphous TaO_x structure are the main probable cation defects. After removing the electric field, the metallic bridge was deteriorated and the Ta atoms were re-oxidized even at high

Memristive Devices Based on 2D Materials

vacuum conductions. The STM image of the surface of a 2 nm thick TaO_x ultrathin layer films is depicted in Figure 6.4b [72]. The formation of highly conductive region with thickness of 40 nm is the main microscopic evidence confirming the formation and expansion mechanisms of conductive channels in oxide film. This observation is a highly important evidence of credibility of newly born STM redox-based lithography techniques [72]. The later I-V measurements were performed to investigate the effect of development of metallic bridge between the STM tip and TaO_x film. The I-V measurements in LRS region confirmed the semiconducting characteristics for TaO_x film (Figure 6.4b1), while the I-V measurement curves after three consecutive cyclic I-V curves shows the formation of highly conductive metallic bridge (Figure 6.4b3). The observation of semiconducting characteristics in early stage of measurements can also confirm the possibility of reduction-oxidation (redox) reactions. It is estimated that the most of oxidation reactions at the tip of STM electrode is attributed to the oxidation of oxygen ions of TaO_x, whereas at the counter electrode (TaO_x nanofilms) the reduction of both metallic cations and residual oxygen were possible [72]. It was found that the current density at the bottom electrode was much smaller than that of the top electrode, which is attributed to the typical geometrical characteristics of bottom electrode with its disk like geometry and atomic layer thickness. It was concluded that the switching mechanism is not merely attributed to the filamentary switching mechanism due to redox reaction [72]. Therefore, there should be another mechanism facilitating the cation motions. The STM studies of HfO_x and TiO_x nano-metric films yielded to the comparable results to those obtained for ultrathin TaO_x film. It was found that mobile cations could considerably participate in RS performance of HfO_x and TiO_x 2D films to finalize the effect of oxygen vacancies on RS phenomenon. The cation mobility was also observed for TiO_{2-x} film in tracer diffusion experiments at the high temperatures [73]. In addition, the study of Ti/TiO_2 system has demonstrated the presence of ECM-type behavior and metallic LRS at the room temperature [74].

There is a connection between VCM and ECM switching mechanism for devices made by TaO_x, HfO_x and TiO_x ultrathin oxide films [72]. Figure 6.5 depicts how different mechanisms control the RS behavior of these memristor devices. It is observed that the RS performance of devices is attributed to the contribution of the mobile ionic species, which represents oxygen and metallic cations. The bipolar counter-eight-wise-switching behavior is also observed when the cyclic potentiodynamic (I-V Sweep) measurements were performed for Ta/TaO_2-Ta_2O_5/Pt. cells. The device was reset to HRS at the positive voltage and was set to LRS when the negative bias was imposed. It was shown that the device has two values of resistance in LRS mode stimulating the capability for the multilevel switching operations. By increasing the maximum sweeping voltage in the set direction, a complementary resistive switching (CRS) behavior was demonstrated when the unit was set at the lower voltage and again reset at the higher voltage at the same polarity (Figure 6.5b). At LRS the resistance value were about $100\ \Omega$. Such a low resistance for 2D based memristor device in LRS mode exhibits the formation of metallic-type filamentary electrical conduction behavior [72]. The typical thermochemical switching mechanism was excluded from the RS mechanism of the above-mentioned 2D memristor, since two Set events occurred at the different absolute voltages. On the other hand, considering that the two reset events happened at the same absolute voltages, the engagement of

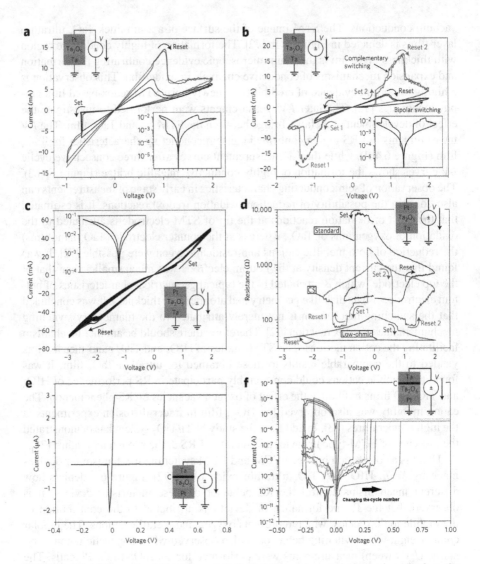

FIGURE 6.5 The cyclic *I-V* curves of Pt/TaO$_x$/Ta based memristor devices. Before the measurement the device was formed into LRS by applying a voltage of −5 V at the current compliance of 1 mA. a. The typical bipolar switching when the sweeping voltage is low. The logarithmic scale of bipolar switching is shown in the inset. b. CRS curves of the same device when the same device when the maximum negative voltage is increased. The bipolar switching is also presented for comparison. c. The *I-V* curve for highly conducting RS behavior of the same device when a high negative sweeping voltage was employed. A low sweep rate is necessary to prevent the device breakdown. d. The comparison of resistance values calculated for switching *I-V* curves. The curves show different levels of conductance for standard, LRS and CRS cases. e. The ECM-type switching of Ta/C/Ta$_2$O$_5$/Pt cell. f. The cyclic sweeping *I-V* curves of Ta/C/Ta$_2$O$_5$/Pt cell. Reprinted with permission from Wedig, A., Luebben, M., Cho, D. Y., Moors, M., Skaja, K., Rana, V., ... Valov, I. (2006). Nanoscale cation motion in TaO$_x$, HfO$_x$ and TiO$_x$ memristive systems. *Nature Nanotechnology*, 11(1), 67–74.

thermally RS mechanism was highly possible. Therefore, the fabricated Ta/TaO$_2$-Ta$_2$O$_5$/Pt cells showed the CRS behavior attributing either to the mobility of oxygen vacancies or to the movement of cation defects [72]. The combination of both RS mechanisms is also predicted. The versatility of device performance was further confirmed that the CRS has originated from the mobility and movement of both oxygen vacancy and metal cations [72]. The increases of negative applying voltage in sweeping *I-V* curves was accompanied by the shift of RS mechanisms from CRS to bipolar RS with highly conducting LRS and HRS events and much smaller hysteresis loops [72]. The ultralow resistance (40–100 Ω) in bipolar RS process is indication of the development of highly conductive metallic channels (Figure 6.5d). The same metallic conductivity type was observed in the study of resistive performance of ultrathin TiO$_2$ film-based devices. It was further demonstrated and practically confirmed that the employment of thin-intermediate amorphous carbon layer between the Ta and Ta$_2$O$_5$ ultrathin films resulted in the ECM behavior during RS with asymmetric set/reset voltage and driven currents (Figure 6.5e, f) [72]. It was additionally found that the set/reset currents are highly dependent on the materials system. Generally, it was demonstrated that the RS mechanism in ultrathin nanoscale TiO$_x$, TaO$_x$ and HfO$_x$ films could be originated from both diffusion of oxygen vacancies and migration of metallic cations. Both mechanisms can lead to the formation of highly conductive filamentary channels by the interconnection of reduced ionic species [72]. Furthermore, it was found that the introduction of ultrathin amorphous carbon layer altered the interface dynamic during the RS process of Ta/Ta$_2$O$_5$ devices [72]. The ECM, VCM-type switching was highly dependent on the boundary conductions. Hence, the employment of intermediate rutile TaO$_2$ film at the Ta/Ta$_2$O$_5$ interface altered the charge transfer mechanisms and improved the movement of both metallic and oxygen vacancy ions [72].

6.4.2 CRS in Heterostructured Memristor Devices

The improvement of energy efficiency and density performance of the cross-bar memristors are the main technological challenges in front of the development of this type of memory devices. This problem arises from the structure of cross-bar memristors, which suffers from the problem of sneak paths [75]. To reduce the power consumption and tackle the challenge of sneak path problem, the complementary resistive switching (CRS) mechanisms were introduced [75]. The schematic structure of the bipolar resistive switching device is depicted in Figure 6.6a. By employing the one-step plasma oxidation of TiN film at the room temperature, a partially oxidized titanium oxynitride (TiN$_x$O$_y$) was fabricated over the TiN film. The deposition of TiO$_2$ layer over TiN$_x$O$_y$ film was followed by fabrication of Pt top electrode. The final structure of device was Pt/TiO$_{2-x}$/TiN$_x$O$_y$/TiN sandwiched structure (Figure 6.6b and c). The device showed complementary resistive switching behavior within the low operation voltage of 1 V. The corresponding *I-V* curve for 20 consecutive cycles is presented in Figure 6.6d demonstrating well-defined clockwise (CW) bipolar RS with high degree of repeatability for Pt/TiO$_{2-x}$/TiN$_x$O$_y$/TiN memristor. When the applied voltage was swept at the sequence of 0 V$\rightarrow$ −1 V$\rightarrow$ 0 V $\rightarrow$1 $\rightarrow$0 V, a quadrupled switching behavior with high repeatability was observed. The first current

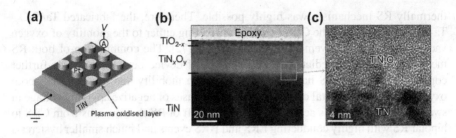

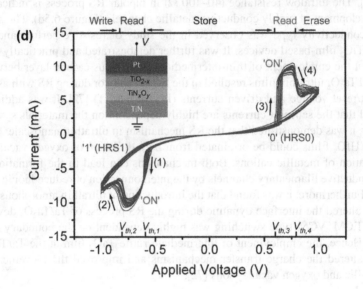

FIGURE 6.6 (a) The schematic of Pt/TiO$_{2-x}$/TiN$_x$O$_y$/TiN cell device. (b) The cross-sectional TEM image of the Pt/TiO$_{2-x}$/TiN$_x$O$_y$/TiN cell and (c) plasma oxidized TiN films. (d) The I-V characteristics of Pt/TiO$_{2-x}$/TiN$_x$O$_y$/TiN cell which shows the CRS behavior of device for 20 consecutive cyclic curves. The inset shows the structure of device. Reprinted and reproduced with permission from Tang, G., Zeng, F., Chen, C., Liu, H., Gao, S., Song, C., ... Pan, F. (2013). Programmable complementary resistive switching behaviours of a plasma-oxidised titanium oxide nanolayer. *Nanoscale*, 5(1), 422–428.

jump discovered at −0.6 V when the applied negative voltage increasing, which is the first threshold voltage (V$_{th.1}$) and then the device turned *on* and shifted to LRS. Afterwards, the increase of negative applied voltage was accompanied by the occurrence of second resistance switching when the device resistance again was altered and switched to HRS^1 at (V$_{th.2}$ = −0.8 V). The device stays in HRS^1 mode while the gate voltage sweeps to positive values. At V$_{th3}$ = 0.6 V the device again turned *on* and switched to LRS^2, and the high resistive state was kept until the voltage reached to V$_{th4}$ = 0.8 V, and again device switched to HRS^0. To explain the memory behavior of device, the HRS^0 and HRS^1 was distinguished by applying the read voltages in the range of V$_{th3}$ < V < V$_{th4}$. The write and erase operation can be respectively operated by application of a voltage value of V < V$_{th2}$ and by application of a voltage value of V > V$_{th4}$. The storage region is the V$_{th1}$ < V < V$_{th3}$. Since both switching process occurred in low voltage values, the Pt/TiO$_{2-x}$/TiN$_x$O$_y$/TiN sandwiched memristor guarantees the capability of suppressing the crosstalk and power consumption problems in the cross-bar array devices [75].

The RS behavior of Pt/TiO$_{2-x}$/TiN$_x$O$_y$/TiN can be explained by the oxygen vacancy migration under bias voltage. Furthermore, the TiO$_{2-x}$ layer consist of two individual layers with different resistance states, which are connected in series to each other. Pt/TiO$_{2-x}$ top interface (R$_{top}$) and the other layer at the TiO$_{2-x}$/TiN$_x$O$_y$ bottom interface (R$_{bot}$), as shown in Figure 6.7a. The upper Pt/TiO$_{2-x}$ is always in LRS and plays a vital role in modulation of oxygen vacancy concentration in the TiO$_{2-x}$/TiN$_x$O$_y$ bottom interface. Imposing a negative bias voltage, a significant number of oxygen vacancies are introduced into the TiO$_{2-x}$ nano-layer by electroforming, and consequently, oxygen-deficient conduction channels are formed which facilitates the LRS switching. It is appeared that simultaneously formed oxygen are physically absorbed and stored in TiN$_x$O$_y$ layer acting as the oxygen reservoir layer (Figure 6.7b). By applying a high positive voltage the oxygen vacancies pushed toward the TiN$_x$O$_y$ interface. It results in the decrease of oxygen vacancy carrier at the top layer accompanied by the subsequent switching from LRS to HRS in top TiO$_2$ layer while the

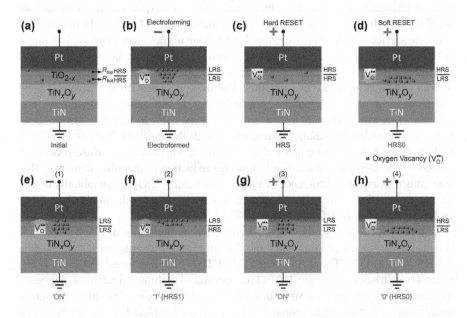

FIGURE 6.7 The proposed switching model for CRS behavior of Pt/TiO$_{2-x}$/TiN$_x$O$_y$/TiN cell device which also shows the oxygen vacancy distribution in active layer of memristor. (a) When the device is in initial stage. (b–d) show the electroforming, hard and soft reset switching process respectively. (e–h) depict the four switching stages. The oxygen vacancies are shown by red globes. The active layer of memristor consists of a top TiO$_x$ nano-layer separated from TiN$_x$O$_y$ bottom layer. The red-dashed lines represent the border between HRS and LRS modes. Marked arrows show the direction of migration of oxygen vacancies. Reprinted with permission from Tang, G., Zeng, F., Chen, C., Liu, H., Gao, S., Song, C., ... Pan, F. (2013). Programmable complementary resistive switching behaviours of a plasma-oxidised titanium oxide nanolayer. *Nanoscale*, 5(1), 422–428.

bottom layer (TiO_{2-x} layer adjacent to TiN_xO_y film) is in LRS state. Furthermore, the annihilation of oxygen vacancies occurs in TiN_xO_y film resulting in the resistant switching from LRS to HRS state in the bottom layer TiN_xO_y film. The observations confirmed that the resistance of the top and bottom layers can be switched between HRS/HRS and LRS/LRS mode while the intermediate layer is in different resistive states. It facilitated the conventional bipolar RS behavior in $Pt/TiO_{2-x}/TiN_xO_y/TiN$ device as presented in Figure 6.7. The detailed explanations are included in the caption of Figure 6.7.

6.4.3 RS Phenomenon in TiO_2 Thin Films

6.4.3.1 Filamentary RS in TiO_2 Ultrathin Film

The microscopic observation and characterization of RS mechanism are highly important methods assisting the understanding of physical aspects of resistive switching in the active layer of memristors during switching process. From the microscopic point of view, the RS is broadly categorized into two following mechanisms. In the valance change mechanism, the electro-migration of oxygen vacancies or cations alters the distribution of charge carriers and valence states of cations. In this mechanism, the device usually shows the bipolar RS. One of the famous reported example is the migration of oxygen vacancies through the dislocation networks of $SrTiO_3$ structure [76]. The second mechanism is mostly attributed to the combined effects of both joule heating and redox reaction. The interplay between the thermal effects and redox reaction in filamentary area facilitates the RS behavior of devices. The mechanism is also known as a fuse-antifuse mechanism. In fuse process the electroforming creation of metallic filaments through the insulator body is facilitated by both joule heating and employed electric field, whereas the antifuse process is caused by the thermal assisted (joule heating) reduction of metallic filaments. This mechanism assists the occurrence of unipolar switching. TiO_2 is undoubtedly one of the most promising semiconductor materials demonstrating both unipolar and bipolar RS mechanisms in TiO_2 memristors [77].

The microstructural study of the cross-section of memristor device by HRTEM shows the formation of Magnéli nanofilaments as the conductive channels during RS process [77]. The as-deposited ALD TiO_2 film was found to have a semi-amorphous pristine structure [77]. The thermal switching might have triggered the transition from pristine structure to more stable phases. The high-resolution transmission electron microscope (HRTEM) studies depicted the development of Ti_nO_{2n-1} or so-called Magnéli phases in conductive filaments during RS of TiO_2 ultrathin film [77]. A typical filament starts growing from the top cathode pole and then grows similar to a conical pillar toward the anode pole. Figure 6.8a demonstrates well-connected conductive filament of Magnéli phases in the TiO_2 active layer [77]. Besides the well-grown nanofilaments, there are several cases of disconnected nanofilaments with Magnéli structure (Figure 6.8b). In fact, during the RS process several nanofilaments started to grow, but when the first nanofilament reaches to the opposite electrode, the current flows between two electrodes and the growth of the other nanofilaments are stopped. The study on TiO_2 film observes that

Memristive Devices Based on 2D Materials

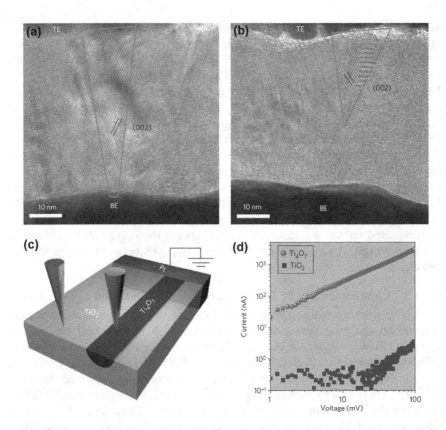

FIGURE 6.8 (a) The high-resolution TEM image of Ti$_4$O$_7$ Magnéli nanofilament with pillar shape. The filament has grown completely and has connected the top and bottom electrodes to each other. (b) The defective Ti$_4$O$_7$ Magnéli nanofilament developed in TiO$_2$ active layer of memristor. (c) Schematic shows the experimental set-up for measurement of I-V characteristics of the TiO$_2$ and Ti$_4$O$_7$ phase on the Set device. (d) The blue line is the characteristic of TiO$_2$ and the red dots show the characteristics of Magnéli nanofilament. Reprinted and reproduced with permission from Kwon, D. H., Kim, K. M., Jang, J. H., Jeon, J. M., Lee, M. H., Kim, G. H., ... Hwang, C. S. (2010). Atomic structure of conducting nanofilaments in TiO$_2$ resistive switching memory. *Nature Nanotechnology*, 5(2), 148–153.

the diameter of a typical filament is around 5 to 10 nm and the distance between them is between 0.1 to 5 μm. This parameter determines the ultimate packing density of resistive random access memory device based on TiO$_2$ ultrathin film. The density and distance between the nanofilaments are vital when the device size is decreased to the nanoscale range. Technically, from the RS point of view, the lower number of CFs would be better for device performance and energy consumption. The multifilament formation will deteriorate the reproducibility of RS behavior of the device [77]. In the present case it was shown that the size of memristor device could be scaled down to the tens of nanometers, considering that the filament diameter is in the range of 5 to 10 nm in fabricated Pt/TiO$_2$/Pt device. To confirm that the Magnéli phases are responsible for the resistive switching, the conductivity of nanofilaments was investigated. The measurement set-up and the corresponding in-situ *I-V* curves of the memristor device are presented in Figure 6.8c and d, respectively. The device was set and then the STM tip was employed operating at the conductive atomic force microscope (AFM) mode. By adjusting the STM tip on TiO_2 and Ti_4O_7, the *I-V* characteristics of the main-body and Magnéli phase of filament were measured. The measurements confirmed that the local conductivity of Magnéli filament is about 1000 times higher (Figure 6.8d) than that in amorphous TiO_2 [77]. It further established the correctness that the reason for the high conductivity of set device returns into the high conductivity of Magnéli phase and the RS is facilitated by their corresponding Ti_4O_7 phase structure [77]. The *I-V* measurements further verified that the high conductivity is not attributed to the oxygen-deficient TiO_2 phase or even is not contributed by the metallic atoms in TiO_2 film [77]. The in-situ observation of reset process confirmed that the Magnéli filaments were partially ruptured and the electrical connections between the joint conductive filaments had been broken, consequently the electrical connection between the top and bottom electrodes has been cut.

6.4.3.2 Complementary RS in In-Doped TiO_2 Ultrathin Film

Ion doping is one the main strategies to alter the optical and bandgap characteristics of semiconductor materials for different functional applications. The doped ions can also affect the RS mechanism of fabricated devices, since the memristive characteristics are also tangibly affected by the presence of other foreign ionic species inside active materials of memristors. For instance, memristor device was made based on ultrathin 7 nm thick TiO_2 film by using ALD technique [78]. To understand the effect of Indium (In) intercalation in active layer of memristor (TiO_2), an ion intercalation set-up was made. The ion intercalation in this set-up was facilitated by imposed driving voltage on the ion-containing solution covering the surface of 2D TiO_2 film. The intercalation mechanism of In-ions into 2D TiO_2 sublayer requires that the In-ions overcome the surface diffusion barrier and bulk diffusion barrier at TiO_2 interface. By applying an external V_G, In-ions are pushed to accumulate at the surface of 7.0 nm think TiO_2 film. By further increase of V_G, In-ions would overcome the surface diffusion barrier and the bulk diffusion barrier to inject into TiO_2 sublayer [78]. The potentiodynamic measurements (*I-V* sweep) of Pt/TiO_2/Au and Pt/In-doped TiO_2/Au memristor devices under the dark condition show different RS mechanisms for In-ion doped and nondoped TiO_2 film. Pt/TiO_2/Au device shows bipolar switching behavior (Figure 6.9a). The set and reset process were measured on ALD TiO_2 2D layered film sandwiched between Au and Pt films (Figure 6.9b). Before switching, all samples were formed into their low resistive state (LRS) by applying −3 V at a current compliance of 1 mA. The typical bipolar switching curves obtained during sweeping from $0{\rightarrow}1{\rightarrow}0{\rightarrow}{-}1{\rightarrow}0$ in the case of Pt/TiO_2/Au samples (Figure 6.9a) with V_{Set} =0.94 and V_{Reset} =−0.98 during *I-V* sweeping measurement. The expected performance was observed during *I-V* sweeping in which the cell was set to LRS at the positive voltage and reset to HRS at the negative voltage, well known as counter-eight-wise-switching mechanism [78]. Bipolar switching can be explained by the ionic drift of oxygen vacancies owning to employed voltage on the electrochemically inert Au and Pt electrodes. This device is categorized as valence change memory

Memristive Devices Based on 2D Materials

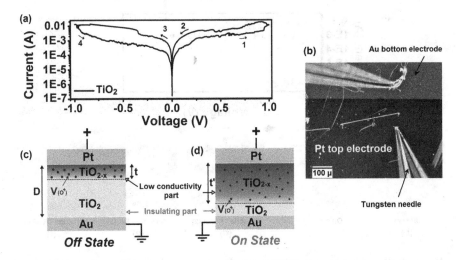

FIGURE 6.9 (a) Typical logarithmic scale of the *(I-V)* sweeping curves for Pt/TiO$_2$/Au. The arrows show the direction of sweeping *I-V* curves. (b) The optical image shows the actual view of memristor device when the tip-electrodes are connected to the bottom Au and top Pt electrodes. (c) and (d) show the gradual movement of oxygen vacancies after imposing the driving gate voltage. Reprinted with permission from Karbalari Akbari, M., & Zhuiykov, S. (2019). A bioinspired optoelectronically engineered artificial neurorobotics device with sensorimotor functionalities. *Nature Communications*, 10(1), 3873.

(VCM) cells. The schematic representation of oxygen vacancies distribution for both *off* and *on* states are shown in Figure 6.9c and d. The dielectric member of memristor with thickness of *D* has two parts: low conductive TiO$_{2-x}$ with the thickness of *t* and insulating part with the thickness of *D-t*. The applied voltage on the electrode caused the ionic drift of oxygen vacancies, which is accompanied by the border shift between low conductivity component and insulating part of active layer of memristor. In this condition, the *I-V* curve of device is not further Ohmic and will be nonlinear. In this proposed model the thickness of insulating layer should be small enough (few nanometer) to let the electric field develops and promotes the ionic drift of defects [78].

A complementary resistive switching (CRS) behavior is observed during the studies of RS of Pt/In-doped TiO$_2$/Au device (Figure 6.10a) where device is set at the higher voltage and again reset at the lower voltage of the same polarity. The cell was set for the first time at $V_{Set1} = 0.732$ V and then reset again in the same polarity at $V_{Reset1} = 0.1277$ V. In the negative voltage, the second LRS to HRS switching occurred at $V_{Set2} = -0.7055$ V and again reset to HRS at $V_{Reset2} = -0.0901$ V. Since two set and reset events occurred almost at the same absolute voltage, thermally assisted mechanism is one of the most possible theory, which can explain the memristor behavior. The unipolar switching confirms the filamentary nature of resistive switching. The filamentary resistive switching is mostly based on the electrochemical metallization (ECM) of the mobile ionic species caused by the migration of metallic cations. It facilitates the development of filamentary metal bridges (*on* state), which

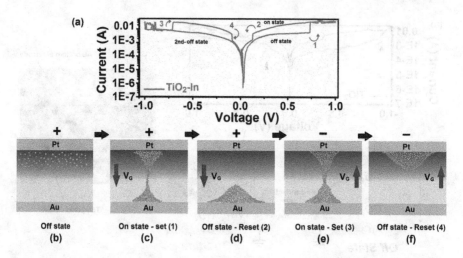

FIGURE 6.10 (a) Typical logarithmic scale depiction of *I-V* sweeping graph of Pt/In-doped TiO₂/Au memristor. (b) Device is set to *off*, the metallic cations are distributed in top layer of In-doped TiO₂ film. (c) The formation of filamentary conductive channels, the device is set to *on* (Set, stage 1). (d) The depletion of metallic cations adjacent to the bottom electrode. The device is set off again (reset, stage 2). (e) By imposing the reverse gate voltage, the metallic cations again move toward the cathode electrode (top Pt electrode) and form a metallic cation bridge and the device is set On (Set-stage 3). (f) By depletion of metallic cations, at Pt electrode the device is set to *off* again (reset, stage 4). Reprinted with permission form Karbalari Akbari, M., & Zhuiykov, S. (2019). A bioinspired optoelectronically engineered artificial neurorobotics device with sensorimotor functionalities. *Nature Communications*, *10*(1), 3873.

can be ruptured later (*off* state). The filamentary mechanism is the characteristic of memristor devices, which are fabricated with electrochemical active electrodes, while the In-doped TiO₂ memristor device fabricated by inert Au and Pt electrodes. Generally, CRS behavior can be explained by both mechanisms, that is, vacancy drift and migration of In-ions. The versatile switching behavior of the In-doped TiO₂ device indicates that both anion and cation components are involved in CRS mechanisms. The proposed model suggested that the redistribution of oxygen vacancies and Indium ions inside TiO₂ layer can affect, alter and facilitate the resistive switching. In the heterostructured oxide stacks, the conjunction of an oxygen-deficient oxide layer (LRS mode) with another oxygen rich layer (HRS mode) trigger the CRS of heterostructured memristor which is caused by the movement of oxygen vacancies. The analytical XPS measurements confirmed that the oxygen concentration in 25 Å thick top layer of In-doped TiO₂ film (7.0 nm) is less than the bottom layer of TiO₂, creating a junction between deficient (LRS) and rich oxygen (HRS) components of In-doped TiO₂ film. This gradual change of oxygen concentration can trigger occurrence of CRS of memristor due to the difference in resistance of TiO₂ stack.

From another point of view, the formation of filamentary conductive channels assist the RS and increase the conductance of devise. At the initial state or *off* state (Figure 6.10b), the top In-doped layer (which also is an oxygen-deficient layer) is in LRS mode while the bottom layer in HRS. By applying the threshold voltage, the ionic species migrate into lower insulating layer resulting in the formation of conductive filaments and paths (Figure 6.10c) and then device reaches to its *on* state (Set1). At higher positive voltage, the ionic species in upper layer (mostly In-ions) are depleted (Figure 6.10d) and again device will reach to its *off* state (reset 2) and demonstrates HRS behavior. The similar phenomenon is expected to occur at the negative biases in which the first *on* state (Figure 6.10e) switches the memristor to new LRS, following another *off* state after depletion of the ionic species (Figure 6.10f). The versatility of CRS behavior of all samples was repeatedly tested several times and the same behavior was observed.

6.4.3.3 Bipolar RS in Pt/TiO$_2$/Ti/Pt 2D Memristors

Two opposite polarities are reported in the bipolar resistive switching (BRS) behavior or Pt/TiO$_2$/Ti/Pt memristor devices [79]. The device has a stacked structure with TiO$_2$ film with ultrathin thickness. The cross-bar device has 60 × 100 nm^2 area with a stacked structure. The top view, cross-sectional and graphical scheme of device is presented in Figure 6.11 a, b, and c. The nanocross-bar structure consists of the 5 nm Ti and then the TiO$_2$ film with either 3 or 6 nm thickness, which are stacked between two top and bottom Pt electrodes. The thermal ALD technique was employed to deposit TiO$_2$ films. The voltage was imposed on Pt bottom electrode while the Pt top electrode is grounded. The Ti ultrathin film is deposited between the TiO$_2$ active layer and top Pt electrode.

The *I-V* characteristics hysteresis loops of Pt/TiO$_2$ (6 nm)/Ti/Pt device is shown in central part of Figure 6.12. It was found that the device shows two different types of RS mechanism as the VCM memristor. The typical eight-wise (8w) and counter-8-wise (c8w) RS behavior is observed during the sweeping *I-V* measurements, which are respectively shown by red and black colors in Figure 6.12. The c8w (black curve) and 8w (red curve) are two different RS mechanisms in VCM-type ReRAMs [79]. The c8w is in fact the manifestation of regular filamentary VCM type of resistive switching of BRS process. The graph is called 8w since the curves evolution is

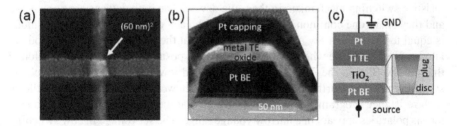

FIGURE 6.11 The Pt/TiO$_2$/Ti/Pt stacked cross-bar cells: (a) the SEM top view of (60 nm)^2 cross-point, (b) The TEM cross-sectional image of 100 nm^2 Pt/TiO$_2$/Ti/Pt device with (c) its corresponding schematic graph of the plug and disc region in the TiO$_2$ layer. Reprinted permission from Zhang, H., Yoo, S., Menzel, S., Funck, C., Cüppers, F., Wouters, D. J., ... Hoffmann-Eifert, S. (2018). Understanding the coexistence of two bipolar resistive switching modes with opposite polarity in Pt/TiO$_2$/Ti/Pt nanosized ReRAM devices. *ACS Applied Materials and Interfaces*, *10*(35), 29766–29778.

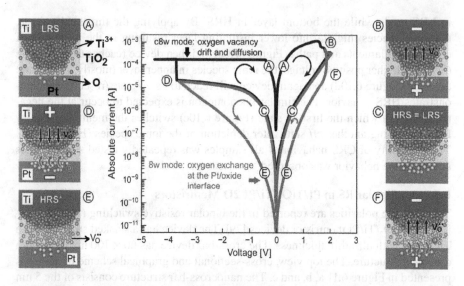

FIGURE 6.12 The bipolar resistive switching (BRS) modes in the Pt (bottom electrode)/TiO$_2$/Ti/Pt (Top electrode) nanocross-bar cell with the area of 100 nm^2. The arrows show the switching orientation. The state and switching events are shown by capitalized words. The occurrence of BRS is attributed to the competition of oxygen vacancy drift and diffusion process with the transfer reaction at the interface of TiO$_2$ and high barrier Pt electrode (without Ti interlayer). The left and right graphical schemes correspond to the dominant ionic process. The color codes inside the *I-V* graphs depict the corresponding atoms, ions or electrodes in memristor device. Reprinted with permission from Zhang, H., Yoo, S., Menzel, S., Funck, C., Cüppers, F., Wouters, D. J., ... Hoffmann-Eifert, S. (2018). Understanding the coexistence of two bipolar resistive switching modes with opposite polarity in Pt/TiO$_2$/Ti/Pt nanosized ReRAM devices. *ACS Applied Materials and Interfaces, 10*(35), 29766–29778.

similar to a written "8" number in English. In the similar way, the graph is called c8w, since the graph shape is similar to written "8" in opposite direction.

The Pt/TiO$_2$/Ti/Pt memristor device reveals the coexistence of a c8w and 8w resistive switching mechanism in the same device. The both devices are nonvolatile and they share one common resistive state with each other. The HRS of c8w mode is equal to the LRS of 8w mode. It was found that the transition from one mode to another one is possible when the driving voltage is positive. Noteworthy, it was found that the 8w BRS in the TiO$_2$ cell is a self-limited low energy and also nonvolatile switching process. Furthermore, the 8w device showed the capability of multilevel high resistance programing. It was established that the transition between two hysteresis polarities appears in a narrow voltage range. The set voltage in c8w switching mechanism is changing proportional to the thickness of TiO$_2$ active layer. It was discovered that the 3 nm TiO$_2$ film provided more stable 8w BRS in Pt/TiO$_2$/Pt device, while the device with higher thickness of active TiO$_2$ layer did not worked as stable as the device with thinner TiO$_2$ film [79]. A comprehensive model has explained the coexistence of BRS model by using the competition between the VCM-type oxygen vacancy drift and diffusion process and the oxygen exchange reaction at the interface between active layer and Pt electrode. It was understood that the as-deposited

ALD TiO_2 film consists of multiple region of conductive reduced TiO_{2-x} phases inside the insulator TiO_2 matrix of active layer. Moreover, the existence of Magnéli-type phases (Ti_nO_{2n-1}) enables an easy path with high conductivity characteristics inside the TiO_2 matrix, which facilitates the formation of highly conductive filaments [79]. The other main important factor in the study of the BR behavior of $Pt/TiO_2/Pt$ memristors is attributed to the switching Schottky interface at Pt/TiO_{2-x} phase at the bottom of the device stack. Because of the geometrical arrangement, the oxygen exchange between TiO_{2-x} and atmosphere is excluded from calculations. Instead, the oxygen diffused and segregated to the Pt grain boundaries. The graphical schemes in Figure 6.12 explain the proposed model for BRS of the $Pt/TiO_2/Pt$ memristor. The cross-sectional graphical scheme from the formed filament in Figure 6.12 a, shows the formation of conductive Ti^{3+} cation filaments and the depletion of ionized oxygen vacancies at the interface of Pt electrode. The standard VCM-type c8w switching behavior is described graphically by the following transition "A→B→C→D".The 8w loop is characterized by the following "C→D→E→F" sequences. It was proposed that the full switching scenario consist of a combination of c8w and 8w switching mechanisms. The transition between c8w and 8w mechanisms and the shared intermediate states are determined and controlled by the amplitude of negative voltage imposed on Pt bottom electrode. The depletion zone in c8w LRS A state is the smallest one among all samples. In c8W reset B, the oxygen vacancies are drifted toward the Ti cathodes. There is the possibility of oxygen transfer through high barrier interface between TiO_2 and Pt bottom electrode. It was confirmed that the oxygen vacancy concentration decreased in active conductive region after the drift of oxygen vacancies through the barrier [79]. Hence the successful transfer from c8w reset (LRS A) to the c8w (HRS C) was facilitated. The study on BRS of $Pt/TiO_2/Ti/Pt$ memristor demonstrated that the exchange reaction at all interfaces of memristor device had tangible effect on RS behavior of memristor device.

6.4.4 RS IN EXFOLIATED 2D PEROVSKITE SINGLE CRYSTAL

The overwhelming need to the low energy consumption devices in neuromorphic technology and computing is undeniable fact in the modern technologies. The RS is the basic physical and electronic phenomenon in performance of the resistive switching devices. The ionic transfer is the basic characteristic of the memristor devices that controls the programmable resistivity of memristor units. *High-k* materials (HfO_x, AlO_x, TaO_x, TiO_x) are one of the main groups of the materials with memristive characteristics [80]. However, their ionic transport activities are not sufficiently high to facilitate the required ionic transport. To tackle the deficiency of *high-k* material for the lack of ionic transport, the defects are introduced into them to enable the oxygen movement in them. Nevertheless, the leakage current is still high in conventional RS, which originates from the defeat-mediated leakage paths [80]. It results in unacceptable level of leakage current. Thus, the resistive devices that can be programmed and works in low energy and low current (PA) level remain elusive. In regard to all challenges mentioned above, the lack of defect-free materials with high-quality resistive characteristics and sufficient ionic transport are highly required [80]. Organic-inorganic hybrid perovskites are one of the advanced materials with mixed ionic-electronic conductivity. While the ionic transport mechanisms are still not well understood in these materials, their capability for significant ionic transport and memory applications are well observed and confirmed. The idea of employment of exfoliated 2D single crystal perovskite as the active layer of memristor is highly interesting, since the lack of grain boundaries and pin-holes in single 2D crystals enable to tackle the problem of leakage current.

An extremely low operating current memristor device is introduced based on the exfoliated organic-inorganic 2D layered perovskite single crystal of PEA_2PbBr_4 [80]. PEA is the abbreviation of phenethylammonium, which is the organic component of 2D single crystal. 2D perovskite layered (Figure 6.13a) single crystals were synthesized by using a modified anti-solvent vapor assisted crystallization method, with nano-metric thickness (1.0 nm) and millimeter-seized lateral dimensions (Figure 6.13b) with bandgap of 2.9 eV. The fabricated memristor unit consists of a 2D perovskite single crystals sandwiched between Au (gate) and graphene (grounded) electrodes. To understand the switching mechanism, the 10 V applied voltage for 60 ms was employed on the device. Alloy distribution was subsequently monitored by the scanning tunneling electron microscope (STEM). It was found that the positive voltage at Au gate has attracted Br^- ions toward the Au/2D perovskite interfaces, which is the main evidence confirming the ion migration in 2D single crystal film of perovskite (Figure 6.13c). By applying a larger (12 V) and longer (120 ms) voltage pulse the resistance switched and high conductivity was attained (Figure 6.13d). The TEM observation confirmed the development of conductive filaments inside the active layer of 2D perovskite (Figure 6.13e). It was established that the dimeter of filament close to graphene electrode is about 30 nm, which is wider than that of the filaments adjacent to Au gate (positive voltage) electrode. It confirms that Br^- ions have been driven from graphene electrode to the Au electrode. By stopping the applied voltage, the filaments were partially broken and disappeared (Figure 6.13f). The proposed mechanism for filamentary RS (Figure 6.13g, h and i) suggested that the migration of ionic Br^- toward the Au gate leads to the formation of Br^- vacancies inside active 2D perovskite. It finally assists the partial formation of filaments inside the active layer. By imposing the higher voltage the complete filament is formed and facilitates the RS process. Moreover, by the formation of highly conductive filaments inside the active layer, a high current is induced, which finally results in the generation of joule heating and rising the temperature of filamentary area. The high-temperature causes the subsequent decomposition of filaments and resistive switching from LRS to HRS.

The characterization of 2D perovskite memristor shows that the power consumption of device is tangibly lower than that of the oxide counterparts. The typical forming curve demonstrates the initial forming voltage of 7.6 V when the compliance current was set to 10 pA. It was found that the compliance current is changing proportional to the thickness of 2D perovskite film [80]. When thickness of perovskite films passed the two-dimensionality limits and became thicker, the forming voltage increased considerably (Figure 6.14b). The measurements discovered the relationship between the thickness and formation voltage of 2D memristor. It was observed that the 2D perovskite film requires 0.24 V/nm to initially form, which is lower than the typical required formation voltage of common memristor devices. The initial formation voltage is typically higher than the other coming formation voltages.

Memristive Devices Based on 2D Materials 203

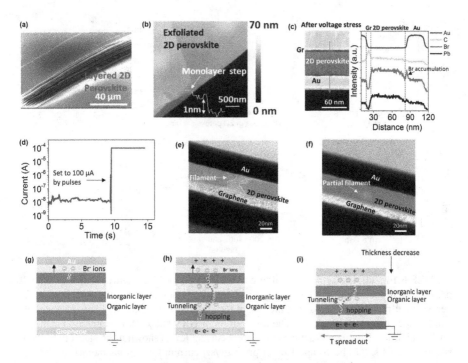

FIGURE 6.13 The (a) SEM image of layered (PEA)$_2$PbBr$_4$ single crystal. (b) The AFM image of layered of 2D perovskite film with a monolayer structure with 1 nm thickness. (c) The cross-sectional STEM image of 2D based perovskite memristor device accompanied by the EDX map analysis, showing the accumulation of Br element at 2D perovskite/Au interface after applying a high voltage long-term pulse. (d) The sudden increase of current for memristor device. (e) The cross-sectional TEM demonstrating the formation of a filament inside the active 2D layer of memristor device. The filament diameter adjacent to graphene electrode is higher and wider than that of close to Au electrode. (f) The filament breakdown and formation of partial filament after inducing the joule heating. (g) The schematic graph of 2D perovskite device when the Br$^-$ ions migrated toward the Au gate (stage i), (h) the formation of a nanofilament by applying a larger pulse voltage. (i) The decomposition of conductive filament induced by joule heating. Reprinted and reproduced with permission from Tian, H., Zhao, L., Wang, X., Yeh, Y. W., Yao, N., Rand, B. P., & Ren, T. L. (2017). Extremely low operating current resistive memory based on exfoliated 2D perovskite single crystals for neuromorphic computing. *ACS Nano*, *11*(12), 12247–12256.

Based on this fact the initial formation of a completed conductive filament requires higher driving force, while after the formation of conductive filaments, the reset process only partially breaks the conductive filaments. Thus, the re-building of a partially decomposed filament requires much less energy than that of the first one. It was observed that the forming voltage of 2D based perovskite device shifted from 7.6 V to 2.6 V from initial to second formation (Figure 6.14c). In the reset negative sweep the current level reaches even to 0.1 pA. It confirms the promising capability of the 2D perovskite memristors as the capable devices for operating at the extremely low energy computation and processing. The reliability test demonstrated

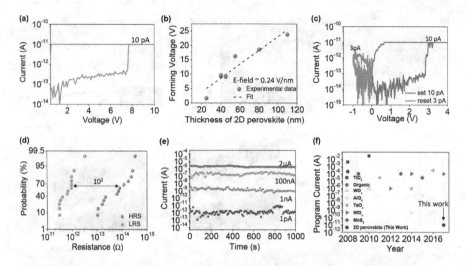

FIGURE 6.14 (a) The typical forming curve of 2D perovskite memristor device with 10 pA compliance current. (b) The variation of forming voltage versus the thickness of 2D perovskite film. (c) Typical logarithmic sale of forming (I-V) curve for 2D perovskite with the compliance current of 10 pA. (d) The distribution of HRS and LRS for operation at 10 pA. (e) The multilevel storage of device at different current levels showing the stability of memristors for 1000s. (f) The comparison of program/set current for various memristor devices. Reprinted and reproduced with permission from Tian, H., Zhao, L., Wang, X., Yeh, Y. W., Yao, N., Rand, B. P., & Ren, T. L. (2017). Extremely low operating current resistive memory based on exfoliated 2D perovskite single crystals for neuromorphic computing. *ACS Nano*, *11*(12), 12247–12256.

that the distribution of HRS and LRS probability in Figure 6.14d. An average of 10^2 memory window was obtained (Figure 6.14d). The capability of multilevel RS and storage is shown for this device by setting the unit at the different current levels. This capability arises from the low energy operation of device (Figure 6.14e). A comparative graph expresses the essential feature that the energy consumption of the 2D perovskite memristor is tangibly lower than the other resistive memory devices. The set/programming current for the other memristors are in the range of 10 mA to 1 µA, while the programming current of the introduced device was in the range of 10 pA (Figure 6.14f) [80].

6.4.5 Memristor Devices Based on 2D MoS$_2$

While the MoS$_2$ is one the first and the most investigated layered semiconductor materials among all of the other 2D TMDs, it did not initially established the considerable RS characteristics. Only the composite structures of 2D MoS$_2$ nanosheets and other materials showed improved resistive characteristics. The mechanically exfoliated MoS$_2$ nanosheets do not demonstrate the RS behavior hence the most of MoS$_2$ nanosheets are made via solution-based methods. Furthermore, there are several cases where the memristive phenomena in CVD-grown monolayer-based memristor devices are reported. The next subsections will review several fabricated devices based on 2D MoS$_2$ memristor materials.

6.4.5.1 Memristor Device Based on Oxidized MoS$_2$ Nanosheets

A memristor unit based on oxidized MoS$_2$ [81] was fabricated with a stacked structure of Ag/MoO$_x$/MoS$_2$/Ag. MoS$_2$ nanosheets were deposited over Ag electrodes and then the samples were thermally annealed to fabricate ultrathin 2D MoO$_x$ over MoS$_2$ film with thickness of less than 3.0 nm. This ultrathin MoO$_x$ film has the main role as the resistive switching layer in device. The graphical scheme of memristor and its cross-sectional scanning electron microscope image are demonstrated in Figure 6.15a and b, respectively. The obtained memristive characteristics were quite interesting, when the device demonstrated an unprecedented high *on/off* ratio of 10^6

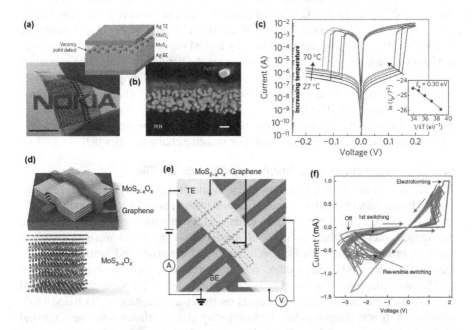

FIGURE 6.15 The schematic graph of memristor device based on oxidized MoS$_2$. (a) The graph shows the memristive array on flexible PEN substrate with vertical MoO$_x$/MoS$_2$ stacked structure. (b) The cross-sectional SEM image of the device with Ag nanowires as the top electrode and PEN as the conductive substrate. (c) The *I-V* curves of devices for different operating temperatures. The inset depicts the activation energy of device. (d) The graphical scheme depicts the configuration of graphene/MoS$_{2-x}$O$_x$/graphene (GMG) memristor device and the bottom image depicts the crystalline structure of active layer of memristor device. (e) The optical microscopic image of measurement set-up. (f) The bipolar switching curves of the corresponding GMG device. Reprinted and reproduced with permission from: (a, b and c) ref. Bessonov, A. A., Kirikova, M. N., Petukhov, D. I., Allen, M., Ryhänen, T., & Bailey, M. J. (2015). Layered memristive and memcapacitive switches for printable electronics. *Nature Materials*, *14*(2), 199, and (d, e and f) ref. Wang, M., Cai, S., Pan, C., Wang, C., Lian, X., Zhuo, Y., ... Miao, F. (2018). Robust memristors based on layered two-dimensional materials. *Nature Electronics*, *1*(2), 130–136.

and the low programming voltage of 0.1–0.2 V. The low programming voltage is a promising characteristic for the potential application at the low energy computational engineering (Figure 6.15c). The proposed model for resistive switching indicates on the effect of ionic transport including ion vacancy migration at the MoO_x/MoS_2 active layers and charge trapping/detrapping effects at the MoO_x/Ag interface [81].

The similar strategy was implemented for oxidation of MoS_2 flakes and obtaining $MoS_{2-x}O_x$ nanosheets [82]. To this aim, MoS_2 flakes were oxidized at 160°C for 1.5 h under the ambient atmosphere to control the oxidation of nano flakes. The fabricated memristor possesses the following heterostructured design: graphene/$MoS_{2-x}O_x$/graphene. The graphical scheme of heterostructured memristor is demonstrated in Figure 6.15d and e. The outstanding performance of memristor device was the excellent bipolar switching with high endurance (10^7), the consistent performance even after 1000 cycles of bending of the device under mechanical loads, and reliable thermal stability up to 340°C. The excellent performance was related to the formation of sharp interface between the graphene electrode and $MoS_{2-x}O_x$ 2D materials. The RS mechanism of device was found by STEM observation. Based on proposed model a conductive channel was formed inside the active materials. The migration of oxygen ions near to the channel region and occupation of empty sulfur vacancies by migrated oxygens lead to the considerable alteration of chemical composition of conductive channel. The increase of oxygen concentration inside the conductive channel region set the device *off*. Thus the oxygen migration was discovered as the main mechanism in alteration of the device resistive characteristics [82].

6.4.5.2 Memristor Device Based on CVD MoS_2 2D Film

The advanced memristors, based on metal-insulator-metal (MIM) structures were limited by the lack of control over the filament formation and external control of the switching voltage. Thus the control of filamentary memristive behavior of device can indeed alter the characteristics of memristors. The CVD method is one of the most promising methods for deposition of 2D materials and fabrication of the 2D-based devices [83]. This method was employed to fabricate a memristor based on 2D MoS_2 material. The control of grain boundaries in 2D MoS_2 film fabricated by CDV method has considerable impact on the characteristics of 2D based MoS_2 memristors. It was observed that the memristive characteristics were merely found in the device with grain boundaries. Hence, the device is referred as intersecting-GB memristors. The switching mechanism of units with different grain boundary orientations was analyzed. Four different groups of the devices were fabricated with various grain boundary orientations [83]. The first group did not contain any grain boundaries, and consequently, did not show the memristive characteristics. The second group contains the grain boundaries connected to one of the electrodes and intersecting at a vertex within the channel. In such device, the Au electrode on the 2D MoS_2 acted as the memristor channel. Another Si gate electrode is employed to control the set voltage of the unit (Figure 6.16a). It was observed that the best-performing device had the grain boundaries that were connected to only one of the two electrodes. The intersecting grain boundary memristor behaved at the HRS mode at $V_G=0V$. The HRS shifts to LRS at $V_G=8V$ accompanied by abrupt increase of device current (Figure 6.16b). The resistance ratio between HRS and LRS state was

Memristive Devices Based on 2D Materials 207

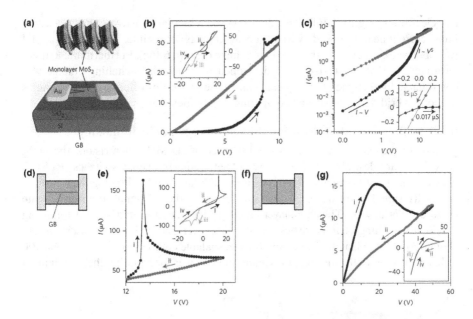

FIGURE 6.16 *I-V* characteristics of 2D MoS$_2$ memristor devices. (a) Graphical scheme of the intersecting grain boundary memristor device with two grain boundaries ended up to one electrode and intersecting at a vertex within the channel. (b) Partial *I-V* characteristics of an electroformed intersecting-GB memristor and (c) its corresponding Log-scale of the same graph. (d) Schematic and (e) partial *I-V* characteristics of bridge-based grain boundary memristor device. (f) Schematic and (g) *I-V* characteristics of electroformed bisecting-GB memristor. Reprinted with permission from Sangwan, V. K., Jariwala, D., Kim, I. S., Chen, K. S., Marks, T. J., Lauhon, L. J., & Hersam, M. C. (2015). Gate-tunable memristive phenomena mediated by grain boundaries in single-layer MoS$_2$. *Nature Nanotechnology*, *10*(5), 403–406.

10^3 (Figure 6.16c). Thus the dynamic negative differential resistance (NDR) behavior confirmed the memristive characteristics of this device [83].

The third group was based on bridge-grain boundary memristor that contained a grain boundary parallel to the channel, which worked as the bridge channel between two electrodes (Figure 6.16d). The electroforming of device was accompanied by the rapid increase of conductivity (G > 1 mS) followed by the NDR phenomenon larger than that of intersecting grain boundary memristor (Figure 6.16e). In contrast to the intersecting grain boundary memristor, the bistable resistance behavior was not observed in the bridge-grain boundary memristor at V=0 V. Highly conductive filaments were formed in bridge-grain boundary memristor which was driven by the sulfur segregation near the grain boundaries. The following joule heating let to the thermal rupture of the conductive filament inducing the switching behavior of device. The proposed RS mechanism of CVD monolayer MoS$_2$ based device is based on the migration of anions and the presence of conductive channel through the grain boundaries. This mechanism is different from the proposed RS mechanism for the solution-based processed memristor. The fourth group of memristors had the grain boundary perpendicular to the channels of the unit and did not contain any connection between two electrodes (Figure 6.16f). This device posed a bipolar resistive switching with a broad current peak followed by a slow decay in the NDR regime. The R_{HRS}/R_{LRS} of 4 was observed for device at the zero bias voltage (Figure 6.16g). Furthermore, a three-terminal 2D MoS_2 memristor was developed in which the switching voltage was controllable by the gate electric field. Thus these properties offer new opportunities for both hybrid data storage and computing capabilities [83].

6.4.5.3 2D MoS_2 Memristor for Efficient-Energy Radio Frequency

Because of the extremely low *on state* resistance of 2D MoS_2 memristors, the MoS_2 atomristors are desirable for the energy efficient radio-frequency switches (RFS). A device structure was executed on monolayer and bilayer MoS_2 allowing the transmission of RF signals if the device is in *on* state or rejects the RF signals if it is in *off* state (Figure 6.17a and b). The monolayer-based device demonstrated several following excellent switching characteristics: zero static power dissipation, low insertion loss about 0.3 dB, high isolation value of 50 GHz, scalable cutoff frequencies beyond 100 THz referring to 0.01 μm² area, maximum input power for handling the switch in *on*

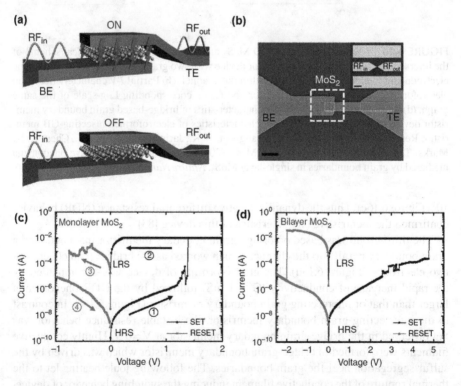

FIGURE 6.17 The (a) device schematic and the (b) optical micrograph of memristor device for radio-frequency switching based on the monolayer MoS_2 film. (c) The *I-V* curve of the bipolar RS effects in monolayer and (d) bilayer MoS_2 based RF switch. Reprinted and reproduced with permission from Kim, M., Ge, R., Wu, X., Lan, X., Tice, J., Lee, J. C., & Akinwande, D. (2018). Zero-static power radio-frequency switches based on MoS_2 atomristor. *Nature Communications*, 9(1), 2524.

Memristive Devices Based on 2D Materials   209

state to above the 20 dBm and the self-switching power level of about 11 dBm [84]. Typical *I-V* curves of RFS unit based on the MoS$_2$ monolayer memristor are presented in Figure 6.17c and d, respectively. The bilayer MoS$_2$ device performed similar to the monolayer-based MoS$_2$ memristor with bipolar resistive switching characteristics. Generally, the switching voltage in the bilayer unit is higher than that of the monolayer device, since the thicker film requires higher driving voltage to turn *on*.

2D MoS$_2$ monolayer film was already employed in the vertical memristor based on two MoS$_2$ monolayers [85]. It was observed that a single monolayer MoS$_2$ does not produce the reliable memristive effects. The electrochemical metallization memory device was observed when and Cu top electrode was employed in the memristor with the double monolayer structure [85] with an active Cu top electrode and an inert Au bottom electrode, as depicted in Figure 6.18a. It was found that the Cu ions diffuse

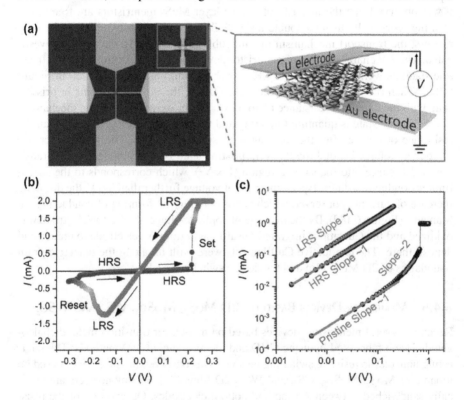

FIGURE 6.18 (a) The optical image of Cu/MoS$_2$ double-layer/Au memristor device with scale bar of 100 μm accompanied by the graphical scheme of device with top Cu electrode and bottom Au electrode. The memristor's cross-bar area is 2 × 2 μm^2. (b) The *I-V* curve of memristor device with bipolar RS with the 0.25 V Set voltage and −0.15 V reset voltage. (c) The logarithmic scale *I-V* curve of device for the pristine state before and after forming process, and for LRs and HRs state after the forming process. Reprinted and reproduced with permission from Xu, R., Jang, H., Lee, M. H., Amanov, D., Cho, Y., Kim, H., ... Ham, D. (2019). Vertical MoS$_2$ double-layer memristor with electrochemical metallization as an atomic-scale synapse with switching thresholds approaching 100 mV. *Nano Letters*, *19*(4), 2411–2417.

through the MoS_2 double monolayer to form atomic scale conductive filaments. Thus, the device falls in the category of electrochemical metallization memory (ECM) device, whereas the memristive performance of other 2D MoS_2 devices are based on the migration of sulfur vacancies to make a conductive filaments, thence they are valence change memories (VCM) [85].

The two-dimensional capabilities accompanied by the electrochemical metalliza-tion effects of Cu cations have resulted in the decrease of switching voltage down to 0.2 V (Figure 6.18b). This is because of the lower migration barrier for Cu cations compared with that of sulfur ions. It is caused by lower diffusion activation energy of Cu cations in MoS_2 active layer compared with that of sulfur vacancies in 2D MoS_2 film [85]. The measured I-V curve of the memristor device clearly showed the bipolar resistive switching with a set voltage at 0.25 V and a reset voltage at −0.15 V. It was observed that the majority of double-layer MoS_2 memristors are free of the forming process. By using a double-layer MoS_2 memristor that exhibits the forming process, the transport mechanism in Cu/double-layer MoS_2/Ag device was investi-gated. The I-V characteristics of memristor device before the forming process (pris-tine state) as well as for HRS and LRS (after forming) are demonstrated in Figure 6.18c. When the applied voltage is less than 0.3 V, the device performs in pristine state with the extracted resistance of 16.6 kΩ. The proposed conduction mechanism in the pristine state is quantum tunneling and the resistance is much larger than the resistance of device after the forming process of memristor. The further increase of applied voltage beyond the 0.3 V in pristine state is accompanied by the transi-tion in I-V curve into the slope 2 region (I $\propto$ V^2), which corresponds to the space-limited conduction states. By the increase of voltage further than 0.6 V, the dramatic increase of current is observed which is attributed to the forming of conducting Cu filaments (Figure 6.18d). By the increase of applied voltage, the Cu top electrode was oxidized and the Cu cations migrated toward the bottom Au electrode to reduce and deposit there. The conductive Cu filaments were built up along the defective grain boundaries of 2D MoS_2 film and the device switched to LRS.

6.4.6 Memristor Devices Based on 2D MoS_2, $MoSe_2$, WS_2, and WSe_2

Recently, several memristive devices based on monolayer transition metal dichalco-genides were fabricated by using CVD and metal organic CVD methods. The inter-esting nonvolatile resistive switching behavior was observed on the devices based on monolayer MoS_2, $MoSe_2$, WS_2, and WSe_2 2D films [86]. The monolayers are verti-cally sandwiched between Au top and bottom electrodes. Observation of the resis-tive behavior in 2D atomic-scaled films turns these atom-resistor devices into highly interesting options for applications in memristors. The deposited nanosheets do not contain any grain boundaries or any oxidized layer. Thus, the interface between nanosheet and Au electrode was highly sharp and clean. The graphical scheme of the 2D memristor devices, the top view of device and the cross-sectional TEM image of Au/MoS_2/Au memristor device are presented in Figure 6.19a, b, and c, respectively. The MoS_2 monolayer has grown on Au foil. The characterization of 2D MoS_2 by atomic resolution STM (Figure 1d) demonstrated relatively high quality and addi-tionally showed local sulfur vacancy defects in the MoS_2 film. The device based on

Memristive Devices Based on 2D Materials 211

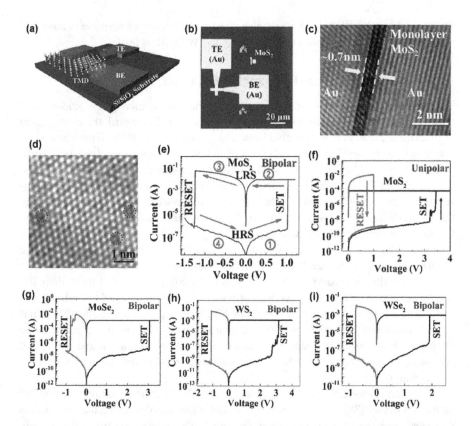

FIGURE 6.19 The (a) schematic graph of metal/insulator/metal atomristor device based on TMD monolayers, (b) the corresponding optical image of device, and the (c) cross-sectional TEM image of Au-Monolayer MoS_2-Au device. (d) The atomically resolved STM image of monolayer MoS_2 film deposited on Au electrode. The sulfur vacancies indicated by dashed lines. (e) The typical *I-V* curve of atomristor device based on MoS_2 monolayer device. The typical bipolar RS behavior is observed. (f) The occasional unipolar RE behavior of device in monolayer MoS_2 device. (g), (h), and (i) graphs respectively demonstrate the typical bipolar RS behavior of atomristor devices based on monolayer $MoSe_2$, WS_2, and WSe_2 films. Reprinted and reproduced with permission from Ge, R., Wu, X., Kim, M., Shi, J., Sonde, S., Tao, L., ... Akinwande, D. (2018). Atomristor: Nonvolatile resistance switching in atomic sheets of transition metal dichalcogenides. *Nano Letters*, *18*(1), 434–444.

monolayer MoS_2 film verified high *on/off* ratio above 10^4 and the corresponding retention time as high as 10^6 s. The electroforming was not observed in the resistive behavior of MoS_2 monolayer memristor, while the abrupt increase of the device current was the indication of dielectric breakdown and formation of CFs in the active layer of memristor (Figure 6.19e). Generally, it was understood that although the electroforming can be avoided by scaling down the thickness of active memristor layer into the few nanometer range, the main drawbacks of this strategy is the excessive leakage current originated from the trap-assisted device tunneling [86]. Occasionally, the unipolar RS was observed in the certain single-layer MoS_2 metal/insulator/metal units.

212　Ultrathin 2D Semiconductors

The typical bipolar RS behavior was also shown in the study of resistive behavior of atomristors based on monolayer $MoSe_2$, WS_2 and WSe_2 films [86].

6.4.7 MEMRISTOR DEVICES BASED ON 2D INSULATING h-BN

It is a common knowledge that the active materials in memristors should comply with the requirements of effective resistive switching. In this regard, the employment of insulating 2D materials has also been considered as one of the interesting alternative options to develop the capabilities of the current memristors based on active insulating layers. The CVD-grown 2D h-BN based units were introduced for the first time at 2016 [87]. The memristors are fabricated on PET substrates and have two different vertical configurations. In the first group of devices, the stacked MIM structure of Ag/h-BN/ITO was fabricated. In the second group of devices, the Au/h-BN/Cu structure was developed and examined (Figure 6.20 a, b). In fabricated h-BN memristor the *on/off* ratio was found to be about 10^2 and the switching endurance for 500 cycles was recorded [84]. The corresponding device retention time was over 10^3 s. Hence the unit worked efficiently with operation capability of more than 700 bending cycles. The *I-V* characteristics of two devices are presented in Figure 6.20c and d, respectively. As it was demonstrated both device performed mostly similar to each other with the same memristive characteristics (Figure 6.20c and d). The dynamic TEM studies of memristors confirmed that the formation and rupture of Ag filaments and conductive bridges between two electrodes are the main switching mechanism of this unit [87].

Moreover, it was recently attempted to improve the existing capabilities in another memristor device [88]. For such purpose the unit is fabricated by using the h-BN flakes deposited over the PET/ITO substrate (Figure 6.20e and f). The h-BN flakes were passivated with the PVOH polymer matrix and then deposited over the substrate. The device showed improved *on/off* ratio and the altered resistive behavior during imposing the negative bias voltages (Figure 6.20g). The durability of memristor was doubled compared with that of the previous 2D h-BN based instruments. It was demonstrated that the device was working efficiently and endured more than 1500 bending cycles. The above-mentioned CVD-grown 2D h-BN flexible device confirmed the low switching voltage (0.4 V) with high *on/off* ration of 10^6. However, as the main drawback retention time of about 10 h was reported, which is not suitable for the practical applications. Nonetheless, the explored unit was highly interesting from the technical point of view since the control of grain boundaries and also the quality of interfaces between h-BN and conductive electrodes as well as polymer provided valuable technological knowledge to the engineers and researcher. The microstructural TEM observations (Figure 6.20h) verified that the RS behavior is driven by the grain boundaries and movement of boron vacancies in the polycrystalline structure of boron nitride. Both mechanisms facilitated the migration of metallic ions from the device electrodes [89].

In Section 6.4, the functionalities of the electrical triggered memristor devices are discussed. Their performances are summarized in Table 6.3. It is worthwhile to mention that the majority of introduced 2D memristors are in their initial stages of development and thus the attentions of researchers and engineers are focused on finding the best materials and device configurational designs. Despite the fact that

Memristive Devices Based on 2D Materials 213

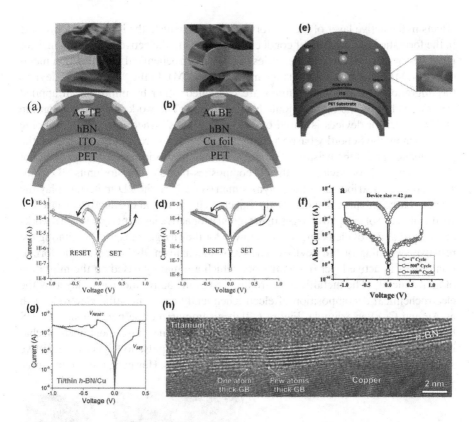

FIGURE 6.20 The memristor devices based on 2D *h*-BN with metal-insulator-metal structures (a) with Ag and (b) Au top electrodes and their corresponding (c) *I-V* curves for Ag/h-BN/ITO and (d) *I-V* curves for Au/*h*-BN/Cu memristor devices. (e) The Ag/*h*-BN-PVOH/Ag memristor device and (f) its corresponding *I-V* curve. (g) The *I-V* curve of Ti/h-BN/Cu device (h) The cross-sectional TEM image of device shows the defective paths through the ultrathin 2D *h*-BN memristor-based device. Reprinted and reproduced with permission from: (a, b, c, and d) ref. Qian, K., Tay, R. Y., Nguyen, V. C., Wang, J., Cai, G., Chen, T., ... Lee, P. S. (2016). Hexagonal boron nitride thin film for flexible resistive memory applications. *Advanced Functional Materials*, 26(13), 2176–2184, and (e and f) ref. Siddiqui, G. U., Rehman, M. M., Yang, Y. J., & Choi, K. H. (2017). A two-dimensional hexagonal boron nitride/polymer nanocomposite for flexible resistive switching devices. *Journal of Materials Chemistry C*, 5(4), 862–871, and (g and h) ref. Pan, C., Ji, Y., Xiao, N., Hui, F., Tang, K., Guo, Y., ... Lanza, M. (2017). Coexistence of grain-boundaries-assisted bipolar and threshold resistive switching in multilayer hexagonal boron nitride. *Advanced Functional Materials*, 27(10), 160481.

2D memristors so far demonstrated a great potential for the high performance computing, still tremendous attentions should be devoted for improvement of the device speed, greater retention, and properties enhancement. This stage is highly important and critically challenging for the 2D-based memristors to pass the infancy era and to enter into industrial stage.

The resistive switching mechanisms in 2D-based memristors are generally divided into three main categories [21]. The first category or model is based on ionic effects in the active layer of memristors. In this mechanism the ionic transfer results in the formation and rupture of conductive filaments in the active layer of memristor materials in thermochemical memories (TCM), electrochemical metallization memories (ECM) and the valence change memories (VCM). In the "ion effects" devices the migration of cations and anions are facilitated either by imposing an applied electric field or by the thermal joule heating effects. The work of the second group of 2D memristor devices is based on "electron effects" where the charge trapping phenomenon, the Schottky barriers and grain boundaries are highly engaged on the performance of 2D memristors. The phase change memories are the other group of 2D based memristor devices. In the 2D composites-based memristor units, 2D nanostructures are distributed in the polymer matrixes where the 2D materials play the role of charge trapping centers for the injected charges from conductive electrodes. By the increase of trapped charges in the memristor active layer, percolation networks are formed and provided conductive pathway for the charge transport facilitating the resistive switching of the devices. The ECM mechanism also refers to the formation of CFs in active layer of memristors, which are mainly based on the migration of active metals. The metallic ions are injected to the active memristor layer by the electrochemical decomposition of electrochemically active metallic electrode such as Ag and Cu. For example, Figure 6.21 has summarized the mechanisms for RS behavior of 2D-based memristors [21]. Noteworthy, all the above-mentioned mechanisms refer to the employment of the applied electrical pulse to trigger the resistive switching and to facilitate the memristive characteristics. However, the driving force

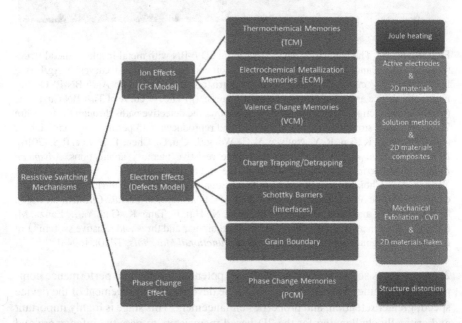

FIGURE 6.21 The classification of resistive switching mechanisms in 2D material-based memristor devices. Reprinted with permission from Zhang, L., Gong, T., Wang, H., Guo, Z., & Zhang, H. (2019). Memristive devices based on emerging two-dimensional materials beyond graphene. *Nanoscale*, *11*(26), 12413–12435.

Memristive Devices Based on 2D Materials 215

is not merely restricted to the electrical pulses. The next section of this chapter will discuss and introduce one of the most recently introduced memristor materials and devices employing the optical signals to facilitate the resistive switching behavior.

6.5 OPTOELECTRONIC MEMRISTOR DEVICES BASED ON ULTRATHIN 2D MATERIALS

6.5.1 NONVOLATILE OPTICAL RESISTIVE MEMORIES VERSUS OPTICAL SENSORS

Optoelectronic memories offer outstanding opportunities for multifunctional integration of optical sensing and data storage capabilities to fabricate a single device with combined capabilities [8]. Contrary to the traditional memory units, the optoelectronic memories utilize the optical stimuli as another factor to manipulate the memristive characteristics of memory devices and facilitate the storage of informative pulsed lights. Furthermore, the combination of electrical readout voltage and optical stimuli has outstanding potential for enlarging the ratio between LRS and HRS, hence enabling the multiple data storage capabilities and enhancing the data storage capacities [8]. One of the main advantages of using optical pulses for triggering the memristor device returns to the reduction of electrical loss during sensory data transmission. Moreover, combining the electrical pulses and optical stimuli together provides promising opportunity to facilitate the digital processing functions for application in arithmetic and logic operation and coincide detections. The suitable integration of the optical sensory functions and memory capabilities opens up new potential opportunities in the data storage, image sensing, and processing. This strategy offers the possibility of direct storage of optical data and the following processing of optical inputs for the image processing technologies [8].

2D materials based on memory devices offer significantly improved capabilities for the next generation of memristors. The capability of charge trapping, the ultrafast processing, and several various privileges put these units in the important position for the next generation of memory devices. The shift from three-terminal structured memristors in field effect transistor design to the two-terminal device tangibly decreased the limitation of integration density of large programming current [8]. The significant optical characteristics of 2D materials with broadband response from ultraviolet region to the infrared corner of light spectrum is highly interesting for various optoelectronic applications including sensing, image processing, artificial visionary systems, optical modulators, logic gates, synaptic devices, and even data storage systems [90].

Technically, both photosensor and optical memristors convert the light signals into the electric responses. However, there are tangible structural distinctions between the optical sensor and optical memristor. In photosensor, the optically generated charge carriers recombine immediately after the removal of optical stimulation source, accompanied by the immediate current drop from the photocurrent (I_p) to the dark current, while in the optical memristors the photo-generated charge carriers are stored in trapping sites [8, 90]. In fact, the working mechanisms of the optically triggered memristor are mainly based on providing the charge trapping sites for the optically generated electron/hole pairs. Thus, the optical memristors have the capability of storage of informative optical signals and reproducing them. The tuning of optical pulses can directly affect the generation of charge carriers and then the amount or the rate of data storage in optical memristor devices. Thus, the light wavelength corresponding to the different light energies, the frequency of light corresponding to the rate of charge generation and charge controller, and also the intensity of light can individually or together alter and control the charge trapping and interfacial phenomena in optical memristors [8, 90]. Thus, using different wavelengths of light combined with different frequencies and intensities can control the charge trapping/detrapping physics and facilitate the multiple resistance states. The next lines will introduce several practical examples of optical memristor devices.

6.5.2 CASE STUDY OF OPTICAL MEMRISTOR DEVICES

6.5.2.1 Optical Memories Based on 2D MoS$_2$

The design concept of optical memories is based on building up the potential wells with an energy band bending between 2D material channels and metal electrodes. Despite the fact that the 2D materials exhibit proper photoresponsivity, their short retention time and low *on/off* ratio limit their applications for multilevel nonvolatile memory devices. A MoS$_2$/SiO$_2$ optical memristor was fabricated with charge trapping contortability and optical triggered memristive capabilities [91]. The SiO$_2$ substrate was firstly treated by oxygen plasma and then the MoS$_2$ monolayer was developed over the SiO$_2$ treated substrate (Figure 6.22a). The band alignments of initial states, under light exposure condition and during *on* and *off* states are graphically depicted in Figure 6.22b [91]. At the initial state, the metal-semiconductor Schottky barrier creates a potential well at the semiconductor-metal interface when a gate voltage is

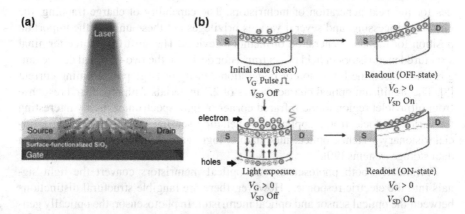

FIGURE 6.22 (a) The schematic illustration of monolayer MoS$_2$-based memory device with functionalized MoS$_2$/SiO$_2$ interface. (b) The corresponding band alignments of the device in its initial state, up on light illumination and reading stage. Reprinted and reproduced with permission from Lee, J., Pak, S., Lee, Y., Cho, Y., Hong, J., Giraud, P., ... Kim, J. M. (2017). Monolayer optical memory cells based on artificial trap-mediated charge storage and release. *Nature Communications*, 8, 14734.

applied. By applying a gate voltage pulse the device is programmed to the *off* state by driving the injected electron charges and trapping them in interfacial trap states at the Schottky barrier. It results in the generation of low off current about 4.0 pA when the applied gate voltage was 3.0 V. After illumination (450 nm), the generated photoelectrons are accumulated at the potential well and then the hole escapes from the electrode through the generated upward band energy bending [89]. When the light illumination source turned off, one of the Schottky barriers is eliminated by the read out process leading to the release of stored electrons. The intentional introduction of charge traps into MoS_2 based optical memory device leads to the improvement of *on/off* ratio (4700) and improvement of retention time of optical memories to 6000 s. The main drawback of this device though is related to its relatively large degradation after the removal of light exposure [91].

Toward the development of the design and performance of optical memories, researchers have chosen another strategy to control the charge trap phenomenon to enhance the memory devices and to additionally facilitate the multilevel and the high responsive nonvolatile 2D optical memories. An optical memory is fabricated by using nanoparticles as the charge trapping components and also as the floating gate between the blocking layer and tunneling dielectric (Figure 6.23a) to store the photo-generated charge carriers in the MoS_2 memory [92]. The fabricated MoS_2 memory unit was capable of performing logic operations and coincidence detection. The device has shown a high *on/off* ratio of 10^7, a long retention time of 10^4 s and high nonvolatile photoresponsivity of 8000 A/W at the illumination power of 0.1 μW [92]. The device structure and the graphs of optical programming are schematically presented Figure 6.23a, and b, respectively. To operate the programming stage, a negative gate voltage is imposed to induce the transmission of generated electrons from the Au nanoparticles into MoS_2 layer. It was found the transition mechanism was Fowler–Nordheim tunneling [92]. The main working principle of device is the transfer of photo-generated electrons from the Au nanoparticles into the main active layer of memristor device (MoS_2) and then controlling and preventing the recombination of photo-generated electrons and holes in MoS_2 2D film (Figure 6.23c). The device is triggered by the photo-illumination source with the wavelength of 655 nm. The visible light-assisted photo programming of the fabricated 2D MoS_2 optical memory unit has facilitated the multilevel resistance obtained by a two-terminal design of photo memory device (Figure 6.23d).

6.5.2.2 Heterostructured 2D-Based Nonvolatile Optical Memory Devices

The performance of 2D-based optical memristors can be affected by the several different factors. Some of the 2D materials can be sensitive to environmental conditions and their long-term endurance can therefore be affected, such as the oxidation of 2D sulfide based nanomaterials. Furthermore, the charge trapping mechanisms can be controlled by various strategies, including interface controlling of the charge transfer and charge transfer suppression. By integrating, combining and designing heterostructured devices, it is possible to extend the optical memristors capabilities for the wide bandgap absorption, increase their light stability and control the charge trapping/detrapping mechanism. In this view, optical memory device was introduced based on WSe_2/h-BN heterostructured 2D device (Figure 6.24a) with high

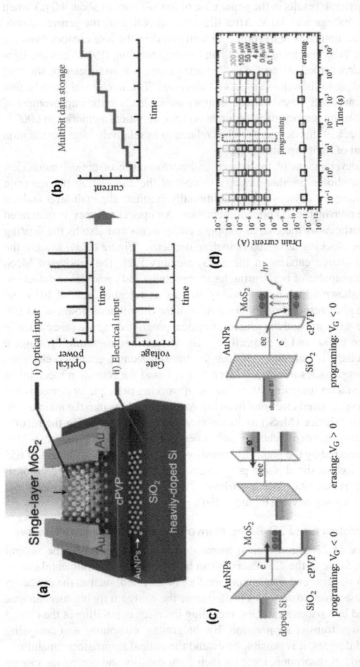

FIGURE 6.23 (a) The schematic illustration of optical 2D MoS_2 memory device with Au nanoparticle as the charge trapping sites. (b) shows the multilevel light-controlled resistance process of optical memristor device. (c) Depicts the band alignment and charge transfer mechanisms in device during programing and erasing processes. (d) Demonstrate the light-controlled multilevel resistance state of optical 2D MoS_2 device at different light intensities. Reprinted and reproduced with permission from Lee, D., Hwang, E., Lee, Y., Choi, Y., Kim, J. S., Lee, S., & Cho, J. H. (2016). Multibit MoS_2 photoelectronic memory with ultrahigh sensitivity. *Advanced Materials,* 28(41), 9196–9102.

Memristive Devices Based on 2D Materials

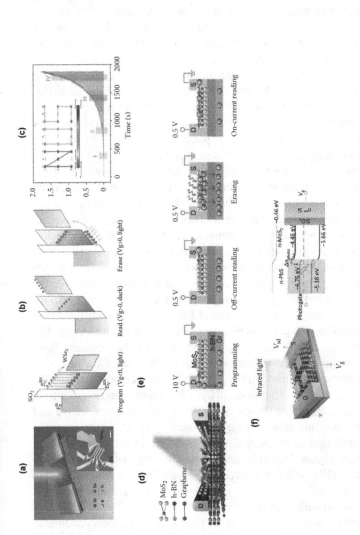

FIGURE 6.24 (a) The graphical scheme of optical memory device based on WSe₂/h-BN heterostructures. (b) The corresponding band alignments and charge transfer mechanism under visible light illumination and using applied gate voltage for programming, read and erasing process. (c) The 130 storage states obtained by applying different numbers of applied light pulses with 405 nm wavelength with duration of 0.5 s and the intensity of 210 mW/cm². (d) The schematic graph of MoS₂/h-BN/graphene heterostructured optical memory devices with (e) the corresponding programming, erasing and reading operations. (f) Schematic of PbS/MoS₂ heterostructured optical memory device with capability of reaction to infrared lights and the mechanism of charge trapping/detrapping at the heterointerfaced optical memory device. Reprinted and reproduced with permission from: (a, b, and d) ref. Xiang, D., Liu, T., Xu, J., Tan, J. Y., Hu, Z., Lei, B., … Chen, W. (2018). Two-dimensional multibit optoelectronic memory with broadband spectrum distinction. *Nature Communications*, 9(1), 2966, and (d and e) ref. Tran, M. D., Kim, H., Kim, J. S., Doan, M. H., Chau, T. K., Vu, Q. A., … Lee, Y. H. (2019). Two-terminal multibit optical memory via van der Waals heterostructure. *Advanced Materials*, 31(7), 1807075, and (f) ref. Wang, Q., Wen, Y., Cai, K., Cheng, R., Yin, L., Zhang, Y., … He, J. (2018). Nonvolatile infrared memory in MoS₂/PbS van der Waals heterostructures. *Science Advances*, 4(4), eaap7916.

photoresponsivity capabilities (1.1×10^6 A/W) [93]. These devices demonstrated more than 128 distinct resistance states and considerable retention time of 4.5×10^4 s [93].

Upon the illumination of WSe_2/h-BN heterostructured 2D devices by 450 nm laser light the electrons are generated in h-BN 2D film. Simultaneously, by applying a negative gate voltage the photo-generated electrons in h-BN 2D film are transmitted to WSe_2. The nonvolatile switching characteristics are observed in the device originating from the remained photo-generated holes in h-BN structure, even after the removal of light illumination and electric field (Figure 6.24b). To erase the memory device, the light illumination accompanied by a positive gate voltage drive the holes from the h-BN to WSe_2 (Figure 6.24b). The unit was capable of showing multiple resistance states by changing the number of applied light pulses. Figure 6.24c shows 130 resistance states in the WSe_2/h-BN heterostructured 2D optical memory device gained by using different numbers of applied optical pulses. It shows the capability of this unit for design of the smart image sensors.

Nevertheless, the high programming voltage still is the main drawback of the three electrode optical memory devices. A recent report has clearly demonstrated that the heterostructured optical memristor based on MoS_2/BN/graphene van der Waals heterostructured unit was fabricated [94]. This unit expressed the essential characteristics such as high nonvolatile responsivity (2×10^4) A/W, extreme low dark current of 10^{-14} A and programming voltage of -10 V [94]. It was found that by applying -10 V negative voltage, the electrons tunnel through the BN layer comes into floating graphene gate. The electrons are accumulated in graphene and then the graphene itself serves as the negative gate bias to program the optical memory device with ultralow dark current value. This stage is the LRS of memristor. Furthermore, the visible light pulse induces the photo-generated holes in the top MoS_2 layer. The photo-generated holes then tunnel the MoS_2/BN barrier into the graphene layer and combine there with the stored electrons. It results in the reset of the device to HRS. The schematic demonstration of these processes is sequentially depicted in Figure 6.24 d and e, respectively. The MoS_2/BN/graphene unit demonstrated high *on/off* ratio of 10^6. Furthermore, the device is capable of the tunable resistive and storage stages by controlling and adjusting the light intensity [94].

Most of the developed optical memory devices work within the ultraviolet and visible light illumination region. However, one of the most important illumination light sources is infrared (IR), which massively employed for data communication and information transfer, wireless networks, night visionary instrument, image sensory, medical diagnosis and variety of telecommunication devise. The development of optical memory devices with sensitivity to IR light sources provide undeniable opportunity for data storage and conversion in photonic circuits, artificial visual intelligence, and designing infrared optical devices by integration of low bandgap 2D materials in memory units. In this view, heterostructured MoS_2/PbS van der Waals optical memory devices are introduced, which response efficiently under wide wavelength of 880 nm to 1940 nm (Figure 6.24f) [95]. The mechanism of heterostructured optical memory unit is based on the generation of photo-generated holes in PbS layer close to the interface of heterostructures. The injection of photo-generated carriers from PbS to the MoS_2 film modulates the conductivity of MoS_2 2D device

Memristive Devices Based on 2D Materials

TABLE 6.4

Comparison of Properties of Optical Memory Devices Based on 2D Materials

Active layer	Controlling gate	RS mechanism	Light wavelength	Programming voltage (V)	Retention time (s)
Graphene/MoS$_2$	SiO$_2$	Charge trapping	635 nm	50	10^5
MoS$_2$	Functionalized SiO$_2$	Charge trapping	450, 650 nm	80	10^4
CuIn$_7$Se$_{11}$	SiO$_2$	Charge trapping	543 nm	80	50
MoS$_2$/PVP/ AuNPs	SiO$_2$/Au NPs	Charge trapping	655 nm	10	10^4
MoS$_2$/PZT	SiO$_2$	Ferroelectric	Halogen	6	10^4
WSe$_2$/BN	SiO$_2$	Charge trapping	410-750 nm	20	4.5×10^4
BP/BN	SiO$_2$	Charge trapping	410-500 nm	15	4.0×10^4
MoS$_2$/h-BN/ graphene	Graphene	Charge trapping	458, 638, 725, 811 nm	10	4.5×10^4
BP/ZnO	–	Oxygen filament	380, 532, 633, 785 nm	5	–
MoS$_2$/PbS	SiO$_2$	Charge trapping	850, 1310, 1550 nm	Not required	10^4
CH$_3$NH$_3$PbI$_3$.$_x$Cl$_x$	–	Charge trapping	White	0.1	10^4
CeO$_{2-x}$/AlOy	–	Charge trapping	499, 560, 638 nm	Not required	10^4

Source: Reprinted and reproduced with permission from [8].

(Figure 6.24f). It was reported that the high photoresponsivity of 1.2×10^4 A/W and the retention time of 10^4 s was observed with multilevel storage for this units [95].

The optical resistive switching of several 2D based optical memory devices and their heterostructures are presented in Table 6.4 [8]. Generally, the nonvolatile optical memory devices based on the 2D materials have responded efficiently to the visible, infrared and UV light sources. The optical energy consumption of device is low and the *on/off* ratio of optical memory devices is considerable. However, the programming voltage of most of the 2D-based memristor devices are still high for application in ultralow programming devices to compete with highly efficient-energy brain programming procedures.

CONCLUSION

Because of the considerable resistive characteristics, the strong light-matter interactions, unique physical as well as chemical properties, electronic and optoelectronic tunability, and progressing synthesis and fabrication techniques, ultrathin 2D materials and their heterostructures have been widely employed to fabricate the memristive devices and optical memory instruments. The integration of 2D materials with their specific resistive characteristics into processing devices, optical sensors and receptors has open up new opportunities for data storage, optoelectronic communication, image processing, computation, neural network systems, and optical data transfer. Further improvement of 2D memristor devices based on 2D materials has facilitated the fabrication and development of low energy data processing systems and neuromorphic devices. As quite recently introduced technology, the integration of optical sensing capabilities by charge trapping/detrapping mechanisms in 2D-based memristor devices exhibited various advantages over other nanostructured materials for the image sensory and visual data processing. However, there are still several challenges which should be addressed in order to reduce and decrease the programming voltage and energy consumption of electrical and optoelectronic memristors and memory units. There are several strategies taken to alter and improve the charge trapping and transfer characteristics of electrical memristors and optical memory devices. These approaches are based on the heterointerfaces design in heterostructured memristors, the control of Schottky barrier height at electrode and semiconductor interface, employment of nanostructured component for charge trapping in the active layer of memristors and several other innovations to improve the increasing requirements for the state-of-the-art performance of memristor devices. The field of 2D memristors is still in its early growing stage and further developments likely expected as well as new outstanding capabilities of the 2D-based memory devices are highly envisaged to be observed in coming years.

REFERENCES

1. Yang, J. J., Strukov, D. B., & Stewart, D. R. (2013). Memristive devices for computing. *Nature Nanotechnology*, *8*(1), 13–24.
2. Beck, A., Bednorz, J. G., Gerber, C., Rossel, C., & Widmer, D. (2000). Reproducible switching effect in thin oxide films for memory applications. *Applied Physics Letters*, *77*(1), 139–141.
3. Sakamoto, T., Lister, K., Banno, N., Hasegawa, T., Terabe, K., & Aono, M. (2007). Electronic transport in Ta_2O_5 resistive switch. *Applied Physics Letters*, *91*(9), 092110.
4. Laughlin, S. B., de Ruyter van Steveninck, R. R., & Anderson, J. C. (1998). The metabolic cost of neural information. *Nature Neuroscience*, *1*(1), 36–41.
5. Merolla, P. A., Arthur, J. V., Alvarez-Icaza, R., Cassidy, A. S., Sawada, J., Akopyan, F., ... Modha, D. S. (2014). A million spiking-neuron integrated circuit with a scalable communication network and interface. *Science*, *345*(6197), 668–673.
6. Likharev, K. K. (2011). CrossNets: Neuromorphic hybrid CMOS/nanoelectronic networks. *Science of Advanced Materials*, *3*(3), 322–331.
7. Tan, C., Liu, Z., Huang, W., & Zhang, H. (2015). Non-volatile resistive memory devices based on solution-processed ultrathin two-dimensional nanomaterials. *Chemical Society Reviews*, *44*(9), 2615–2628.
8. Zhou, F., Chen, J., Tao, X., Wang, X., & Chai, Y. (2019). 2D materials based optoelectronic memory: Convergence of electronic memory and optical sensor. *Research*, *2019*, 9490413.
9. Zhou, F., Zhou, Z., Chen, J., Choy, T. H., Wang, J., Zhang, N., ... Chai, Y. (2019). Optoelectronic resistive random access memory for neuromorphic vision sensors. *Nature Nanotechnology*, *14*(8), 776–782.

Memristive Devices Based on 2D Materials

10. Zhu, X., & Lu, W. D. (2018). Optogenetics-inspired tunable synaptic functions in memristors. *ACS Nano, 12*(2), 1242–1249.

11. Tan, H., Liu, G., Zhu, X., Yang, H., Chen, B., Chen, X., ... Li, R. W. (2015). An optoelectronic resistive switching memory with integrated demodulating and arithmetic functions. *Advanced Materials, 27*(17), 2797–2703.

12. Yang, J. J., Pickett, M. D., Li, X., Ohlberg, D. A., Stewart, D. R., & Williams, R. S. (2008). Memristive switching mechanism for metal/oxide/metal nanodevices. *Nature Nanotechnology, 3*(7), 429–433.

13. Sun, W., Gao, B., Chi, M., Xia, Q., Yang, J. J., Qian, H., & Wu, H. (2019). Understanding memristive switching via in situ characterization and device modeling. *Nature Communications, 10*(1), 3453.

14. Chang, T. C., Chang, K. C., Tsai, T. M., Chu, T., & Sze, S. M. (2015). Resistance random access memory. *Materials Today, 19*(5), 254–264.

15. Sawa, A. (2008). Resistive switching in transition metal oxides. *Materials Today, 11*(6), 28–36.

16. Strukov, D. B., Snider, G. S., Stewart, D. R., & Williams, R. S. (2008). The missing memristor found. *Nature, 453*(7191), 80–83.

17. Balatti, S., Larentis, S., Gilmer, D. C., & Ielmini, D. (2013). Multiple memory states in resistive switching devices through controlled size and orientation of the conductive filament. *Advanced Materials, 25*(10), 1474–1478.

18. Yang, J. J., Zhang, M.-X., Strachan, J. P., Miao, F., Pickett, M. D., Kelley, R. D., ... Williams, R. S. (2010). High switching endurance in TaO_x memristive devices. *Applied Physics Letters, 97*(23), 232102.

19. Nowotny, J., & Rekas, M. (1989). Defect structure of cobalt monoxide: I. The ideal defect model. Journal of the American Ceramic Society, 72(7), 1199–1207.

20. Waser, R., Dittmann, R., Staikov, G., & Szot, K. (2009). Redox-based resistive switching memories-nanoionic mechanisms, prospects, and challenges. *Advanced Materials, 21*(25–26), 2632–2663.

21. Zhang, L., Gong, T., Wang, H., Guo, Z., & Zhang, H. (2019). Memristive devices based on emerging two-dimensional materials beyond graphene. *Nanoscale, 11*(26), 12413–12435.

22. Yim, J. W. L., Chen, D., Brown, G. F., & Wu, J. (2009). Synthesis and ex situ doping of ZnTe and ZnSe nanostructures with extreme aspect ratios. *Nano Research, 2*(12), 931937.

23. Sato, Y., Kinoshita, K., Aoki, M., & Sugiyama, Y. (2007). Consideration of switching mechanism of binary metal oxide resistive junctions using a thermal reaction model. *Applied Physics Letters, 90*(3), 033503.

24. Strachan, J. P., Strukov, D. B., Borghetti, J., Yang, J. J., Medeiros-Ribeiro, G., & Williams, R. S. (2011). The switching location of a bipolar memristor: Chemical, thermal and structural mapping. *Nanotechnology, 22*(25), 254015.

25. Karg, S. F., Meijer, G. I., Bednorz, J. G., Rettner, C. T., Schrott, A. G., Joseph, E. A., ... Caimi, D. (2008). Transition-metal-oxide-based resistance-change memories. *IBM Journal of Research and Development, 52*(5), 481–492.

26. Russo, U., Ielmini, D., Cagli, C., & Lacaita, A. L. (2009). Self-accelerated thermal dissolution model for reset programming in unipolar resistive-switching memory (RRAM) devices. *IEEE Transactions on Electron Devices, 56*(2), 193–200.

27. Jeong, D. S., Schroeder, H., & Waser, R. (2007). Coexistence of bipolar and unipolar resistive switching behaviors in a $Pt/TiO_2/Pt$ stack. *Electrochemical and Solid-State Letters, 10*(8), 51–53.

28. Rohde, C., Choi, B. J., Jeong, D. S., Choi, S., Zhao, J., & Hwang, C. S. (2005). Identification of a determining parameter for resistive switching of TiO_2 thin films. *Applied Physics Letters, 86*(26), 262907.

29. Choi, B. J., Jeong, D. S., Kim, S. K., Rohde, C., Choi, S., Oh, J. H., ... Tiedke, S. (2005). Resistive switching mechanism of TiO_2 thin films grown by atomic-layer deposition. *Journal of Applied Physics, 98*(3), 033715.

30. Miao, F., Strachan, J. P., Yang, J. J., Zhang, M. X., Goldfarb, I., Torrezan, A. C., ... Williams, R. S. (2011). Anatomy of a nanoscale conduction channel reveals the mechanism of a high-performance memristor. *Advanced Materials, 23*(47), 5633–5640.

31. Kim, K. M., Choi, B. J., Shin, Y. C., Choi, S., & Hwang, C. S. (2007). Anode-interface localized filamentary mechanism in resistive switching of TiO_2 thin films. *Applied Physics Letters, 91*(1), 012907.

32. Munjal, S., & Khare, N. (2017). Valence change bipolar resistive switching accompanied with magnetization switching in $CoFe_2O_4$ thin film. *Scientific Reports, 7*(1), 12427.

33. Valov, I., Waser, R., Jameson, J. R., & Kozicki, M. N. (2011). Electrochemical metallization memories fundamentals, applications, prospects. *Nanotechnology, 22*(25), 254003.

34. Ielmini, D., Bruchhaus, R., & Waser, R. (2011). Thermochemical resistive switching: Materials, mechanisms, and scaling projections. *Phase Transitions, 84*(7), 570–602.

35. Strukov, D., Alibart, F., & Williams, R. S. (2012). Thermophoresis/diffusion as a plausible mechanism for unipolar resistive switching in metal-oxide-metal memristors. *Applied Physics. Part A, 107*(3), 509–518.

36. Jiang, W., Kamaladasa, R. J., Lu, Y. M., Vicari, A., Berechman, R., Salvador, P. A., ... Skowronski, M. (2011). Local heating-induced plastic deformation in resistive switching devices. *Journal of Applied Physics, 110*(5), 054514.

37. Yao, J., Zhong, L., Natelson, D., & Tour, J. M. (2012). In situ imaging of the conducting filament in a silicon oxide resistive switch. *Scientific Reports, 2*, 242.

38. Lu, H. L., Liao, Z. M., Zhang, L., Yuan, W. T., Wang, Y., Ma, X. M., & Yu, D. P. (2013). Reversible insulator-metal transition of LaAlO3/SrTiO3 interface for nonvolatile memory. *Scientific Reports, 3*, 2870.

39. Chang, S. H., Chae, S. C., Lee, S. B., Liu, C., Noh, T. W., Lee, J. S., ... Jung, C. U. (2008) Effects of heat dissipation on unipolar resistance switching in Pt/NiO/Pt capacitors. *Applied Physics Letters, 92*(18), 183507.

40. Chang, S. H., Lee, J. S., Chae, S. C., Lee, S. B., Liu, C., Kahng, B., ... Noh, T. W. (2009). Occurrence of both unipolar memory and threshold resistance switching in a NiO Film. *Physical Review Letters, 102*(2), 026801.

41. Pickett, M. D., Borghetti, J., Yang, J. J., Medeiros-Ribeiro, G., & Williams, R. S. (2011). Coexistence of memristance and negative differential resistance in a nanoscale metal-oxide-metal system. *Advanced Materials, 23*(15), 1730–1733.

42. Alexandrov, A. S., Bratkovsky, A. M., Bridle, B., Savel'ev, S. E., Strukov, D. B., & Stanley Williams, R. (2011). Current-controlled negative differential resistance due to Joule heating in TiO_2. *Applied Physics Letters, 99*(20), 202104.

43. Yang, Y., Gao, P., Gaba, S., Chang, T., Pan, X., & Lu, W. (2012). Observation of conducting filament growth in nanoscale resistive memories. *Nature Communications, 3*, 732.

44. Johnson, S. L., Sundararajan, A., Hunley, D. P., & Strachan, D. R. (2010). Memristive switching of single-component metallic nanowires. *Nanotechnology, 21*(12), 5.

45. Oka, K., Yanagida, T., Nagashima, K., Kanai, M., Xu, B., Park, B. H., ... Kawai, T. (2012). Dual defects of cation and anion in memristive nonvolatile memory of metal oxides. *Journal of the American Chemical Society, 134*(5), 2535–2538.

46. Kumar, M., Abbas, S., & Kim, J. (2018). All-oxide-based highly transparent photonic synapse for neuromorphic computing. *ACS Applied Materials and Interfaces, 10*(40), 34370–34376.

Memristive Devices Based on 2D Materials

47. Henkel, C., Zierold, R., Kommini, A., Haugg, S., Thomason, C., Aksamija, Z., & Blick, R. H. (2019). Resonant tunneling induced enhancement of electron field emission by ultra-thin coatings. *Scientific Reports*, *9*(1), 6840.
48. Février, P., & Gabelli, J. (2018). Tunneling time probed by quantum shot noise. *Nature Communications*, *9*(1), 4940.
49. Simmons, J. G. (1967). Poole-Frenkel effect and Schottky effect in metal-insulator-metal systems. *Physiological Reviews*, *155*(3), 657.
50. Inoue, I. H. M., & Rozenberg, M. J. (2008). Taming the Mott transition for a novel Mott transistor. *Advanced Functional Materials*, *18*(16), 2289–2292.
51. Hirose, Y., & Hirose, H. (1976). Polarity-dependent memory switching and behavior of Ag dendrite in Ag-photodoped amorphous As_2S_3 films. *Journal of Applied Physics*, *47*(6), 2767–2772.
52. Russo, U., Kamalanathan, D., Ielmini, D., Lacaita, A. L., & Kozicki, M. N. (2009). Study of multilevel programming in programmable metallization cell (PMC) memory. *IEEE Transactions on Electron Devices*, *56*(5), 1040–1047.
53. Lu, W., Jeong, D. S., Kozicki, M., & Waser, R. (2012). Electrochemical metallization cells-blending nanoionics into nanoelectronics? Materials Research Bulletin, *37*, 124–130.
54. Guo, X., Schindler, C., Menzel, S., & Waser, R. (2007) Understanding the switching-off mechanism in Ag+ migration based resistively switching model systems. *Applied Physics Letters*, *91*(13), 133513.
55. Kever, T., Bottger, U., Schindler, C., & Waser, R. (2007). On the origin of bistable resistive switching in metal organic charge transfer complex memory cells. *Applied Physics Letters*, *91*(8), 083506.
56. Hui, F., Grustan-Gutierrez, E., Long, S., Liu, Q. K., Ott, A. K., Ferrari, A. C., Lanza, A. (2017). 2D resistive switching memories: Graphene and related materials for resistive random access memories. *Advanced Electronic Materials*, *3*, 1600195.
57. Bertolazzi, S., Bondavalli, P., Roche, S., San, S., Choi, S. Y., Colombo, L., Bonaccorso, F., & Samorì, P. (2019). Nonvolatile memories based on graphene and related 2D materials. *Advanced Materials*, *31*, 1806663.
58. Standley, B., Bao, W., Zhang, H., Bruck, J., Lau, C. N., & Bockrath, M. (2008). Graphene-based atomic-scale switches. *Nano Letters*, *8*(10), 3345–3349.
59. Sangwan, V. K., Jariwala, D., Kim, I. S., Chen, K. S., Marks, T. J., Lauhon, L. J., & Hersam, M. C. (2015). Gate-tunable memristive phenomena mediated by grain boundaries in single-layer MoS_2. *Nature Nanotechnology*, *10*(5), 403.
60. Guo, X., Schindler, C., Menzel, S., & Waser, R. (2007). Understanding the switching-off mechanism in Ag+ migration based resistively switching model systems. *Applied Physics Letters*, *91*(13), 133513.
61. Yuan, F., Zhang, Z., Liu, C., Zhou, F., Yau, H. M., Lu, W., ... Chai, Y. (2017). Real-time observation of the electrode-size-dependent evolution dynamics of the conducting filaments in a SiO_2 layer. *ACS Nano*, *11*(4), 4097–4004.
62. Bonaccorso, F., Bartolotta, A., Coleman, J. N., & Backes, B. (2016). 2D-crystal-based functional inks. *Advanced Materials*, *28*(29), 6136–6166.
63. Cai, Z., Liu, B., Zou, X., & Cheng, H. M. (2018). Chemical vapor deposition growth and application of two-dimensional materials and their heterostructures. *Chemical Reviews*, *118*(13), 6091–6033.
64. Karbalaei Akbari, M., & Zhuiykov, S. (2019). Tailoring two-dimensional semiconductor oxides by atomic layer deposition. In S. Walia (Ed.), *Low power semiconductor devices and processes for engineering applications in communications, computing and sensing* (pp. 117–156). Boca Raton, FL: CRC Press Press/Taylor & Francis.

65. Mahjouri-Samani, M., Gresback, R., Tian, M., Wang, K., Puretzky, A. A., Rouleau, C. M., ... Geohegan, D. B. (2014). Pulsed laser deposition of photoresponsive two-dimensional GaSe nanosheet networks. *Advanced Functional Materials, 24*(40), 6365–6371.

66. Magda, G. Z., Pető, J., Dobrik, G., Hwang, C., Biró, L. P., & Tapasztó, L. (2015). Exfoliation of large-area transition metal chalcogenide single layers. *Scientific Reports, 5*, 14714.

67. Sun, Z., Liao, T., Dou, Y., Hwang, S. M., Park, M. S., Jiang, L., ... Dou, S. X. (2014). Generalized self-assembly of scalable two-dimensional transition metal oxide nanosheets. *Nature Communications, 5*, 3813.

68. Poh, S. M., Tan, S. J., Wang, H., Song, P., Abidi, I. H., Zhao, X., ... Loh, K. P. (2018). Molecular-beam epitaxy of two-dimensional In_2Se_3 and its giant electroresistance switching in ferroresistive memory junction. *Nano Letters, 18*(10), 6340–6346.

69. Wang, J., & Hu, W. (2017). Recent progress on integrating two-dimensional materials with ferroelectrics for memory devices and photodetectors. *Chinese Physics. Part B, 26*(3), 3.

70. Kapitanova, O. O., Panin, G. N., Cho, H. D., Baranov, A. N., & Kang, T. W. (2017). Formation of self-assembled nanoscale graphene/graphene oxide photomemristive heterojunction using photocatalytic oxidation. *Nanotechnology, 28*(20), 204005.

71. Waser, R., & Aono, M. (2007). Nanoionics-based resistive switching memories. *Nature Materials, 6*(11), 833–840.

72. Wedig, A., Luebben, M., Cho, D. Y., Moors, M., Skaja, K., Rana, V., ... Valov, I. (2006). Nanoscale cation motion in TaO_x, HfO_x and TiO_x memristive systems. *Nature Nanotechnology, 11*(1), 67–74.

73. Davies, J. A., Domeij, B., Pringle, J. P. S., & Brown, F. (1965). The migration of metal and oxygen during anodic film formation. *Journal of the Electrochemical Society, 112*(7), 675–680.

74. Hu, C., McDaniel, M. D., Posadas, A., Demkov, A. A., Ekerdt, J. G., & Yu, E. T. (2014). Highly controllable and stable quantized conductance and resistive switching mechanism in single-crystal TiO_2 resistive memory on silicon. *Nano Letters, 14*(8), 4360–4367.

75. Tang, G., Zeng, F., Chen, C., Liu, H., Gao, S., Song, C., ... Pan, F. (2013). Programmable complementary resistive switching behaviours of a plasma-oxidised titanium oxide nanolayer. *Nanoscale, 5*(1), 422–428.

76. Szot, K., Speier, W., Bihlmayer, G., & Waser, R. (2006). Switching the electrical resistance of individual dislocation in single-crystalline $SrTiO_3$. *Nature Materials, 5*(4), 312–320.

77. Kwon, D. H., Kim, K. M., Jang, J. H., Jeon, J. M., Lee, M. H., Kim, G. H., ... Hwang, C. S. (2010). Atomic structure of conducting nanofilaments in TiO_2 resistive switching memory. *Nature Nanotechnology, 5*(2), 148–153.

78. Karbalari Akbari, M., & Zhuiykov, S. (2019). A bioinspired optoelectronically engineered artificial neurorobotics device with sensorimotor functionalities. *Nature Communications, 10*(1), 3873.

79. Zhang, H., Yoo, S., Menzel, S., Funck, C., Cüppers, F., Wouters, D. J., ... Hoffmann-Eifert, S. (2018). Understanding the coexistence of two bipolar resistive switching modes with opposite polarity in $Pt/TiO_2/Ti/Pt$ nanosized ReRAM devices. *ACS Applied Materials and Interfaces, 10*(35), 29766–29778.

80. Tian, H., Zhao, L., Wang, X., Yeh, Y. W., Yao, N., Rand, B. P., & Ren, T. L. (2017). Extremely low operating current resistive memory based on exfoliated 2D perovskite single crystals for neuromorphic computing. *ACS Nano, 11*(12), 12247–12256.

81. Bessonov, A. A., Kirikova, M. N., Petukhov, D. I., Allen, M., Ryhänen, T., & Bailey, M. J. (2015). Layered memristive and memcapacitive switches for printable electronics. *Nature Materials, 14*(2), 199.

82. Wang, M., Cai, S., Pan, C., Wang, C., Lian, X., Zhuo, Y., ... Miao, F. (2018). Robust memristors based on layered two-dimensional materials. *Nature Electronics, 1*(2), 130–136.

83. Sangwan, V. K., Jariwala, D., Kim, I. S., Chen, K. S., Marks, T. J., Lauhon, L. J., & Hersam, M. C. (2015). Gate-tunable memristive phenomena mediated by grain boundaries in single-layer MoS_2. *Nature Nanotechnology, 10*(5), 403–406.

84. Kim, M., Ge, R., Wu, X., Lan, X., Tice, J., Lee, J. C., & Akinwande, D. (2018). Zero-static power radio-frequency switches based on MoS_2 atomristor. *Nature Communications, 9*(1), 2524.

85. Xu, R., Jang, H., Lee, M. H., Amanov, D., Cho, Y., Kim, H., ... Ham, D. (2019). Vertical MoS_2 double-layer memristor with electrochemical metallization as an atomic-scale synapse with switching thresholds approaching 100 mV. *Nano Letters, 19*(4), 2411–2417.

86. Ge, R., Wu, X., Kim, M., Shi, J., Sonde, S., Tao, L., ... Akinwande, D. (2018). Atomristor: Nonvolatile resistance switching in atomic sheets of transition metal dichalcogenides. *Nano Letters, 18*(1), 434–444.

87. Qian, K., Tay, R. Y., Nguyen, V. C., Wang, J., Cai, G., Chen, T., ... Lee, P. S. (2016). Hexagonal boron nitride thin film for flexible resistive memory applications. *Advanced Functional Materials, 26*(13), 2176–2184.

88. Siddiqui, G. U., Rehman, M. M., Yang, Y. J., & Choi, K. H. (2017). A two-dimensional hexagonal boron nitride/polymer nanocomposite for flexible resistive switching devices. *Journal of Materials Chemistry C, 5*(4), 862–871.

89. Pan, C., Ji, Y., Xiao, N., Hui, F., Tang, K., Guo, Y., ... Lanza, M. (2017). Coexistence of grain-boundaries-assisted bipolar and threshold resistive switching in multilayer hexagonal boron nitride. *Advanced Functional Materials, 27*(10), 160481.

90. Leydecker, T., Herder, M., Pavlica, E., Bratina, G., Hecht, S., Orgiu, E., & Samorì, P. (2016). Flexible nonvolatile optical memory thin-film transistor device with over 256 distinct levels based on an organic bicomponent blend. *Nature Nanotechnology, 11*(9), 769–775.

91. Lee, J., Pak, S., Lee, Y., Cho, Y., Hong, J., Giraud, P., ... Kim, J. M. (2017). Monolayer optical memory cells based on artificial trap-mediated charge storage and release. *Nature Communications, 8,* 14734.

92. Lee, D., Hwang, E., Lee, Y., Choi, Y., Kim, J. S., Lee, S., & Cho, J. H. (2016). Multibit MoS_2 photoelectronic memory with ultrahigh sensitivity. *Advanced Materials, 28*(41), 9196–9102.

93. Xiang, D., Liu, T., Xu, J., Tan, J. Y., Hu, Z., Lei, B., ... Chen, W. (2018). Two-dimensional multibit optoelectronic memory with broadband spectrum distinction. *Nature Communications, 9*(1), 2966.

94. Tran, M. D., Kim, H., Kim, J. S., Doan, M. H., Chau, T. K., Vu, Q. A., ... Lee, Y. H. (2019). Two-terminal multibit optical memory via van der Waals heterostructure. *Advanced Materials, 31*(7), 1807075.

95. Wang, Q., Wen, Y., Cai, K., Cheng, R., Yin, L., Zhang, Y., ... He, J. (2018). Nonvolatile infrared memory in MoS_2/PbS van der Waals heterostructures. *Science Advances, 4*(4), eaap7916.

7 Artificial Synaptic Devices Based on Two-Dimensional Semiconductors

7.1 INTRODUCTION TO ARTIFICIAL SYNAPTIC FUNCTIONALITIES

7.1.1 CHALLENGES OF DATA PROCESSING

Both the concept of program storage and that of binary principles were proposed for the first time by von Neumann in 1945 [1]. This concept was later used as the fundamental basis for design and development of a new generation of computers, known as von Neumann–type computers [1], in which the central processing unit (CPU) was separated from the memory [2]. Although the von Neumann system attained unprecedented achievements in processing and computer knowledge, it faced a fundamental problem stemming from its structural design. The separated structure in von Neumann system is the Achilles' heel of this design limiting further development of this computing system. The separated structure in von Neumann systems demands a huge data transfer process between CPU and memory device. When the CPU needs to execute and operate huge data, the management of data traffic turns into a serious limitation. As the speed of data transfer increases, the longer time idle has spent by the processor waiting for the data to be fetched from the memory unit. In this case the CPU becomes idle which is known as the bottleneck of von Neumann systems [3]. No matter how fast a processor can operate, the system performance is restricted by the rate of data transfer between CPU and memory. In the current design of von Neumann computers, the faster CPU is equal to the longer idle time of the system. Furthermore, while the speed of processors has increased significantly during last decade, the technological advancements in novel memory devices mostly focus on the increased capacity and data storage density, thus it does not solve the problem of data transfer traffic between CPU and memory unit [4]. The other challenge is attributed to the power consumption in von Neumann system, which is highly challenging for complex computation and processing operation in modern computers [5].

All technological challenges mentioned above have triggered the motivation of engineers to find new approaches to overcome the von Neumann bottleneck challenge. There are several strategies, which can be taken to tackle the limitations of von Neumann devices. They are as follows:

(1) *Catching* (the storage of frequently used data in a specific area of RAM),
(2) *Prefetching* (moving the data to *Catch* before it is requested for speed access),
(3) *Multithreading* (simultaneous management of multiple request in separate threads),
(4) Using *Random access memories* (RAM) to activate outputs on both the rising and falling edge of the systems,
(5) Employment of *RAMBUS*, which is a memory subsystem with a RAM, RAM controller and the path (bus) which connects the RAM to microprocessor of computer.
(6) Integration of processor and memory in a single device [6].

To fulfill the information technology requirements, a huge amount of data should be processed, transferred and stored. It is therefore required to have new processing systems with the capability of high-speed information transfer and processing to facilitate the visionary light processing, pattern recognition, sound localization, device automation, artificial intelligence, and machine learning. The von Neumann computing and other replacement methods are failed to overcome the present challenges of computer engineering. Alternatively, mimicking the human brain functionalities can be the most promising strategy to tackle the deficiencies of von Neumann computers [7]. The human brain operation is highly energy efficient. The data storage and processing are functioned in brain with the lowest possible energy consumption. This ultralow energy consumption is inspired by the functionalities of biological nervous system. Based on the brain performance, a novel type of computing and processing system was proposed by Alan Turing in 1948, when he introduced the idea of using a computing device with the interconnected network of artificial neurons [8]. It is envisaged that the interconnect network of artificial synapses can effectively solve and overcome the restrictions of the traditional van Neumann computing.

7.1.2 Biological Synapse versus the Artificial Synapse

7.1.2.1 Biological Synapse

Biological synapses, as one of the key members of neural system, are the information channels ensuring short-term computation, long-term learning, and memorization by tuning the synaptic weights [9]. As a biological structure, a chemical synapse is a gapped connection between two neurons through which the communication between two individual axons is created by biochemical reactions. Uniqueness of the brain as the main intelligent organ arises from the highly-energetically efficient synaptic data processing. A typical neuron has up to several thousands of synapses where the presynaptic neurons are connected to the postsynaptic neurons via dendrites. A graphical interpretation of the synapse is demonstrated in Figure 7.1 [10]. The synapse is a biological junction where the electrochemical waves (action potentials)

Synaptic Devices Based on 2D Semiconductors

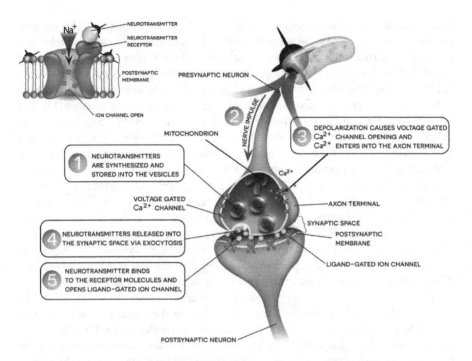

FIGURE 7.1 The graphical scheme of biological synapse when the presynaptic pulses are transferred from axons to synaptic junctions (Reprinted with permission from John, R., Liu, F., Chien, N. A., Kulkarni, M. R., Zhu, C., Fu, Q., ... Mathews, N. (2018). Synergistic gating of electro-iono-photoactive 2D chalcogenide neuristors: Coexistence of Hebbian and homeostatic synaptic metaplasticity. *Advanced Materials*, *30*(25), 1800220).

are transmitted through the axons of neuron. At the end of the axon, the presynaptic action potential signal reaches the presynaptic terminal and triggers the channel to release the chemical neurotransmitters, which are usually Na⁺ and K⁺ ions[11]. The released ionic neurotransmitters travel the synapse distance and bind to the molecular receptors in membrane of the opposite neuron junction. The delivered ionic species generate postsynaptic current on the opposite side of the synaptic cleft to finalize the signal transition process, which is briefly called interneural signaling. The postsynaptic response to a presynaptic signal is classified as being either excitatory or inhibitory. The excitatory postsynaptic current (EPSC) and inhibitory postsynaptic current (IPSC) are two famous postsynaptic outputs [12].

The energy consumption of a biological synaptic event is tangibly lower than that of any other similar analogous device. It is believed that a synaptic event in brain requires 1 to 10 fj energy [13]. The synaptic performances assist the occurrence of some fundamental human capabilities and activities. Learning, remembering, processing, recognition, biological reactions in body, conscious and unconscious actions/reaction and voluntary and non-voluntary movements are some of the well-known examples operating by the nervous system. There are some fundamental concepts and definitions for biological synaptic events that must be introduced to further understand the synaptic properties and behavior of artificial synapses. The learning process in brain occurs by the strengthening and weakening of synaptic weights and connections through the neural system networks which is termed as the functional synaptic plasticity [14]. This learning can be explained based on *short-term plasticity* (STP) and *long-term plasticity* (LTP). STP is actually described as the rapid change of the synaptic strength for the time scale from milliseconds to few minutes [14]. The STP of synapses can enhance the certainty of synaptic transition and regulate the balance between the excitation and inhibition of cerebral cortex. It is believed that STP is involved in realization of advanced functional capabilities of nervous system such as learning, memorization, sleep rhythm, and attention [15]. Paired pulsed facilitation (PPF)/depression (PPD) are the dynamic enhancement/reduction of the neurotransmitter release [15]. It was observed that the IPSC and EPSC are changing by the increase or decrease of interval time between two consecutive presynaptic pulse signals. Both inhibitory and exhibitory postsynaptic currents are intensified (Increased) by the decrease of time interval. For exhibitory synapse, the magnitude of the second postsynaptic current (A_2) is higher than the first synaptic current (A_1). The inhibitory synapse further inhibits the magnitude of the current (A) where the $A_2 < A_1$ by the decrease of time intervals. This process corresponds to PPF/PPD [15].

Compared with STP, the time scale of LTP is in the range of few minutes to several hours. There are two other characteristics attributed to the LTP: long-term potentiation and long-term depression or inhibition. The long-term potentiation is the strengthening of synaptic events over the considerable time span causing the storage of events and information in human memory. It is the persistent enhancement of the transition between two neurons originating from the simultaneous stimulation of two neurons [16]. The ability of synapse for changing its intensity and weight is synonym to capabilities of synapses for memorization process. Since the memory is encoded by the alteration or change of synaptic weights, the long-term potentiation or depression are widely considered as the main molecular mechanism underlying the learning and memorization process of the human brain [17]. On the opposite side, long-term depression is equal to the weakening the connections and disconnecting of the synaptic paths to forget and remove the unwanted memories, so it is also called long-term inhibition [18]. There are several reasons which may cause the long-term depression. Strong synaptic stimuli or long-term weak stimuli can cause long-term depression in the biological synapse [19]. From a biological point of view, the long-term depression is also equal to the decrease of receptor intensity or change in the release of presynaptic neurotransmitters in synaptic junction [20].

The Hebbian theory (1949) described the synaptic plasticity as a continuous and repetitive stimulation of the postsynaptic neurons by the presynaptic neurons which finally results in the increase of efficiency of synaptic transition. The Hebbian theory emphasizes that the association learning occurs by the successive excitation of neuron B by the neuron A. To be precise, the neuron A should contribute the excitation of neuron B and this process should not occur simultaneously [21]. This theory is the bases of the *spike-time-dependent-plasticity* (STDP). Considering the plasticity concept in synaptic behavior, the STDP refers to the combination of long-term potentiation and long-term depression events, which are triggered by spike-time correlations [22]. Besides, the STDP proposes that the synaptic plasticity is not merely attributed to the delay time but also to the frequency of presynaptic spike signals. For example, it was observed that high frequency presynaptic signals (200 Hz) caused long-term potentiation, while the low frequency signals (5 Hz) resulted in long-term depression [23]. These synaptic functionalities executed by the network of synapses in the neural system. In the human brain, the arrived information is processed and then orders are delivered to the corresponding autonomic and somatic nervous systems. Nowadays, the performance of the artificial neural networks and synaptic devices are evaluated based on the same biological concepts. In this context, the STP, LTP, STDP, long-term potentiation, and long-term depression, PPF and the other properties of artificial synapses are measured and evaluated.

7.1.2.2 Artificial Synapse

As it was discussed, a biological synapse mostly works based on the biochemical phenomena where the synaptic events facilitate the ultralow energy and efficient human brain processing. Analogously, an artificial synapse is a device which mimics the behavior and performance of the biological synaptic junction to transfer the synaptic signals. From a technical point of view, memristors are the best devices capable of successful emulation of the characteristics of a biological synapse (Figure 7.2a and b) [24]. The resistive switching mechanism is the basic principle of the memristor devices (Figure 7.2b). In memristor the charge transfer between two conducting electrodes are controlled by the alteration of resistance values of the interlayer material between the conductive electrodes [25]. A sandwiched insulator or

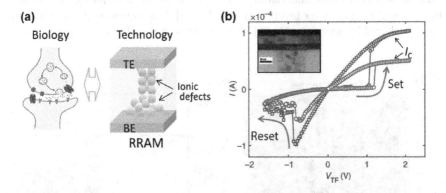

FIGURE 7.2 (a) Biological synaptic characteristics can be represented and emulated by a memristor device similar to RRAM, where the conductivity changes by voltage-induced ion migration and corresponding formation and dissociation of conductive filaments. (b) Typical current-voltage (I/V) curve of a RRAM device where a positive voltage are applied to the top electrode to trigger the transition from HRS to LRS in memristor device. The bipolar resistive switching mechanisms are observed. The transition from HRS to LRS is controlled by the compliance current I_c. A larger I_c resulted in a higher conductance, thus enabling the occurrence of time-dependent potentiation and depression in the RRAM synaptic device (Reprinted and reproduced with permission from Wang, W., Pedretti, G., Milo, V., Carboni, R., Calderoni, A., Ramaswamy, N., ... Ielmini, D. (2018). Learning of spatiotemporal patterns in a spiking neural network with resistive switching synapses. *Science Advances*, 4(9), eaat4752).

transition metal oxide between two conductive electrodes is the typical structure of an artificial synaptic device (Figure 7.2a) [24]. However, the device configuration is not merely restricted to the horizontally stacked structures, since the synaptic behavior was also observed in the vertically stacked memristors or even was found at the field effect transistor design [26]. The neural synaptic electronics are observed in variety of materials with different resistive and synaptic mechanisms. In fact, the diversity of materials with synaptic characteristics experiences a rapid progress, thus the fundamental understanding of related synaptic mechanisms are currently progressing. Metal oxide resistive random access memories (RRAMs) (such as TiO_2, AlO_x, HfO_x, WO_x, TaO_x), conductive bridge synapses (Ag- and Cu-based electrodes), phase-change materials-based synapses (PC synapse), Ferro electric materials-based synapses, Magneto resistive (MR)–based synapses, carbon nanotube–based synapses, and organic nanostructured-based synapses are some of active memristor materials [27–35].

The von Neumann computing technology is divided into software and hardware implementation. The implementation of software is still highly dependent on the basis of von Neumann structures, such as the google AlfaGo. However, this software technology still suffers from the high power consumption and low integration density. The hardware devices mainly include resistive random access memories (RRAM), which is a type of nonvolatile resistive memories (memristors), static RAM and phase-change memory. The memristor crossbar arrays configure the artificial neural network (ANN) system (Figure 7.3a) [36]. It was demonstrated that the ionic motion in insulator material assisted the creation and annihilation of defects in the sandwiched memristor layer (Figure 7.3b) [36]. Consequently, the device resistance is altered by applying external electric pulses. It is highly promising approach

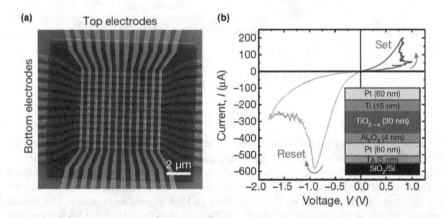

FIGURE 7.3 The top view image of an interconnected network of synaptic device for application in ANN system. (b) Synaptic memristor device consist of TiO_{2-x}/Al_2O_3 active layer sandwiched between Pt electrodes (Reprinted and reproduced with permission from Prezioso, M., Merrikh-Bayat, F., Hoskins, B. D., Adam, G. C., Likharev, K. K., & Strukov, D. B. (2015). Training and operation of an integrated neuromorphic network based on metal-oxide memristors. *Nature*, *521*(7550), 61–64).

to design series of interconnected memristors ($\sim 10^5$ bits cm^{-1}) to achieve ultra-high density ANN systems. The ANN device architecture can perform similar to the human cerebral cortex system which contains almost 10^{15} synapses. While the synaptic characteristics are well recognized in memristors, their development is also the subject to key factors for improvement of their performance. Insufficient device performance, the lack of stability and reliability and high-energy consumption are the main challenging factors limiting the progress of integrated ANN systems based on artificial synapses [37]. To fulfill the requirements of artificial intelligence technology for new electronic era, tremendous attention and urgent promotion are required to develop the neuromorphic/brainlike computing. For hardware engineering, new materials and new structures are always investigated to implement neuromorphic networks and facilitate the efficient brain like computing. More recently, 2D materials have attracted considerable attention of scientific communities for application in memristors to simulate the synaptic functionalities. 2D materials with their unique characteristics, distinguished charge transfer mechanisms, ultrathin thickness, high quality, and peculiar physical properties and the capability of phase transformation from metal to insulator are intriguing options for employment in the synaptic devices as active memristor layers. Graphene was the first 2D material introduced and practically integrated in electronic instruments. However, its application in electronic and optoelectronic fields is still restricted by the lack of bandgap in pristine form [38]. Although the development of grapheme is progressing, noteworthy that the worldwide attention has been shifted to research, synthesis, characterization, and employment of novel 2D materials for variety of electronic and optoelectronic application. It was observed that 2D materials from transition metal dichalcogenide, black phosphorus (BP), hexagonal boron nitride (h-BN) and ultrathin 2D films of oxide materials can effectively show the synaptic characteristics and tangibly reduce the energy consumption for synaptic events [39]. The short-channel effect phenomenon is the main properties of 2D materials, which can ensure the balance between the energy consumption and synaptic device efficiency. Another main property of 2D material is their outstanding compatibility for running electric features. The bandgap engineering and alteration of electronic properties of 2D materials can be achieved by changing the number of layers in 2D nanostructues. It consequently facilitates the alteration of resistance state and conductivity values of 2D materials helping to control the synaptic phenomenon and to achieve ultralow energy consumption of artificial synapses [40]. It is well-known that the band gap values of 2D materials cover a wide range of energies ranging from ultimate conductive to insulating properties [39]. Such energy modulation creates a variety of electronic properties in 2D materials, including semiconducting, metallic, semimetallic, insulating and superconducting characteristics. The high carrier density and mobility put 2D materials in a strong position for replacement of common Si-based channel materials. The bandgap engineering and the control of mobility characteristics of 2D materials make it possible to realize the synaptic plasticity and neuromorphic properties in 2D based synaptic devices [40]. This is achieved either by the control of the carrier migration dynamics on the surface of 2D materials or by the internal control of carrier motion dynamics inside of 2D structures. Furthermore, the decreased size of 2D materials lets the engineers to further scale down the dimensions of artificial synapses and increase the integration density of present artificial synapses. This is highly important factor for engineers to move closer to the capacity and capability of human brain. Generally the two-dimensionality based properties open up a whole new pathway for fabrication of new artificial synaptic devices with highly efficient working mechanisms.

7.2 SYNAPTIC ELECTRONICS BASED ON 2D MATERIALS

7.2.1 DESIGN AND STRUCTURE OF DEVICES

The library of 2D materials is expanding rapidly since the advent of exfoliated graphene. The ultrabandgap diversity of 2D materials provides a flexible platform to choose 2D materials with desired property for designed electronic applications. The electronic properties of 2D materials cover a broad range of electrical characteristics from insulating to conducting properties. For example, due to the lack of bandgap and high carrier mobility, 2D graphene film is one of the attractive electrode materials for electrical measurement of properties of 2D-based devices [41]. The lack of dangling bond in 2D graphene avoids the formation of Fermi-level pining effect at graphene/2D materials heterostructure. Consequently, the van der Waals 2D graphene heterostructures are famous for their ultralow resistance. Furthermore, the semimetallic characteristics (TMDCs: WTe_2, $TiSe_2$, $SnSe_2$), the semiconducting properties (TMDCs: MoS_2, WS_2, $MoSe_2$, WSe_2, TiO_x, ZrO_2) and the insulating states (HfO_2, h-BN) of 2D materials are well documented [42–56]. The RS property is another intriguing characteristics of 2D materials, which can be employed for the development of artificial synaptic materials [57]. Thus, 2D materials can be classified according to their switching mechanisms. Based on the RS mechanisms, the memristor synapses can be categorized into electrostatic/ electron ion synapses, electrochemical metallization/conductive bridge synapses, phase-change synapses, thermochemical synapses, and oxidation/reduction chemical valence synapse [40].

Generally, the semiconductor and insulator materials contain some level of defect sates and impurities. The occupation or the release of the impurities and defect states by the electron or ions affect the electron/ion mobility in 2D material. The alteration of electron/ion mobility inside 2D material channel facilitates the modulation of synaptic plasticity. There are several typical designs for synaptic devices including, the sandwiched structure with top and bottom electrode, the two (or more) terminal electrodes deposited on top of the 2D materials with lateral structure, and the gate transistor synapses with tunable conductivity. However, the structural design of artificial synapses is not merely restricted to these structures. It includes and covers any other innovative design that facilitates the occurrence of charge-trapping phenomena. Table 7.1 provides a detailed comparison for the different structural types of artificial synaptic instruments with their individual characteristics [58]. The next sections will provide the detailed information on various types of artificial synaptic instruments based on 2D materials. The devices are categorized based on the mechanism of resistive switching and charge-trapping.

TABLE 7.1

Comparison of Artificial Synapses Respecting to Materials and Structure

Device type	Active layer	Drawbacks	ON current	Forming voltage	Size	Depiction
Visible light-responsive, fast ≤ 20 ms response	$CH_3NH_3PbI_3$ (MAPbI$_3$)	Lateral structure, non-CMOS	3 µA	0.3 V	500 nm	
Photo-electro, monolayer, self-rectify, volatile, low current	n-MoS$_2$	Nonvolatile operation, limited UV range	10 nA	NA	50 µm	
High mobility, photo responsive, monolayer	Black phosphorus	Four terminal, high voltage, double- layered	10^{-7} A	10 V	2 µm	

(Continued)

TABLE 7.1 (CONTINUED)

Comparison of Artificial Synapses Respecting to Materials and Structure

Device type	Active layer	Drawbacks	ON current	Forming voltage	Size	Depiction
Flexible, transparent, biocompatible, biodegradable	Natural organic polymer – Chitosan	Large-sized, lateral structure	0.5 mA	1.5 V	1 mm	
Flexible biocompatible, biodegradable	Natural organic polymer – Lignin	Large-sized, high current	0.5 mA	NA	0.1 mm	
Flexible biocompatible, transparent	Fish collagen	High current	0.5 mA	1.5 V	0.1 mm	
2D, low power, high reliability, 300 ms switching	Graphene	Side gate terminal	50 pA	Electrochemical potential	15 µm	

(Continued)

TABLE 7.1 (CONTINUED)

Comparison of Artificial Synapses Respecting to Materials and Structure

Device type	Active layer	Drawbacks	ON current	Forming voltage	Size	Depiction
2D FET	MoS_2 with PVA proton conductor	Side gate terminal	20 μA	NA	10 μm	
Humidity sensitive, very low current	MoO_3 with ion liquid	Liquid phase electrolyte	20 nA	NA	10 μm	
2D, high thermal stability, high endurance, flexible	Graphene/$MoS_{2-x}O_x$/graphene	High current, complex structure	1 mA	2 V	10 nm	

(Continued)

TABLE 7.1 (CONTINUED)

Comparison of Artificial Synapses Respecting to Materials and Structure

Device type	Active layer	Drawbacks	ON current	Forming voltage	Size	Depiction
FET, flexible, high mobility	SWCNT	High current, complex fabrication	0.5 mA	NA	20 μm	
Small size, FET, CMOS compatible	CNT	Complex fabrication	10 μA	NA	1 μm	
30 ns switching, multilevel operation	$Zr_{0.5}Hf_{0.5}O_2$ graphene oxide quantum dots	High current, complex structure	0.5 mA	0.6 V	100 μm	

Source: Reprinted with permission from [58].

7.2.2 Electronic/Ionic Artificial Synapses

7.2.2.1 Polycrystalline 2D MoS₂ Synaptic Device

The multiterminal hybrid memory-transistors (memtransistors) are capable of showing synaptic plasticity. A memtransistors based on CVD deposited polycrystalline single-layer MoS$_2$ was fabricated with multiterminal structure (Figure 7.4a) [59]. The synaptic memtransistors work based on the induced migration of defects that causes the RS phenomenon. The *I-V* curve of 2D MoS$_2$ memristor is shown in central section

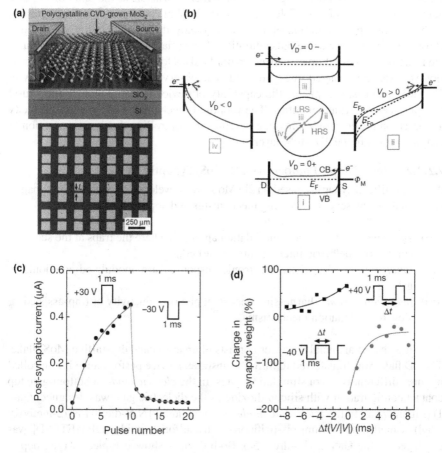

FIGURE 7.4 (a) Graphical scheme of single-layer MoS$_2$ memtransistors device with gate-tunable synaptic functionalities. Optical micrograph of fabricated MoS$_2$ memtransistors with channel length of L. (b) Schematic of energy band alignment corresponding to different resistive states for bipolar resistive switching of synaptic device. The defect migrations are depicted by arrows. (c) The effect of pulse number on the postsynaptic current of single-layer MoS$_2$ memtransistor device. (d) The changes of synaptic weight as the function of the paired pulses with different pulse intervals (Reprinted and reproduced with permission from Sangwan, V. K., Lee, H. S., Bergeron, H., Balla, I., Beck, M. E., Chen, K. S., & Hersam, M. C. (2018). Multi-terminal memtransistors from polycrystalline monolayer molybdenum disulfide. *Nature*, 554(7693), 500–504).

of Figure 7.4b. The schematic configuration of energy band alignment corresponding to the bipolar resistive switching state of MoS_2 memristor is presented in Figure 7.4b. It was observed that the multiterminal structure memtransistors based on single-layer polycrystalline MoS_2 film is capable of emulating the synaptic functionalities. The alternating gate voltages were employed with different intensities to modulate the synaptic weight of artificial synaptic device. The long-term potentiation/depression properties were observed in the behavior of artificial synapses demonstrating that the device is capable of emulating the complex neuromorphic properties. The graphs of variation of postsynaptic current of memtransistors as the function of pulse number are shown in Figure 7.4c. The electrical voltaic pulses with 30 V magnitude at the constant pulse duration of 1 ms was employed. The depression voltage was chosen at −30 V at the same pulse duration. The variation of the synaptic weight as the function of interval times between paired pulses for the potentiation and depression voltages are depicted in Figure 7.4d. The observation of synaptic plasticity in 2D MoS_2 memristor confirms the capability of device for emulation of the neural synaptic learning functionalities. The proposed mechanisms of synaptic plasticity are based on the defect mobility engineering and controlling the dynamic motion of charge carriers in the 2D MoS_2 layer.

7.2.2.2 Electro-Iono-Photoactive 2D MoS_2 Synaptic Device

Another artificial synapses based on 2D MoS_2 was developed to emulate the synaptic functionalities by several following mechanisms and approaches [60]:

(i) *electronic mode*: a defect modulation approach where the traps at the semiconductor–dielectric interface are perturbed;
(ii) *ionotronic mode*: where electronic responses are modulated via ionic gating;
(iii) *photoactive mode*: harnessing persistent photoconductivity or trap-assisted slow recombination mechanisms.

The MoS_2-based artificial synapses consists of an exfoliated crystalline MoS_2 flake. The 2D flake was transferred to Si/SiO_2 substrate. Device performance is controlled by three different types of stimulating gates. In the *electric mode*, a bottom-gate top contact configuration with silicon dioxide as the dielectric gate was designed (Gate 1) (Figure 7.5a). In the *ionotronic mode*, an ionic liquid [N,N-diethyl-N-(2-methoxy ethyl)-N-methylammonium-bis-(trifluoromethylsulfonyl)-imide (DEMETFSI)] was employed as the Gate 2 (Figure 7.5a). Both devices showed typical *n*-type depletion-operation mode with linear mobility. To mimic the ion flux transmission and neurotransmission release in the biological synapses, different trapping possibilities were reutilized. The *optical stimulation* (as the third gate) was used to modulate the photogenerated charge carriers in atomically thin 2D MoS_2 film allowing modulation of the synaptic weights with photonic pulses. This optical capability of artificial synapse is highly important that capitalizes the possibility of high-bandwidth optical communication protocols. The schematic of 2D MoS_2-based synaptic device is depicted in Figure 7.5a. As the main privilege, the incorporation of the optical biases with electrical pulses addressed different charge-trapping mechanisms to precisely

Synaptic Devices Based on 2D Semiconductors 243

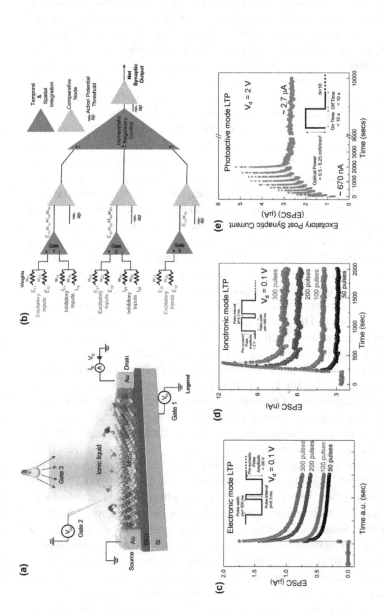

FIGURE 7.5 (a) Graphical scheme of triple modulated architecture of artificial 2D MoS$_2$ synaptic device. (b) The analog circuit depicting the interplay between the various parameters for stimulation of synaptic plasticity. The interplay between the Hebbian and synaptic plasticity in MoS$_2$ base 2D synaptic device, which depicts the enhancement in the strength of long-term potentiation functionalities of device as the function of training by (d) electronic mode, (e) ionotronic mode, and (f) photovoltaic mode. In photovoltaic mode, the light intensity is altered (from 0.5 to 5.25 mW mm^{-2}) to enable training by optical pulses (Reprinted and reproduced with permission from John, R. A., Liu, F., Chien, N. A., Kulkarni, M. R., Zhu, C., Fu, Q., ... Mathews, N. (2018). Synergistic gating of electro-iono-photoactive 2D chalcogenide neuristors: Coexistence of Hebbian and homeostatic synaptic metaplasticity. *Advanced Materials, 30*(25), 1800220).

modulate the synaptic weights. Furthermore, the instrument was capable of neuro-modulation to achieve the plasticity and metaplasticity properties via dynamic control of spike-time-dependent plasticity (Hebbian theory). The schematic of an analog circuit is demonstrated in Figure 7.5b depicting the interplay between Hebbian and homeostatic plasticity. The interplay between Hebbian and homeostatic synaptic plasticity in MoS_2 synapse is provided in Figure 7.5 c, d, and e, respectively. The long-term potentiation strength was achieved by employment of sequential voltaic training pulses in synaptic instrument. It was observed that the synaptic weights increase linearly by increasing the number of training pulses. It was also demonstrated that the application of 100 training pulses resulted in the weight change of synapse up to 117% in electronic mode, while the same number of training pulses in ionotronic mode resulted in the increase of weight up to 223% (Figure 7.5 c and d). In the optically stimulated synaptic device, a strong long-term potentiation was observed in the 2D MoS_2 artificial synapse via using the optical gating function (photoactive mode). The light pulses ($\lambda = 445$ nm) played the role of presynaptic signals. Up on the illumination on the surface of 2D MoS_2 synapse, the conductance increased rapidly and then continued until the gradual saturation when the conductance was one order of magnitude above the value of dark current. A rapid drop of conductance occurred on the termination of optical illumination source, but the conductance still remained in a high values (Figure 7.5e). The band-to-band transition is expected to be the main reason for rapid transition, while the long-lasting high conductance values in opto-electronic mode were attributed to the defect or trap-centered slow recombination of generated electron/holes. The observations are in accordance with the proposed model of random local potential fluctuation (RLFP). This state is called persistent conductivity [60]. By the increase of optical pulses and higher number of photons, the decay time became slower. This is consistent with the RLPF model where higher carrier numbers occupy sites of the local potential minima, and therefore, the rate of election/hole recombination would be slower. The vivid manifestation of this phenomenon is slower degradation of photogenerated current in synaptic device.

The capability of developed electro-ion-photoactive 2D MoS_2 synapse [60] for emulation of the human brain learning process was evaluated by using the classical conditioning Pavlov's dog experiment. In this test the classical conditioning was employed in the 2D MoS_2 thin-film memtransistors by modulation coupling between optical and electrical pulses. To organize the test, the light pulses emulate the food-unconditioned stimuli, which activate the salivation-unconditioned response from postsynaptic terminal. Simultaneously, the voltage pulses are imposed on the back gate to modulate the bell-conditioned stimuli to finally activate the conditioned response of a dog (Figure 7.6a). The graph of EPSC for electronically pulse/conditioned stimulus and optically pulse/unconditioned stimulus are demonstrated in Figure 7.6b and c, respectively. In the test, initially the light pulses (unconditioned stimulus) led to an efficient unconditioned response, which is known as salvation state (state a in Figure 7.6.d). However, the imposed back gate voltage on SiO_2 film did not lead to the efficient conditioned response before the training stage. It was observed when the EPSC was less than 500 nA, the salivation stage was not observed. It was verified that the initial training stage with conditioned and unconditioned stimuli (without sequential training) did not result in the effective association of learning stage as it was depicted in state b in Figure 7.6d. It was understood that a repeated training resulted in strong

Synaptic Devices Based on 2D Semiconductors 245

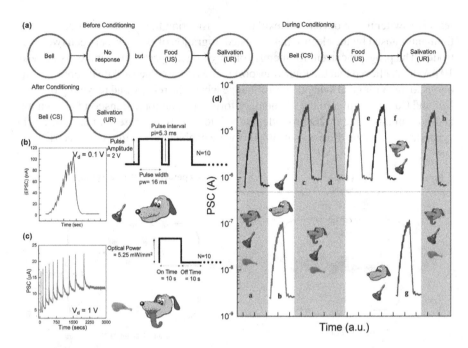

FIGURE 7.6 (a) Schematic outline of classical conditioning experiments famous as Pavlov's experiment. (b) The voltage pulse/conditioned stimulus accompanied by the EPSC reaction of synaptic device. (c) The unconditioned optical pulsed stimulus and their EPSC responses. (d) Different states in learning process of artificial synaptic device by using the Pavlov's experiment protocol (Reprinted with permission from John, R. A., Liu, F., Chien, N. A., Kulkarni, M. R., Zhu, C., Fu, Q., ... Mathews, N. (2018). Synergistic gating of electro-iono-photoactive 2D chalcogenide neuristors: Coexistence of Hebbian and homeostatic synaptic metaplasticity. *Advanced Materials*, *30*(25), 1800220).

association. After the successful training, the conditioned stimulus alone triggered the salivation behavior. The state *c* and *f* depict the condition when the current is higher than 500 nA. When the conditioned stimulus was used alone without accompanying the unconditioned stimulus for 2 h, the EPSC decreased to values below the salivation threshold (state *g*). Finally, the association was recovered back faster by using simultaneous conditioned and unconditioned stimuli (state *h*). The coexistence of these multiple forms of synaptic plasticity elevated the efficiency of memory storage and furthermore enhanced the processing capacity of artificial synaptic system. It finally facilitated the design and great performance of efficient novel neural system with capability of mimicking the learning process.

7.2.2.3 Ionic Transport in 2D Perovskite Synaptic Devices

An ionic-based artificial synapse based on 2D bromide perovskite device with extremely low energy consumption was recently demonstrated [61]. The synapse was capable of emulating the synaptic functionalities of human brain. The artificial synapse was based on a layered 2D perovskite (2D layered PEA_2PbBr_4 crystal) resistive active alloy sandwiched between bottom Au and top graphene electrodes. The resistive switching in device was explained considering the formation of conductive filaments caused by the electric field assisted migration of Br⁻ ions between two electrodes. The program current for memristor based on 2D perovskite single crystal was 10 pA. It is believed that the ionic transport in 2D Br-based perovskite layered is analogous to the biological synaptic performance where the release and recycling of ionic Na⁺ and Ca⁺ are similar to the filamentary formation and ionic transport between top and bottom electrode in 2D bromide perovskite synapse (Figure 7.7a and b). In 2D synaptic devices, the diffusion of Br⁻ ions were enabled by the imposed electric field

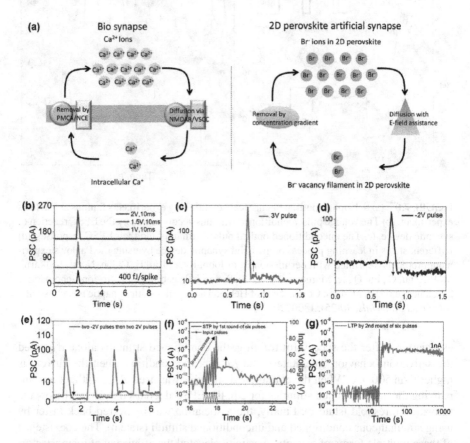

FIGURE 7.7 The synaptic characteristics of 2D Br-based perovskite device. (a) Comparison of functionality of a biological synaptic device with its (b) analogous functions in artificial 2D perovskite synaptic device. The EPSC of device with different amplitudes of electrical potential pulses. (c) The EPSC of 100 pA by applying a 3V pulse for 10 ms. (d) Depression induced by negative pulsed voltage. (e) The short-term depression and potentiation. (f) Short-term potentiation after imposing six voltage pulses. (g) The transform from short-term potentiation to long-term potentiation by applying second round of six pulses. The current level reached to 1na with the retention time of 1000 s (Reprinted with permission from Tian, H., Zhao, L., Wang, X., Yeh, Y. W., Yao, N., Rand, B. P., & Ren, T. L. (2017). Extremely low operating current resistive memory based on exfoliated 2D perovskite single crystals for neuromorphic computing. *ACS Nano, 11*(12), 12247–12256).

Synaptic Devices Based on 2D Semiconductors

on working electrodes and the removal of Br$^-$ ions were caused by the concentration gradient. By the employment of different voltaic pulses, the artificial synapse showed EPSC values with different intensities. The calculated energy consumption for a singular spike voltage of 1.0 V for 10 ms pulse duration was only 400 fj, which is close to the energy consumption of biological synapse. The synaptic weight of device was changed by employment of stronger synaptic pulses (3 V) (Figure 7.7c). It is noted that the positive voltages drive the negative Br ions toward the positive Au electrode to from the partial conduction filaments composed of Br$^-$ vacancies. When a negative pulsed voltage was applied (–2 V), the Br$^-$ ions returns to their initial states resulting in the recovery of conductance value and the drop of postsynaptic current to lower current level than that of the initial state (Figure 7.7d). This behavior is equal to the short-term depression. The short-term potentiation and depression were observed by applying sequential negative and positive voltages (Figure 7.7e). The transition from short-term potentiation (Figure 7.7f) to the long-term potentiation (Figure 7.7 g) was facilitated after applying twelve consecutive voltaic pulses. The required time for recovery from long-term potentiation state to the initial state (retention time) was estimated to be around 1000 s. The energy consumption of fabricated synaptic unit based on 2D perovskite single crystal was tangibly lower than that of artificial synapses based on 3D perovskite polycrystalline film. The presence of grain boundaries and defects are the main reasons for large leakage current in the 2D perovskite polycrystalline structure. The high leakage current resulted in high operation current, and consequently, increased the power consumption. Due to the employment of 2D single crystalline structure the leakage current was tangibly blocked (0.1 pA). It resulted in the low operation current of device (10 pA). Thus, the consumption energy of synapses was considerably lower than the same polycrystalline structure.

7.2.2.4 In-Ion Doped Ultrathin TiO$_2$ Optical Synaptic Devices

Two-terminal optical synapse based on In-doped TiO$_2$ ultrathin film was also fabricated for emulation of the synaptic functionalities [62]. The In-ions were doped into ~7.0 nm thick TiO$_2$ film deposited by atomic layer deposition (ALD) technique. It was observed that the ion doping affected the resistive switching mechanism and also extended the optical absorption edge of ultrathin 2D TiO$_2$ film to the visible light range. It was measured that the set voltage of the TiO$_2$ based synaptic device was shifted to the lower voltages after optical stimulation of In-doped TiO$_2$ memristor. These observations (Figure 7.8a) confirmed the impact of photogenerated electron/hole on occupation of trapping sites in amorphous TiO$_2$ film and thus resulted in the decrease of the set voltage (Figure 7.8b) and the increase of memristors' conductance (Figure 7.8b). Furthermore, the increase of light intensity slightly affected the set voltage of memristors, which is an indication of saturation state in the number of photogenerated charge carriers. Therefore, it does not let the device to achieve the lower V_{Set} values. The emulation of optical synaptic characteristics was verified by the observation of unrested postsynaptic current caused by singular optical pulse (Figure 7.8c). PPF synapse values were demonstrated by two consecutive EPSC spikes with pulse intervals of 0.3 s (Figure 7.8d). The rapid decay of PPF index of device [the ratio of amplitude of the second EPSC (A$_2$) to the first EPSC (A$_1$)] was discovered after the increase of pulse intervals. It shows the sensitivity of short-term

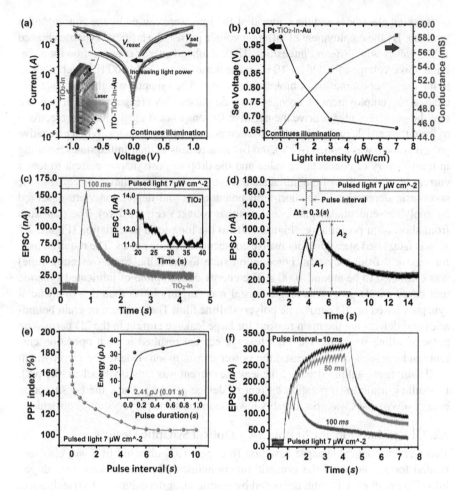

FIGURE 7.8 The optical synaptic characteristics of In-doped ultrathin TiO$_2$ films. (a) The logarithmic scale *I-V* curves of ITO/In-doped/Au synaptic device. (b) Variation of the set voltage and conductance vs changes of light intensity for pulsed lasers. (c) The EPSC of synaptic device induced by 7 μW cm^{-2} laser pulse. (d) The PPF of device. (e) The variation of PPF index versus the pulse intervals. Inset shows the variation of energy consumption versus pulse duration. (f) The variation of EPSC of synaptic device stimulated with the pulsed light with different pulse intervals (Reprinted and reproduced with permission from Karbalari Akbari, M., & Zhuiykov, S. (2019). A bioinspired optoelectronically engineered artificial neurorobotics device with sensorimotor functionalities. *Nature Communications, 10*(1), 3873).

plasticity of optical synapses to the sequence of optical pulses (Figure 7.8e). The calculated energy consumption of In-doped TiO$_2$ optical synaptic device for a singular pulse was estimated to be around 2.41 pJ (pulsed optical signal with 10 ms duration) (Figure 7.8e). The operational energies of selected metal oxide-based synapses are presented in Table 7.2 [62]. It was observed that the energy consumption of In-doped ultrathin TiO$_2$ optical synapse was highly efficient. Moreover, when the successive

Synaptic Devices Based on 2D Semiconductors

TABLE 7.2

The Operation Energy of Selected Metal-Oxide-Based Synaptic Devices

Materials	Explanations	Driving source	Operation energy
Human Synapse	Biological	Ionic pulses	10 fJ
IZO	Electrochemical (Transistor)	Voltage pulse	45 pJ
ZnO_x	Ta_2O_5 as electrolyte (Transistor)	Voltage pulse	35 pJ
α-MoO_3	Ionic liquid electrolyte (Transistor)	Voltage pulse	9.6 pJ
α-MoO_3	$LiClO_4$/PEO electrolyte (Transistor)	Voltage pulse	0.16~1.8 pJ
CuO/Ta_2O_5	Oxide heterostructures	Voltage pulse	2.3 pJ
CMOS	Analog, 180 nm transistor	Voltage pulse	100 pJ
CMOS	Digital, 28 nm transistor	Voltage pulse	25 pJ
CMOS	Digital, programmable	Voltage pulse	10 nJ
TiO_x/HfO_x	Resistive change	Voltage pulse	0.85–24 pJ
TiO_x	Resistive change	Voltage pulse	200 nJ
WO_x	Resistive change	Voltage pulse	40 pJ
PCMO	CMOS-RRAM	Voltage pulse	6–600 pJ
$HfO_x/TiO_x/HfO_x/TiO_x$	Analog	Voltage pulse	40 μJ
$HfO_x/TiO_x/HfO_x/TiO_x$	Various initial resistance, CMOS – 16,348 RRAM device	Voltage pulse	0.85 pJ –24 pJ
In_2O_3/ZnO_2	Oxide heterostructures	Optical pulse	0.2 nJ
TaO_x/TiO_2	CMOS – 3D RRAM device	Voltage pulse	10 fJ
In-Doped TiO_2	Ion-doped memristor	Optical pulse	2.41 pJ

Source: Reprinted with permission from [62].

laser pulses with different frequencies were employed, transition from STP to LTP was emulated by the same optical synaptic unit (Figure 7.8f). These results demonstrated that the optical stimuli with lower pulse intervals are beneficial for facilitation the LTP capabilities. Observations also confirmed that shorter pulse intervals resulted in higher gain values consisting with the effect of residual generated carries on the following pulses. The EPSC and conductance saturation were also detected and observed during the study of synaptic behavior of artificial synapses. This phenomenon was attributed to the saturation of photogenerated electron/holes. It was also shown that the instrument was capable of emulation of bidirectional analog switching.

7.2.3 ELECTROCHEMICAL METALLIZATION/CONDUCTIVE BRIDGE (ECM/CB) ARTIFICIAL SYNAPSES

The electrochemical metallization is the process of formation of a conductive bridge between the counter electrodes of a memristor mainly proposed for solid electrolyte-based synapses. These types of memristors have one electrochemically active electrode (Cu, Ag, Ni) in one side and another auxiliary electrode in another side (Pt,

W, etc.) that plays the role of electrochemical inert electrode of the unit. The mechanism for synaptic plasticity of conductive bridge synapses is the control of oxidation, migration, and reduction of metallic cations (Cu^{2+}, Ag^+, Ni^{2+}) of electrochemically active electrodes. By applying positive voltage on electrochemically active electrode the conductive filaments (conductive bridge) are formed, which is subsequently deteriorated by imposing the positive voltages on inert electrode during the voltage switching. The current state can be altered in these types of artificial synapses by controlling the formation of filamentary conductive channels. An electronic synapse based on 2D hexagonal boron nitride (h-BN) was fabricated [63] and compared with the oxide synapses. It was claimed that the multilayer h-BN stacks showed both volatile and nonvolatile resistive switching depending on the programming state and applied electrical pulses. The volatile and nonvolatile RS allow the implementation of STP and LTP using a 2D h-BN as the active switching layer.

Interestingly, the mechanical exfoliated multilayer h-BN stacks did not show any kind of RS behavior [63]. This is due to the lack of native defects in mechanically exfoliated film; therefore, the dielectric breakdown is accompanied by irreversible physical damage and material removal from the resistive layer. Considering the fabrication technique, the CVD-grown multilayer h-BN film contains a larger amount of native defects consisting of lattice disorders (LD) and thickness fluctuations (TF) (Figure 7.9a and b). The lattice disorders are propagated out-of-the-plane. It generated defect paths across the layered h-BN stacks (Figure 7.9b, c, and d). Despite the fact that the defects are not favorable for the application in gate dielectrics in transistor or anti-scattering substrate, it is useful for application as active layer in memristors. The microstructural studies confirmed that the granular 2D h-BN film contains local defects. The electronic synapse based on Au/Ti/5–7-layer h-BN/Cu device was evaluated. The short-term potentiation was observed when the sequential voltaic pulses are employed on synapses (Figure 7.9e). The paired-pulsed facilitation characteristics of synapses with the relaxation time were observed during the evaluation of synaptic characteristic of artificial synapse (Figure 7.9f). The volatile resistive switching behavior happened when the time interval between the sequential peaks was in the range of several seconds. Through the decrease of time intervals between imposed voltaic pulses, the volatile RS of device changed and shifted to the nonvolatile RS (Figure 7.9g). Since the potentiation, depression and relaxation time of synapse were dependent on the time intervals between the voltaic pulses, this device was able to mimic the synaptic characteristics. Self-relaxation of the dielectric breakdown event during volatile RS operation is related to the disruption of the conductive filaments initiating by the detrapping of metallic ions [63]. These metallic ions were injected into the h-BN layer during the set event. The process is time-dependent and also attributed to the amplitude of injected charge for the formation of conductive filaments. The modulation of plasticity of synapse was successfully attained by the control of the magnitude of voltaic pulses and also by the control of time intervals between imposed pulses. The progressive synaptic potentiation was one of the fundamental properties of the device (Figure 7.9h and i). The current reached to the saturation level and did not increase further even after imposing thousands number of voltaic pulses. The application of sequential voltaic pulses (V = 1.2 V) was accompanied by abrupt potentiation, which later led to the additional sudden

Synaptic Devices Based on 2D Semiconductors   251

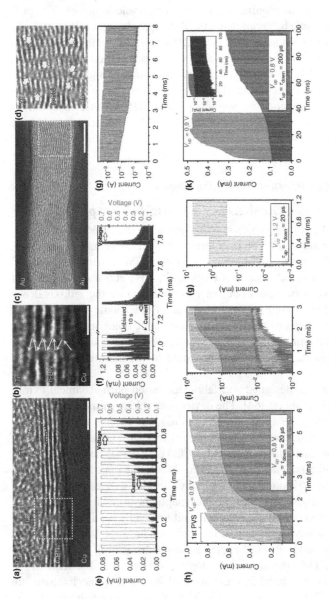

FIGURE 7.9 Cross-sectional TEM image of Au/Ti/h-BN/Cu memristor device. (a) The 5-7 layer h-BN film deposited by CVD film with lattice disorder (LD) and thickness fluctuations (TF). (b) The same film with 15-18 h-BN layer. (c) and (d) show the high magnification image of cross-section with the natural defects. The scale bars are: (a) 2 nm (b) 1 nm, (c), 5 nm, and (d) 2 nm. (e) The synaptic potentiation of synaptic device. (f) The synapse relaxation time. (g) Current signal measured by applying voltaic pulses with $V_{down} = -0.1$ V and $V_{up} = -0.7$ V, showing a progressive reset transition. This demonstrates the nonvolatile RS of device. (h) Progressive synaptic potentiation with applying sequential pulses with two different amplitudes ($V = 0.8$ V and $V = 0.9$ V) with 20 µs pulse intervals. (g) Employment of sequential voltaic pulses accompanied by abrupt increase of current and switching from volatile to nonvolatile RS. (k) The synaptic potentiation by applying voltaic pulses with different amplitudes ($V = 0.8$ V and $V = 0.9$ V) with the pulse interval of 200 µs (Reprinted and reproduced with permission from Shi, Y., Liang, X., Yuan, B., Chen, V., Li, H., Hui, F., ... Lanza, M. (2018). Electronic synapses made of layered two dimensional materials. *Nature Electronics*, 1(8), 458–465).

increase of the current level. At this stage, the synapse resistance was shifted to the nonvolatile RS regime. By applying sequential voltaic pulse with the different intensities, the synaptic potentiation was observed. The increase of EPSc was observed when the time intervals increased from 20 μs (Figure 7.9h) to 200 μs (Figure 7.9k). However, longer pulse interval requires larger amount of voltaic pulses. The proposed mechanism for synaptic behavior of artificial synapses based on 2D h-BN was attributed to the filing of boron vacancies in the 2D h-BN film by the ionic metals separated from Cu electrode. This migration and occupation resulted in the development of conductive channels between two electrodes allowing the alteration of resistance and conductance of synaptic device. The power consumption of synapse was as low as 0.1 fW when the conversion time was less than 10 ns [63]. These observations clearly confirmed outstanding capabilities of the device for emulation of synaptic functionalities of the human brain.

7.2.4 Redox/Valence Changing Artificial Synapses

7.2.4.1 Proton Intercalated Quasi-2D α-MoO$_2$ Artificial Synaptic Transistor

The redox/valence type artificial synapses work similar to the ECM/CB synapses. Both artificial synapses work based on the electrochemical reactions and migration of ionic species. Resistive switching mechanism in the ECM instruments is based on the facile oxidation of metallic cations (Ag^+, Cu^{2+}, Ni^{2+}, etc) and formation of conduction bridge between two electrodes of memristor, while the redox/valence changing synapses perform based on the movement of other ionic species including proton, oxygen vacancies or the other introduced ionic components present in dielectric component of memristors. The migration of ionic species in redox/valance changing synapses alters the chemical valence state of the main elements in 2D dielectric materials.

An artificial synapse was designed and fabricated based on mechanical exfoliated quasi-2D α-MoO$_3$ nanosheets [64]. The device was able to mimic the essential characteristics of biological synapses. Three-terminal transistor is made by using typical layered α-MoO$_3$ nanosheets (double sheet of MoO$_6$). The layered structure facilitated the proton ion intercalation in 2D MoO$_3$ flake. In the fabricated ionic/electronic hybrid three-terminal device based on 2D α-MoO$_3$, the ionic liquid [(1-ethyl-3-methylimidazolium bis-(trifluoromethanesulfonyl)-imide] was employed as the gate material. The ionic liquid serves as the presynaptic neuron which controls the generation and the migration of proton ionic transmitters, while the ultrathin flake of α-MoO$_3$ serves as the postsynaptic neuron in which conductance is controlled and modulated by ionic liquid. Figure 7.10 explains the mechanism of artificial three-terminal synapses [64]. The optical microphotograph and graphical scheme of the unit accompanied by the graphical interpretation of its working mechanisms are demonstrated in Figure 7.10 a~d, respectively. The employed gate voltage on ionic liquid facilitates the repletion of ionic protons toward the interface of ionic layer and 2D α-MoO$_3$ film. The accumulation of protons can intercalate the top-most layer of α-MoO$_3$ film at higher voltage and alter the phase structure of MoO$_3$ film. By protons intercalation into MoO$_3$ film the valence of Mo^{6+} is changed to Mo^{5+} and

Synaptic Devices Based on 2D Semiconductors 253

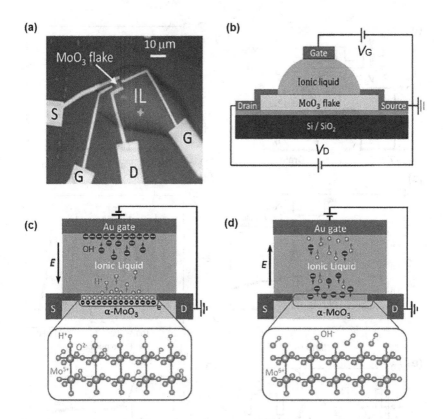

FIGURE 7.10 The color online image of quasi-2D MoS₂ based synaptic neurotransistor device. (a) The optical micrograph of device. (b) The schematic of structure of quasi-2D α-MoO₃ memristor. (c) Graphical scheme from the intercalation of protons and hydroxyl groups in the device under the positive gate voltage and the (d) corresponding graph at a negative gate voltage (Reprinted and reproduced with permission from Wang, S., Zhang, D. W., & Zhou, P. (2019). Two-dimensional materials for synaptic electronics and neuromorphic systems. *Scientific Bulletin, 64*(15), 1056–1066).

(Mo-O)-H bonds are formed. The process is defined as the electrochemical doping process and the altered phase is well-known as the molybdenum bronze (H_xMoO_3). The electrochemical doping process and phase alteration introduces external charge carriers into MoS₂ flake resulting in the increase of electron density in the channel. The chemical doping reaction is described by the following equation [64]:

$$Mo^{6+}O_3 + e^- + H^+ \rightarrow Mo^{5+}(O_2)(OH) \tag{7.1}$$

The gate-controlled reversible electrochemical doping of ionic liquid/2D α-MoO₃ synapses closely resembles to the transmission process of neurotransmitters between the biological presynaptic and postsynaptic neurons. Such device successfully emulates the synaptic characteristics by showing the EPSC, depression, and potential of synaptic weight, PPF and transition from STP to the LTP. Figure 7.11a depicts

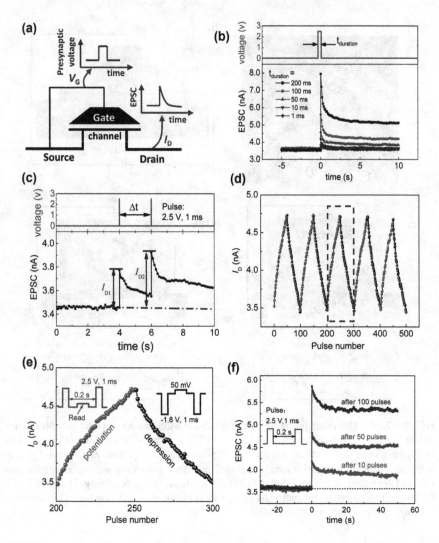

FIGURE 7.11 (a) The schematic diagram of three-terminal synaptic transistor device. (b) The EPSC of device triggered by the series of presynaptic voltaic pulses with different pulse durations. (c) The ESPCs triggered by paired voltaic pulses, which shows the PPF behavior of device. PPF index, defined as (($I_{D2} - I_{D1}$)/I_{D1}) × 100%, (d) The gradual increase and decrease of EPSCs after imposing 50 consecutive positive electric pulses (2.5 V for 1 ms interval and 200 ms duration) accompanied by 50 consecutive negative voltage pulses (−1.8 V for 1 ms interval and 200 ms duration) and (e) one of the enlarged cycles. (f) The EPSC decay after applying 10, 50, and 100 pulses (Reprinted and reproduced with permission from Yang, C. S., Shang, D. S., Liu, N., Shi, G., Shen, X., Yu, R. C., ... Sun, Y. (2017). A synaptic transistor based on quasi-2D molybdenum oxide. *Advanced Materials*, 29(27), 1700906).

Synaptic Devices Based on 2D Semiconductors

the graphical structure of three terminal synapses [64] and Figure 7.11b shows the impact of pulse duration on EPSC of the unit. The increase of voltaic pulses results in the increase of EPSC. The same increasing trend of EPSC values was observed when the pulse duration time was decreased. The smallest calculated energy consumption for a single synaptic event was 9.6 pJ (2.5 V signal and 1 ms duration), which was tangibly lower than the conventional CMOS devices. The PPF behavior of synapses (Figure 7.11c) shows that the amplitude of generated EPSC of artificial synapse by the second presynaptic pulse is considerably larger than the EPSC of first one. The PPF behavior was explained based on the adsorption/desorption behavior of ionic protons into 2D α- MoO_3 flake. When the voltaic pulse interval is shorter than the adsorption time of proton in 2D MoO_3 film, the saturated proton on the ionic liquid/2D MoO_3 interfaces cannot be fully desorbed. Therefore, the employement of the following voltaic pulse resulted in the increase of proton adsorption and the increase of channel current. The long-term potentiation and depression behaviors of 2D α- MoO_3 based memristor were evaluated and confirmed by applying 50 consecutive positive (2.5 V for 1 ms) and then negative voltaic pulses (-1.85 V for 1ms and spaced by 200 ms intervals) (Figure 7.11d,e). It was observed that device did not return to its initial current state after applying 50 sequential positive voltage pulses indicating LTP and long-term memory capabilities of the artificial synapses. The same behavior was also observed after imposing different number of voltaic pulse signals (10, 50 and 100 pulses) with 2.5 V amplitude indicating the long-term memorization characteristics and synaptic plasticity in the device (Figure 7.11f).

7.2.4.2 Li-Ions Intercalated Quasi-2D α-MoO_2 Artificial Synaptic Transistor

An artificial neurotransistor synapse was introduced and developed based on the 2D α- MoO_3 in which the intercalation and extraction of Li^+ ions into the 2D α-MoO_3 film facilitated the synaptic characteristics [65]. In this unit 18 nm thick mechanically exfoliated α- MoO_3 nano flake was employed as the channel materials by three terminals configuration transistor (Similar to Figure 7.10). The solid electrolyte consists of the dissolved $LiClO_4$ in polyethylene oxide (PEO) matrix. By employment of positive gate voltage on solid electrolyte, the Li^+ ions were driven toward the MoO_3 channel and intercalated into 2D α- MoO_3 film. The intercalated Li^+ ions served as dopants in MoO_3 film contributing the free electron carrier and thus increased the conductance of channel. The electrochemical doping process was accompanied by the development of molybdenum bronze xLi^+MoO_3 [65]. The fabricated artificial synapse demonstrated the characteristics similar to a biological synapse. The atomic microscopy image of device is shown in Figure 7.12a where the presynaptic voltaic pulses (gate electrode) pushed the Li^+ ions toward the α-MoO_3/solid electrolyte interface and then intercalated into MoO_3 channel. The effect of voltage amplitude on EPSC of synapse confirmed the effect of magnitude of postsynaptic voltaic pulse on the efficient intercalation of Li^+ ions into MoO_3 film. At the low voltage the accumulation and intercalation of Li^+ ions are not facilitated. Consequently, some of Li^+ ions are trapped at the solid electrolyte/MoO_3 interface. At higher following voltage, a larger number of Li^+ ions are intercalated into MoO_3 film accompanied by the increase of channel conductance. The dependency of EPSC amplitude to the pulse duration was observed during the performance of fabricated artificial synapse where

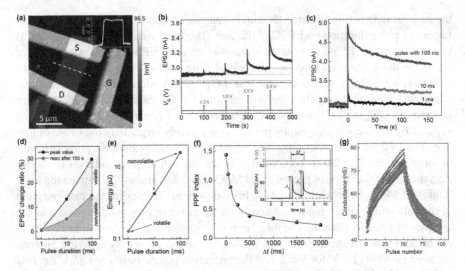

FIGURE 7.12 (a) The atomic force microcopy image of the neurotransistors with the MoO₃ nanoflake channel with 2D structure between source and drain. (b) The EPSC of devices stimulated by voltaic pulses with different amplitude. (c) The effect of pulse duration for three different voltaic pulses with the same amplitude. (d) The effect of pulse duration on the EPSC changes of synaptic device. (e) Pulse duration dependence of the energy consumption of device for a single pulse. (f) The variation of PPF index of device versus the pulse intervals. (g) The condensed plot of LTP and LTD for 15 synaptic device based on 2D α-MoO₃ devices (Reprinted and reproduced with permission from Yang, C. S., Shang, D. S., Liu, N., Fuller, E. J., Agrawal, S., Talin, A. A., LI, Y. Q., Shen, B. G., & Sun, Y. (2018). All-solid state synaptic transistor with ultra-low conductance for neuromorphic computing. *Advanced Functional Materials*, *28*, 1804170).

the portion of nonvolatile characteristics of device increased by the increase of pulse duration and vice versa (Figure 7.12 b and c). This behavior is attributed to dependence of the Li⁺ ion intercalation into the amplitude of voltaic pulse. The energy efficiency of fabricated artificial neurotransmitter demonstrated the minimum energy usage of 0.16 pJ for a nonvolatile conductance phenomenon triggered by 2.5 V pulse while the minimum energy usage for a nonvolatile synaptic event is about 1.8 pJ (Figure 7.12d). The required energy for synaptic events in the metal-oxide-based device is tangibly lower than that of in the conventional CMOS circuits. Table 7.3 provides a comparative data of the synaptic behavior of several electrochemical transistors. Apparently, the energy consumption of Li⁺ ion intercalated 2D α- MoO₃ synapse is one of the lowest value among all samples. However, the proton doped 2D α-MoS₂ neurotransistor was the lowest one among all artificial synapses. The dependency of PPF index of device to the pulse intervals (Δt) is shown in Figure 7.12f, where the longer pulse intervals accompanied by the abrupt decrease of PPF index. The capability of LTP and LTD of the synaptic weight were confirmed, as presented in Figure 7.12g, where the employment of 50 sequential positive voltage (2.0 V) pulses caused potentiation. Then the following negative sequential pulses (−2.0 V) induced depressions. The credibility of device was shown by the measurement of LTP and

TABLE 7.3
Operation Energy of the Selected Metal-Oxide-Based Synaptic Devices

Channel material	Electrolyte	Action ions	Conductance	Operation Energy
p-Si	$RbAg_4I_5$/MEH-PPV	Ag^+	$\approx$130 nS	10 pJ
Indium zinc oxide	SiO_2	H^+	$\approx$26 nS	45 pJ
PEDOT:PSS	KCl	H^+	$\approx$750 μS	–
Carbon nanotube	PEG	H^+	$\approx$45 nS	7.5 pJ
ZnOx	Ta_2O_5	O^{2-}	$\approx$35 nS	35 pJ
WSe_2	$LiClO_4$/PEO	Li^+	$\approx$570 pS	30 fJ
PEO/P3HT	Ion gel	–	$\approx$0.28 nS	1.23 fJ
α-MoO_3	Ionic liquid	H^+	$\approx$95 nS	0.2 fJ
PEDOT:PSS/PEI	KCl	H^+	$\approx$850 μS	10 pJ
$Li_{1-x}CoO_2$	LiPON	Li^+	$\approx$250 μS	–
α-MoO_3	$LiClO_4$/PEO	Li^+	$\approx$75 nS	0.16 pJ

Source: Reprinted and reproduced with permission from [65].

LTD for 15 different synapses with the same technical characteristics. Altogether, the artificial neurotransistor synapse based on Li^+ ion intercalated 2D α-MoO_3 film was capable of emulation of essential functionalities of biological synapse including EPSC, PPF, LTP, and LTP [65].

7.2.5 Phase-Change Synapses

The concept of phase-change memories and the materials with phase-change capabilities have been developed several decades ago (1968) since they were employed for the development of nonvolatile memories [66]. The phase-change materials are usually members of chalcogenides that can be converted from amorphous into crystalline phase. The set process occurs when the amorphous phase transforms into crystalline state by employment of appropriate voltaic pulses. The phase-change transition from amorphous to crystalline structure is accompanied by the switching from HRS to LRS. The crystalline to amorphous structural phase-change requires a high amplitude/short duration (ns) voltaic pulse, while a phase change from amorphous into crystalline structure requires medium amplitude but longer duration pulse (μs) [67–69]. Recently, the phase-change 2D materials have been extensively attracted attention of the researchers for application in artificial synapse and neuromorphic computing.

As an example, heterostructured artificial synapse was introduced in which the phase alteration is the main mechanism of memristive switching [70]. The unit consists of a modifiable synaptic architecture based on the phase-engineered tungsten oxide/selenide monolithic heterostructure. Figure 7.13a depicts the graphical image of three-terminal device, which is in fact a gate-tunable memristor composed of the WO_{3-x} resistor component over the WSe_2/graphene barristor. The WSe_2 layer was mechanically exfoliated and deposited over the graphene layer to arrange a variable-barrier Schottky diode. The employment of van der Waals (vdW) WSe_2 layer is necessary, since it makes

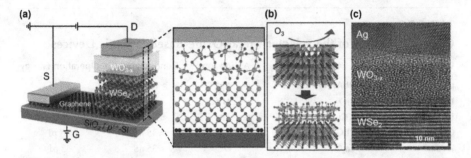

FIGURE 7.13 (a) The graphical demonstration of synaptic device with a vertically integrated WO_{3-x} memristor and WSe_2/graphene barristor. (b) The schematic of UV-ozone assisted oxidation process. (c) The cross-sectional HRTEM image of synaptic device (Reprinted and reproduced with permission from Huh, W., Jang, S., Lee, J. Y., Lee, D., Lee, D., Lee, J. M., ... Lee, C. H. (2018). Synaptic barristor based on phase-engineered 2D heterostructures. *Advanced Materials*, *30*(35), 1801447).

possible to arrange a variable-barrier contact between WSe_2 and graphene film, which does not contain Fermi-level pining. The WSe_2/graphene bottom barristor component enables to electronically control the current flow in the device. By using this controlling gate the switching state in WO_{3-x} layer can be modulated. The WO_{3-x} layer was fabricated by the UV-ozone oxidation of WSe_2 film (Figure 7.13b). The memristive characteristics of heterostructured device were originated from the WO_{3-x} layer in the heterostructured Ag/WO_{3-x}/WSe_2/graphene configuration (Figure 7.13c).

Three-terminal configuration of the artificial synapses successfully emulated the synaptic properties [70]. The synaptic barristor play the role of an additional neuromodulator operating in biological synapse. The device is similar to the biological astrocyte, which plays a critical role in cellular programming to achieve the highest level of data processing. In the fabricated instrument, the gate, drain and source terminals of synapses were defined as the neuromodulator, pre, and post neurons, respectively (Figure 7.14a). Unlike the conventional two-terminal synaptic device, the weight control in this artificial synapse is induced by modulation of the gate voltage, independent from the input signals to the source and drain electrodes. In this instrument, the role of the gate terminal is analogous to the role of a neuromodulator as the synapse gate-keeper (Figure 7.14b). Characteristics of synaptic plasticity including LTP and LTD were demonstrated accelerating by application of negative voltaic pulses with larger amplitude on the gate terminal (Figure 7.14c). This behavior is attributed to the formation of electric field across the WO_{3-x} layer. As the Schottky barrier at WSe_2/graphene decreased, the dynamic variation range of the postsynaptic current is increased. The gate-assisted modulation was known as an applicable characteristic of the device. Without employment of the gate voltage, the artificial synapse was not able to emulate the LTP behavior. For example, the PSC of artificial synapse did not experience an increasing trend, when the V_G was not employed and the artificial synapse was only capable of emulation the STP behavior (Figure 7.14d). On the other hand, by employment of −30.0 V gate voltage the artificial synapse showed the LTP behavior and the current output of device increased continuously (Figure 7.14d).

Synaptic Devices Based on 2D Semiconductors

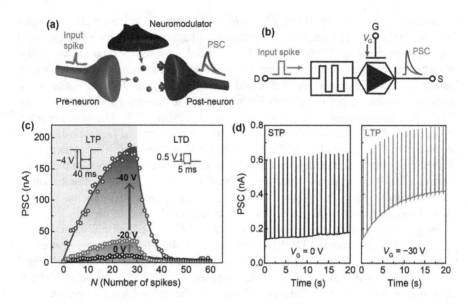

FIGURE 7.14 (a) The schematic illustration of the three-terminal synaptic barristor and its corresponding (b) circuit diagram. (c) The sequential increase and decrease of PSC by applying sequential negative potentiating (−4.0 V) and positive depression spikes (0.5 V) at various V_G of 0, −20 and −40 V. (d) Plot of PSC as a function of amplitude of applying spike voltage. In both graphs the applied V_{DS} is −1.0 V. However, the STP behavior was observed when the $V_G = 0$. By applying the $V_G = -30$ V, the synapses showed LTP states. (Reprinted with permission from Huh, W., Jang, S., Lee, J. Y., Lee, D., Lee, D., Lee, J. M., ... Lee, C. H. (2018). Synaptic barristor based on phase-engineered 2D heterostructures. *Advanced Materials*, *30*(35), 1801447).

Thus, optimization of the electrical characteristics of the artificial synapses with three-terminal configuration offered considerable power saving benefits.

7.2.6 Thermochemical/Joule Heating Synapses

In the RRAM the bipolar resistive switching occurs due to the electric field assisted reaction and migration of oxygen ions. In bipolar RS the set and reset events are observed under opposite voltage polarities. When RS mechanisms are dominated by joule heating effects, the resistance threshold switching behavior controls the RS mechanism of memristor. In this mechanism the conductive filament spontaneously fuses if the amplitude of voltage drop to lower than specific values and the device returns to its HRS. There is always a competition between the electric field driven RS and the joule heating RS mechanisms. All of the same phenomena and RS behavior are expected to be present in the artificial synapse based on 2D materials.

For instance, a single-layer MoS_2 synapse was recently introduced based on the joule heating effect [71]. The unit also uses the privilege of electrostatic doping induced synaptic computation. Therefore, its energy consumption is ultralow and close to biological synapses. The resistive heating of monolayer MoS_2 film allows the increase of residual temperature of 2D film by using a low electrical energy with the magnitude of ~10 fJ [71]. This causes the increase (facilitation) or decrease of (depression) the conductance of the 2D MoS_2 film. The artificial synapse based on 2D MoS_2 film showed variable conductance and continues tunable short-term memory states, which assisted the emulation of synaptic STP and facilitated the synaptic computation in the biological system. A positive correlation between the temperature and conductance of 2D MoS_2 semiconductor device was demonstrated in Figure 7.15a. The device conductance elevated by the resistive heating, which is the manifestation of increased synaptic conductivity. Gate dependent STP behavior of 2D MoS_2-based artificial synapse confirmed the effect of voltage amplitude and polarity (positive or negative) on the device conductance. It was understood that a negative voltage with higher magnitude enabled the electron density in the n-doped 2D MoS_2 artificial synapse, which actually generates a lower conductivity (Figure 7.15b). The device also showed the capability of emulation of synaptic computation including the STP, PPF and paired pulsed depression (Figure 7.15c). The tunable STP characteristics of 2D MoS_2-based synapse ensured the fabrication of a sound localization device (Figure 7.15d). The instrument work based on two major working mechanisms of the sound localization including the coincidence detection by internal time difference (ITD) and by the internal level difference (ILD) as it is shown in Figure 7.15e. The artificial synapse demonstrated a great breakthrough for the complete neuromorphic computing, mimicking the complicated and accurate synaptic information processing and human brain functionalities.

7.2.7 2D HETEROSTRUCTURED-BASED SYNAPSES

The control of charge transfer between heterostructured 2D materials is demonstrated to be a reliable and practical strategy to emulate the synaptic behavior. The PO_x/BP 2D heterostructured device successfully emulated the synaptic functionality realizing by the charge transfer between PO 2D oxide and the BP channel (Figure 7.16a) [72]. The unit is fabricated based on the deposition of a 20 nm thick BP flake on the 2.0 nm thick PO_x film. The main characteristics of biological synapse, including LTP and the spike-timing-dependent plasticity (STDP) were mimicked by the PO_x/BP 2D heterostructured synapse. It was found in this device that both positive and negative synaptic weight changes are anisotropic properties. It means that the changes in the synaptic weight of device along the x (armchair) direction are ultimately different from the synaptic weight along the y (zigzag) direction for BP crystal. Due to anisotropic characteristics for charge transfer in BP crystal, STDP along the x and y axis resembles the STDP of the biological synapse. The device worked at the vacuum conditions. The STDP behaviors of 2D BP based artificial synapse built along the x and y directions of the BP are demonstrated in Figure 7.16b and c, respectively [72]. The potentiation was recognized as the consequence of accumulated trapped charge and strengthening of synaptic connection. The longer positive time intervals resulted in the weaker potentiation. However, it led to the weakening of the synaptic connection (depression) when the presynaptic spike arrives after the postsynaptic pulses. Again, longer negative time intervals ultimately resulted in the reduced depression response of the device. The behavior of the artificial synapse based on PO_x/BP 2D heterostructure is similar to the performance of the biological synapse.

Synaptic Devices Based on 2D Semiconductors 261

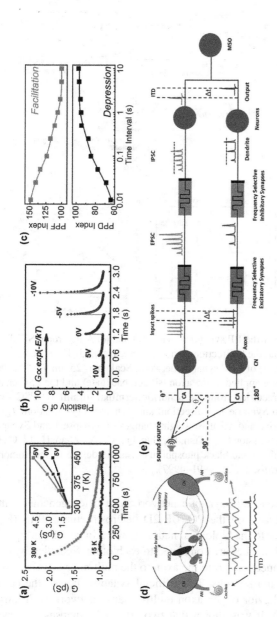

FIGURE 7.15 The properties of artificial synaptic device by Joule heating. (a) Absence of STP at a low temperature of 15 K. The inset shows the changes in temperature-dependent conductance of the MoS$_2$ device with three different gate voltages. (b) Normalized gate-voltage-dependent STP and time constants. (c) PPF and PPD indexes after two consecutive pulses as a function of the interpulse interval. (d) Schematic picture for sound localization with both ITD and ILD. (e) Schematic picture of the working mechanism of synaptic computation for ITD-based sound localization. "CA" means cochlea shown in panel (d). The blue circles represent neurons. The horizontal dashed lines represent the potential threshold for neuronal firing (Reprinted and reproduced with permission from Sun, L., Zhang, Y., Hwang, G., Jiang, J., Kim, D., Eshete, Y. A., ... Yang, H. (2018). Synaptic computation enabled by joule heating of single-layered semiconductors for sound localization. *Nano Letters*, *18*(5), 3229–3234).

262   Ultrathin 2D Semiconductors

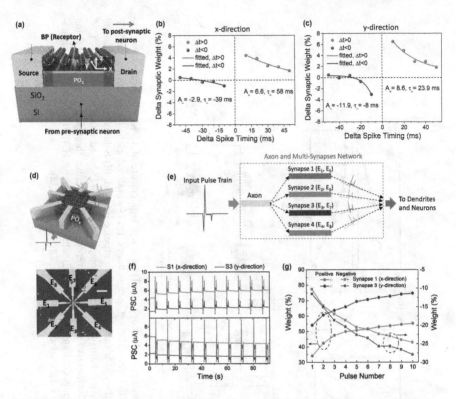

FIGURE 7.16 (a) Schematic of the BP synaptic device. (b) and (c) Anisotropic STDP characteristics in the x-direction and the y-direction devices, respectively. (d) The schematic (left) and an optical micrograph (right) of a BP synaptic network. Scale bar: 25 μm. (e) The device testing scheme. Input pulses are applied to the axon (silicon backgate) and the PSC changes of each synaptic device are recorded from the dendritic terminal (drain). (f) PSC signals simultaneously recorded from Synapse 1 (x-direction) and Synapse 3 (y-direction). (g) The simultaneously recorded positive and negative weight changes of Synapse 1 and Synapse 3. (Reprinted and edited with permission from Tian, H., Guo, Q., Xie, Y., Zhao, H., Li, C., Cha, J. J., ... Wang, H. (2016). Anisotropic black phosphorus synaptic device for neuromorphic applications. *Advanced Materials*, 28(25), 4991–4997).

A compact artificial axon-multi-synaptic network with heterogeneous connection strengths was made based on the PO$_x$/BP 2D heterostructured films (Figure 7.16d). The device is similar to the biological neural network in which multiple synapses with heterogeneous branches are connected to each other. Specifically, Figure 7.16e demonstrates the connections from the axon to the multiple synapses network and then to the dendrites and neurons. The artificial synapse showed the capability of the emulation of behavior of the axon-multi-synapse network with intrinsic connections (Figure 7.16f). It was shown that two artificial synapses can operate simultaneously in response to the pulse trainings applied at the axon electrode. It was also demonstrated that the synaptic weight of device was distinguished in the y-direction. The weight change was significantly larger in the y-direction compared to that in the x-direction for both applied positive and negative responses (Figure

7.16g). Generally it was shown that the artificial anisotropic axon-multi-synaptic network is capable of emulating the performance of the biological network of multi-synaptic junctions. Thus, the BP based artificial synapse is a promising device for the new generation of neuromorphic electronics for computational and neural network engineering [72].

7.2.8 Optical-Based Synapses

7.2.8.1 Optical Synaptic Devices versus Photodetectors

The direct storage of optical information can be arranged by the integration of optical sensing and optical charge trapping in RRAMs [73]. In this view, the optical signals are received by the light sensors and converted into electrical signals for digital applications. Optical RAM are highly desirable instrument since they are capable of bridging the optical sensing and neuromorphic computing to cause the direct processing of optical information. Compared with a photosensor or nonvolatile optical RAMs, the optically stimulated RAMs demonstrated both light assisted and light modulated nonvolatile and volatile switching to finally generate the light-tunable synaptic characteristics. The difference in the signal characteristics of photodetector and STP/LTP of optical synapses are depicted in Figure 7.17 [73]. The sharp and fast reaction of photosensor or image sensor is in clear contrast with the optical RAM synaptic device. In the optical RAM artificial synapse the light-tunable and time-dependent plasticity are the fundamental properties to implicate the potential applications for development of neuromorphic visual systems [73].

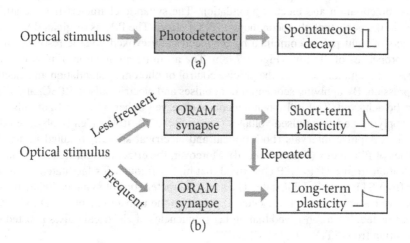

FIGURE 7.17 The comparison between the performance of (a) a photodetector without tunable plasticity, and (b) an optical RRAM synaptic device with light-tunable short-term plasticity and long-term plasticity. (Reprinted with permission from Zhou, F., Chen, J., Tao, X., Wang, X., & Chai, Y. (2019). 2D materials based optoelectronic memory: Convergence of electronic memory and optical sensor. *Research*, *2019*, 9490413).

7.2.8.2 Optical RRAM Synaptic Devices

It is appeared to indicate that the first reported optical synapses are based on the charge-trapping phenomenon at the heterointerfaces between graphene and carbon nanotubes (CNTs). This optical synaptic unit demonstrated different synaptic properties including STP, LTP, STDP, and spatiotemporal learning rule (STLR) [74]. The STP is triggered owning to the charge transfer from CNTs to graphene upon the illumination of heterostructure by the light modulating the built-in field at the CNT/graphene interface. In another case, the 2D heterostructured MoS_2/ perylene-3,4,9,10-tetracarboxylic dianhydride (PTCDA) heterojunction showed considerable changes in synaptic weight and PPF ratio [75]. A hybrid structure based on 2D WSe_2 and boron doped Si nanocrystals was employed to further broaden the absorption spectrum of light sensor and to reduce the power consumption of optical synapse. The artificial synapse showed wide absorption spectrum from ultraviolet to near-infrared region. The power consumption of device was about 75 fJ, which is considered as highly efficient artificial synapse for the low-power performance of visual intelligence system beyond the detection limits of human's eye [76]. As it was discussed in previous section, the synaptic characteristics including the STP and LTP were observed for In-doped ultrathin TiO_2 film. The device worked based on the visible light assisted resistive switching of ultrathin TiO_2 film [62].

An optical synapse is made based on monolayer MoS_2 film. While the bulk MoS_2 is not really an ideal option for optoelectronic devices, the monolayer MoS_2 with 1.8 eV bandgap showed outstanding optical characteristics. An optoelectronic memristor based on monolayer MoS_2 films was fabricated on p-Si substrate and showed the rectification ratio of 4×10^3 [77]. An optical memristor synapse is made based on the monolayer MoS_2 film (Figure 7.18a). The device was capable of mimicking the photonic potentiation and electric habituation. The synaptic characteristics are attributed to the persistent photoconduction (PPC) effect. The PPC is originated from the effects of light pulses combined by the electrically evoked volatile restive switching properties of 2D film (Figure 7.18b). The accurate modulation of photonic and electric parameters assisted the precise control of photonic potentiation and electric depression. By applying sequential light pulses and electric pulses, EPSC, and IPSC can be achieved. The PPF phenomenon is observed either by using light pulses or by applying negative pulsed voltages (Figure 7.18c, d). It was clearly observed that the longer pulse intervals (both optical and electrical stimuli) resulted in smaller values of PPF index (Figure 7.18c, d). Moreover, the effect of light frequency on the transition from STP to LTP confirmed that higher frequencies facilitated the transfer from STP state to LTP state. The same characteristics of synaptic behavior was observed when the effect of electrical pulses on the transition from STD to LTD is investigated. It was observed that higher frequencies of electrical pulses assisted the transition from STD to LTD [77].

A heterostructured-based 2D optical synaptic device was made where the charge transfer was controlled at the heterojunction between 2D inorganic MoS_2 and the 2D organic (PTCDA) semiconductor films [78]. Two device were designed based on optical an electric modulation of synaptic weight (Figure 7.19 a, b). For optical

Synaptic Devices Based on 2D Semiconductors

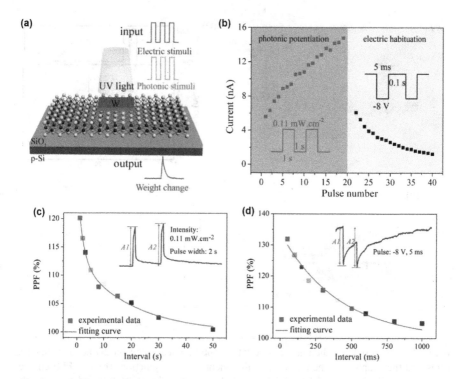

FIGURE 7.18 (a) The schematic of memristive optical synapse based on monolayer MoS$_2$. (b) Photonic potentiation and electric inhibition in the W/MoS$_2$/p-Si memristive synapse. (c) The variation of PPF versus the time intervals between the optical pulses. (d) The variation of PPF index versus time intervals between electrical pulses. Reprinted with permission from He, H.-K., Yang, R., Zhou, W., Huang, H. M., Xiong, J., Gan, L., … Guo, X. (2018). Photonic potentiation and electric habituation in ultrathin memristive synapses based on monolayer MoS$_2$. *Small, 14*(15), 1800079.

modulation of synaptic device, a three-terminal configuration is explored without the gate dielectric and the control gate options. While for the electric modulation, a four-terminal architecture is developed in which an Al$_2$O$_3$ film was employed as the gate dielectric and a top Au electrode modulated the device as the gate electrode. The electron transfer between 2D MoS$_2$ film and 2D PTCDA organic semiconductor was created by using relatively different voltaic pulses in both directions. This distinguished charge transfer property was facilitated by the developed heterointerfaces between two individual 2D films. The optical synaptic heterostructured device demonstrated PPF phenomenon, where the pulses with longer intervals showed lower PPF index values (Figure 7.19c). Almost 60 synaptic weight changes were achieved due to incredible optical modulation of 2D MoS$_2$ synaptic device. Furthermore, in the case of four-terminal device with Al$_2$O$_3$ gate electrode the maximum obtained facilitation was about 500% after imposing 50 relatively negative pulses. On the other hand, through 50 positive pulses, the minimum facilitation of 3% was achieved

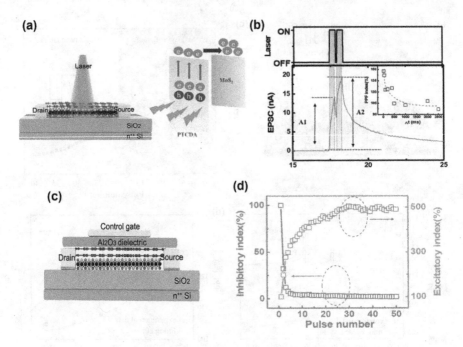

FIGURE 7.19 (a) The schematic of optical synaptic device based on 2D heterostructured films and the graphical scheme of band alignment in heterojunction. (b) The PPF characteristics of optical synaptic device. The inset shows the variation of PPF index versus the pulse interval duration. (c) The schematic of four-electrode electrical synaptic device. (d) Excitatory (blue)/and inhibitory (red) index as the function of pulse number. Reprinted and reproduced with permission from Wang, S., Chen, C., Yu, Z., He, Y., Chen, X., Wan, Q., Shi, Y., Zhang, D. W., Zhou, H., Wang, X., & Zhou, P. (2018). MoS$_2$/PTCDA hybrid heterojunction synapse with efficient photoelectric dual modulation and versatility. *Advanced Materials*, *31*, 1806227.

(Figure 7.19d). Totally, the device exhibited distinguished capability for modulation of synaptic weight through both electric stimuli and optical stimuli [78].

A two-terminal black phosphorus (BP)-based synaptic device was fabricated to achieve the optically stimulated excitatory and inhibitory action potentials without applying external altering electric field. The 2D black phosphorous is an emerging material with broadband absorption, strong light-matter coupling, with facile tunable optoelectronic characteristics [79]. The optical synaptic properties of the device was triggered by using the advantages of oxidation-related defects in BP and their impacts on optoelectronic properties of BP. In this device the few-layered BP nanosheets are transferred on to flexible polyethylene naphthalate (PN) substrate to fabricate a flexible optical synaptic device based on few-layered BP (Figure 7.20a). The optical device demonstrated stable dynamic synaptic plasticity by employment of pulsed light with λ=365 nm wavelength (Figure 7.20b). The stability and credibility of device was confirmed even after 1000 cycles of bending. The device interestingly demonstrated the positive EPSC when pulsed light with λ=280 nm wavelength was employed to stimulate the artificial synapse (Figure 7.20c). However, when the

Synaptic Devices Based on 2D Semiconductors

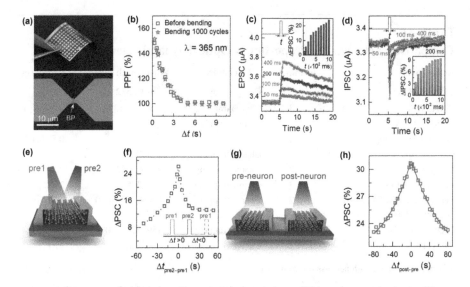

FIGURE 7.20 (a) Photograph of flexible BP optical synaptic device on the PEN substrate accompanied by the magnified image of an individual device. (b) The variation of PPF index of device versus the pulse intervals before bending and after 1000 cycles of bending. The λ = 365 nm pulse laser was employed to trigger the synaptic functionalities device. (c) The λ = 280 nm optical pulsed lights were employed to mimic the EPSC property and (d) the λ = 365 nm pulsed lights were employed to stimulate the IPSC. (e) Schematic of a multineural system where two individual light sources with different excitation wavelengths are employed as the presynaptic 1 and presynaptic 2 pulses. (f) The variation of ΔPSC versus the time intervals between the optical stimulation of presynaptic 1 and presynaptic 2. (g) The schematic design of two individual optical synaptic devices based on BP. The devices were stimulated by the light with the wavelength of 280 nm. (h) The variation of ΔPSC as the function of time difference between pre- and postsynaptic pulses ($\Delta t_{Post-Pre}$). Reprinted and reproduced with permission from Ahmed, T., Kuriakose, S., Mayes, E. L. H., Ramanathan, R., Bansal, V., Bhaskaran, M., ... Walia, S. (2019). Optically stimulated artificial synapse based on layered black phosphorus. *Small, 15*(22), 1900966.

λ=365 nm pulsed light was used, the synaptic device demonstrated the negative current values. This sudden shift from positive to negative EPSC values are attributed to the oxidation-related defects on the surface of few-layered BP film. The optical synaptic device was capable of emulation of synaptic properties at various wavelengths of lights owning to outstanding light sensitivity of BP film. The EPSC response experienced increasing trend when the pulsed light duration was increased. This behavior was similar to human brain learning process. The fabricated multineural system device (Figure 7.20e) was capable of operating the spatiotemporally correlated dynamic logic through the pristine-like 2D BP. (Figure 7.20f). The other device (Figure 7.20g) was stimulated by two individual presynaptic pulsed lights (λ=660 nm for presynaptic 1, and λ=280 nm for presynaptic 2). The symmetric characteristics of ΔPSC versus the $\Delta t_{pre2-pre1}$ are depicted in Figure 7.20g. Both devices were also capable of mimicking the STDP properties. It was shown that the optical synaptic

TABLE 7.4
Comparison between the Properties of Optical Synaptic Devices

Active layer	Terminal	Mechanism	Wavelength nm	Spike energy
Graphene/CNT	3	Charge trapping	405, 532	250 pJ
Si-NC/WSe$_2$	3	Charge trapping	375, 532, 1342	0.000075 pJ
MoS$_2$/PTCDA	3	Charge trapping	532	5.38 pJ
2D Perovskite/graphene	3	–	520	0.592pJ
CH$_3$NH$_3$PbI$_3$	2	Charge trapping	500, 635	–
CdS/CNT	2	Charge trapping	405, 532	–
InGaZnOx/Al$_2$O$_3$	3	Charge trapping	365	2400 pJ
Pentacene/PMMA/CsPbBr$_3$ quantum dots	3	Charge trapping	365, 450, 520, 660	0.76 pJ
ITO/Nb:SrTiO$_3$	2	Charge trapping	459, 528, 630	1178 pJ
C8-BTBT	3	Charge trapping	360	2160 pJ

Source: Reprinted with permission from [73].

device based on 2D BP was capable of emulation of sophisticated neural functionalities merely through the optical modulation of synaptic characteristics [79].

Table 7.4 briefly summarized a comparative data set from the performance of optical artificial synapse based on 2D materials with the different functionalities [73].

7.3 CONCLUSIONS

Because of the outstanding physical and chemical capabilities, electrostatic interaction, tunability, and confined charge-trapping mechanisms, 2D materials and their heterostructured films have been widely employed for different memory and synaptic applications. Most of the previous research works focused on functionalities of ultrathin 2D materials and films to emulate the biological synaptic properties. The synaptic behavior and its mechanisms are mostly based on the resistive switching and charge-trapping phenomena in ultrathin 2D films. Although the capabilities of 2D materials for emulation of synaptic plasticity are proven in several reported devices, the development of integrated synaptic systems with potential of neuromorphic computation is still in fancy stage. Therefore, in addition to the development of novel artificial synapses based on 2D materials, it is highly required to develop novel neuromorphic system in which the synaptic weights are controlled in interconnected network of artificial synapses to emulate the neuromorphic performance of the human brain. On the other hand, fabrication of interconnected synaptic systems is also highly dependent on the advance fabrication techniques. The main current fabrication challenge is the large-scale growth of advanced ultrathin 2D materials with precise controllability over their thickness, chemical composition, and electronic properties. Thus, the fabrication techniques must be improved to ensure uniform integration of the synaptic devices with the same physical and structural characteristics and uniform electrical properties. Furthermore, the reliability, energy

Synaptic Devices Based on 2D Semiconductors

efficiency and long-term stability of the 2D-based artificial synapses and their networks are still considered as challenging targets. Thus, further explorations should to be carried out to meet the requirements of high-performance artificial synapses in neuromorphic engineering.

REFERENCES

1. Von Neumann, J. (1993). First draft of a report on the edvac. *IEEE Annals of the History of Computing*, *15*(4), 27–75.
2. Indiveri, G., & Liu, S. C. (2015). Memory and information processing in neuromorphic systems. *Proceedings of the IEEE*, *103*(8), 1379–1397.
3. Nawrocki, R. A., Voyles, R. M., & Shaheen, S. E. (2016). A mini review of neuromorphic architectures and implementations. *IEEE Transactions on Electron Devices*, *63*(10), 3819–3829.
4. Dawson, J. W. (2007). The essential Turing: Seminal writings in computing, logic, philosophy, artificial intelligence, and artificial life plus the secrets of enigma. *Review of Modern Logic*, *10*, 179–181.
5. Upadhyay, N. K., Joshi, S., & Yang, J. J. (2016). Synaptic electronics and neuromorphic computing. *Science China Information Sciences*, *59*(6), 061404.
6. Efnusheva, D., Cholakoska, A., & Tentov, A. (2017). A survey of different approaches for overcoming the processor-memory bottleneck. *International Journal of Computer Science and Information Technology*, *9*(2), 151–164.
7. Schneider, M. L., Donnelly, C., Russek, S. E., Baek, B., Pufall, M. R., Hopkins, P. F., ... Rippard, W. H. (2018). Ultralow power artificial synapses using nanotextured magnetic Josephson junctions. *Science Advances*, *4*(1), 1701329.
8. Turing, A. M. (1995). *Computing machinery and intelligence* (pp. 11–35). Cambridge, MA: MIT Press.
9. Ho, V. M., Lee, J., & Martin, K. C. (2011). The cell biology of synaptic plasticity. *Science*, *334*(6056), 623–628.
10. John, R., Liu, F., Chien, N. A., Kulkarni, M. R., Zhu, C., Fu, Q., ... Mathews, N. (2018). Synergistic gating of electro-iono-photoactive 2D chalcogenide neuristors: Coexistence of Hebbian and homeostatic synaptic metaplasticity. *Advanced Materials*, *30*(25), 1800220.
11. López, J. C. (2001). A fresh look at paired-pulse facilitation. *Nature Reviews: Neuroscience*, *2*(5), 307.
12. Perea, G., & Araque, A. (2007). Astrocytes potentiate transmitter release at single hippocampal synapses. *Science*, *317*(5841), 1083–1086.
13. Laughlin, S. B., de Ruyter van Steveninck, R. R., & Anderson, J. C. (1998). The metabolic cost of neural information. *Nature Neuroscience*, *1*(1), 36–41.
14. Citri, A., & Malenka, R. C. (2007). Synaptic plasticity: Multiple forms, functions, and mechanisms. *Neuropsychopharmacology*, *33*(1), 18–41.
15. Di Filippo, M., Picconi, B., Tantucci, M., Ghiglieri, V., Bagetta, V., Sgobio, C., ... Calabresi, P. (2009). Short-term and long-term plasticity at corticostriatal synapses: Implications for learning and memory. *Behavioural Brain Research*, *199*(1), 108–118.
16. Costa, R. P., Mizusaki, B., Sjöström,P. J., & van Rossum, M. C. W. (2017). Functional consequences of pre- and postsynaptic expression of synaptic plasticity. *Philosophical Transactions of the Royal Society of London Series B*, *372*, 20160153.
17. Poldrack, R. A., & Packard, M. G. (2003). Competition among multiple memory systems: Converging evidence from animal and human brain studies. *Neuropsychologia*, *41*(3), 245–251.
18. Choquet, D., & Triller, A. (2013). The dynamic synapse. *Neuron*, *80*(3), 691–703.

19. Kandel, E. R., Schwartz, J. H., Jessell, T. M., Siegelbaum, S. S., Hudspeth, A. J., & Mack, S. (2000). *Principles of neural science*. McGraw-Hill, New York.
20. Fioravante, D., & Regehr, W. G. (2011). Short-term forms of presynaptic plasticity. *Current Opinion in Neurobiology*, 21, 269–274.
21. Hebb, D., Martinez, J., & Glickman, S. (1994). The organization of behavior - A neuro-psychological theory. *Contemporary Psychology*, *39*, 1018–1020.
22. Caporale, N., & Dan, Y. (2008). Spike timing-dependent plasticity: A Hebbian learning rule. *Annual Review of Neuroscience*, *31*, 25–46.
23. Rachmuth, G., Shouval, H. Z., Bear, M. F., & Poon, C. S. (2011). A biophysically-based neuromorphic model of spike rate-and timing-dependent plasticity. *Proceedings of the National Academy of Sciences of the United States of America*, *108*(49), E1266–E1274.
24. Wang, W., Pedretti, G., Milo, V., Carboni, R., Calderoni, A., Ramaswamy, N., ... Ielmini, D. (2018). Learning of spatiotemporal patterns in a spiking neural network with resistive switching synapses. *Science Advances*, *4*(9), eaat4752.
25. Yang, J. J., Strukov, D. B., & Stewart, D. R. (2013). Memristive devices for computing. *Nature Nanotechnology*, *8*(1), 13–24.
26. Beck, A., Bednorz, J. G., Gerber, C., Rossel, C., & Widmer, D. (2000). Reproducible switching effect in thin oxide films for memory applications. *Applied Physics Letters*, *77*(1), 139–141.
27. Sakamoto, T., Lister, K., Banno, N., Hasegawa, T., Terabe, K., & Aono, M. (2007). Electronic transport in Ta_2O_5 resistive switch. *Applied Physics Letters*, *91*(9), 092110.
28. Merolla, P. A., Arthur, J. V., Alvarez-Icaza, R., Cassidy, A. S., Sawada, J., Akopyan, F., ... Modha, D. S. (2014). A million spiking-neuron integrated circuit with a scalable communication network and interface. *Science*, *345*(6197), 668–673.
29. Tan, H., Liu, G., Zhu, X., Yang, H., Chen, B., Chen, X., ... Li, R. W. (2015). An opto-electronic resistive switching memory with integrated demodulating and arithmetic functions. *Advanced Materials*, *27*(17), 2797–2703.
30. Russo, U., Ielmini, D., Cagli, C., & Lacaita, A. L. (2009). Self-accelerated thermal dissolution model for reset programming in unipolar resistive-switching memory (RRAM) devices. *IEEE Transactions on Electron Devices*, *56*(2), 193–200.
31. Jeong, D. S., Schroeder, H., & Waser, R. (2007). Coexistence of bipolar and unipolar resistive switching behaviors in a $Pt/TiO_2/Pt$ stack. *Electrochemical and Solid-State Letters*, *10*(8), 51–53.
32. Miao, F., Strachan, J. P., Yang, J. J., Zhang, M. X., Goldfarb, I., Torrezan, A. C., ... Williams, R. S. (2011). Anatomy of a nanoscale conduction channel reveals the mechanism of a high-performance memristor. *Advanced Materials*, *23*(47), 5633–5640.
33. Kever, T., Bottger, U., Schindler, C., & Waser, R. (2007). On the origin of bistable resistive switching in metal organic charge transfer complex memory cells. *Applied Physics Letters*, *91*(8), 083506.
34. Standley, B., Bao, W., Zhang, H., Bruck, J., Lau, C. N., & Bockrath, M. (2008). Graphene-based atomic-scale switches. *Nano Letters*, *8*(10), 3345–3349.
35. Sangwan, V. K., Jariwala, D., Kim, I. S., Chen, K. S., Marks, T. J., Lauhon, L. J., & Hersam, M. C. (2015). Gate-tunable memristive phenomena mediated by grain boundaries in single-layer MoS_2. *Nature Nanotechnology*, *10*(5), 403.
36. Prezioso, M., Merrikh-Bayat, F., Hoskins, B. D., Adam, G. C., Likharev, K. K., & Strukov, D. B. (2015). Training and operation of an integrated neuromorphic network based on metal-oxide memristors. *Nature*, *521*(7550), 61–64.
37. Gruning, A., & Bohte, S. M. (2014). Spiking neural networks: Principles and challenge. European symposium on artificial neural networks, computational intelligence and machine learning, Belgium. Retrieved from www.i6doc.com/fr/livre/?GCOI=280011 00432440. 02/01/2020.

Synaptic Devices Based on 2D Semiconductors

38. Nevius, M. S., Conrad, M., Wang, F., Celis, A., Nair, M. N., Taleb-Ibrahimi, A., ... Conrad, E. H. (2015). Semiconducting graphene from highly ordered substrate interactions. *Physical Review Letters*, *115*(13), 136802.

39. Li, C., Zhou, P., & Zhang, D. W. (2017). Devices and applications of van der Waals heterostructures. *Journal of Semiconductors*, *38*(3), 031005.

40. Wang, S., Zhang, D. W., & Zhou, P. (2019). Two-dimensional materials for synaptic electronics and neuromorphic systems. *Scientific Bulletin*, *64*(15), 1056–1066.

41. Liu, N., Chortos, A., Lei, T., Jin, L., Kim, T. R., Bae, W. G., ... Bao, Z. (2017). Ultratransparent and stretchable graphene electrodes. *Science Advances*, *3*(9), e1700159.

42. Najmaei, S., Amani, M., Chin, M. L., Liu, Z., Birdwell, A. G., O'Regan, T. P., ... Lou, J. (2014). Electrical transport properties of polycrystalline monolayer molybdenum disulfide. *ACS Nano*, *8*(8), 7930–7937.

43. Kang, K., Xie, S., Huang, L., Han, Y., Huang, P. Y., Mak, K. F., ... Park, J. (2015). High-mobility three-atom-thick semiconducting films with wafer-scale homogeneity. *Nature*, *520*(7549), 656.

44. Wang, X., Gong, Y., Shi, G., Chow, W. L., Keyshar, K., Ye, G., ... Ajayan, P. M. (2014). Chemical vapor deposition growth of crystalline monolayer MoS_2. *ACS Nano*, *8*(5), 5125–5131.

45. Chang, Y. H., Zhang, W., Zhu, Y., Han, Y., Pu, J., Chang, J. K., ... Li, L. J. (2014). Monolayer MoS_2 grown by chemical vapor deposition for fast photodetection. *ACS Nano*, *8*(8), 8582–8590.

46. Gao, Y., Liu, Z., Sun, D. M., Huang, L., Ma, L. P., Yin, L. C., ... Ren, W. (2015). Large-area synthesis of high-quality and uniform monolayer WS_2 on reusable Au foils. *Nature Communications*, *6*, 8569.

47. Kobayashi, Y., Sasaki, S., Mori, S., Hibino, H., Liu, Z., Watanabe, K., ... Miyata, Y. (2015). Growth and optical properties of high quality monolayer WS_2 on graphite. *ACS Nano*, *9*(4), 405663.

48. Chen, J., Liu, B., Liu, Y., Tang, W., Nai C. T., Li, L., Zheng, J., Gao, L., Zheng, Y., Shin, H. S., Jeong, H. Y., & Loh, K. P. (2015). Chemical vapor deposition of large-sized hexagonal WSe_2 crystals on dielectric substrates. *Advanced Materials*, *27*, 6722–6727.

49. Yoshida, M., Iizuka, T., Saito, Y., Onga, M., Suzuki, R., Zhang, Y., ... Shimizu, S. (2016). Gate-optimized thermoelectric power factor in ultrathin WSe_2 single crystals. *Nano Letters*, *16*(3), 2061–2065.

50. Keyshar, K., Gong, Y., Ye, G., Brunetto, G., Zhou, W., Cole, D. P., ... Ajayan, P. M. (2015). Chemical vapor deposition of monolayer rhenium disulfide (ReS_2). *Advanced Materials*, *27*(31), 4640–4648.

51. Hafeez, M., Gan, L., Li, H., Ma, Y., & Zhai, T. (2016). Large-area bilayer ReS_2 film/multilayer ReS_2 flakes synthesized by chemical vapor deposition for high performance photodetectors. *Advanced Functional Materials*, *26*(25), 4551–4560.

52. Hafeez, M., Gan, L., Li, H., Ma, Y., & Zhai, T. (2016). Chemical vapor deposition synthesis of ultrathin hexagonal $ReSe_2$ flakes for anisotropic Raman property and optoelectronic application. *Advanced Materials*, *28*(37), 8296–8301.

53. Naylor, C. H., Parkin, W. M., Ping, J., Gao, Z., Zhou, Y. R., Kim, Y., ... Johnson, A. T. (2016). Monolayer single-crystal 1T'-$MoTe_2$ grown by chemical vapor deposition exhibits weak antilocalization effect. *Nano Letters*, *16*(7), 4297–4304.

54. Zhou, J., Liu, F., Lin, J., Huang, X., Xia, J., Zhang, B., ... Liu, Z. (2017). Large-area and high-quality 2D transition metal telluride. *Advanced Materials*, *29*(3), 1603471.

55. Huang, C., Wu, S., Sanchez, A. M., Peters, J. J., Beanland, R., Ross, J. S., ... Xu, X. (2014). Lateral heterojunctions within monolayer $MoSe_2$-WSe_2 semiconductors. *Nature Materials*, *13*(12), 1096.

56. Gong, Y., Lin, J., Wang, X., Shi, G., Lei, S., Lin, Z., ... Ajayan, P. M. (2014). Vertical and in-plane heterostructures from WS_2/MoS_2 monolayers. *Nature Materials, 13*(12), 1135.

57. Zhang, L., Gong, T., Wang, H., Guo, Z., & Zhang, H. (2019). Memristive devices based on emerging two-dimensional materials beyond graphene. *Nanoscale, 11*(26), 12413–12435.

58. Berco, D., & Ang, D. S. (2019). Recent progress in synaptic devices paving the way toward an artificial cogni-retina for bionic and machine. *Advanced Journal of Intelligent Systems, 1*, 1900003.

59. Sangwan, V. K., Lee, H. S., Bergeron, H., Balla, I., Beck, M. E., Chen, K. S., & Hersam, M. C. (2018). Multi-terminal memtransistors from polycrystalline monolayer molybdenum disulfide. *Nature, 554*(7693), 500–504.

60. John, R. A., Liu, F., Chien, N. A., Kulkarni, M. R., Zhu, C., Fu, Q., ... Mathews, N. (2018). Synergistic gating of electro-iono-photoactive 2D chalcogenide neuristors: Coexistence of Hebbian and homeostatic synaptic metaplasticity. *Advanced Materials, 30*(25), 1800220.

61. Tian, H., Zhao, L., Wang, X., Yeh, Y. W., Yao, N., Rand, B. P., & Ren, T. L. (2017). Extremely low operating current resistive memory based on exfoliated 2D perovskite single crystals for neuromorphic computing. *ACS Nano, 11*(12), 12247–12256.

62. Karbalari Akbari, M., & Zhuiykov, S. (2019). A bioinspired optoelectronically engineered artificial neurorobotics device with sensorimotor functionalities. *Nature Communications, 10*(1), 3873.

63. Shi, Y., Liang, X., Yuan, B., Chen, V., Li, H., Hui, F., ... Lanza, M. (2018). Electronic synapses made of layered two dimensional materials. *Nature Electronics, 1*(8), 458–465.

64. Yang, C. S., Shang, D. S., Liu, N., Shi, G., Shen, X., Yu, R. C., ... Sun, Y. (2017). A synaptic transistor based on quasi-2D molybdenum oxide. *Advanced Materials, 29*(27), 1700906.

65. Yang, C. S., Shang, D. S., Liu, N., Fuller, E. J., Agrawal, S., Talin, A. A., LI, Y. Q., Shen, B. G., & Sun, Y. (2018). All-solid state synaptic transistor with ultra-low conductance for neuromorphic computing. *Advanced Functional Materials, 28*, 1804170.

66. Ovshinsky, S. R. (1986). Reversible electrical switching phenomena in disordered structures. *Physical Review Letters, 21*(20), 1450.

67. Lankhorst, M. H., Ketelaars, B. W., & Wolters, R. A. (2005). Low-cost and nanoscale nonvolatile memory concept for future silicon chips. *Nature Materials, 4*(4), 347.

68. Bez, R. (2009). Chalcogenide PCM a memory technology for next decade. *Proceedings of the electron devices meeting (IEDM)*. IEEE International, Baltimore, MD.

69. Servalli, G. (2009). A 45 nm generation phase change memory technology. *Proceedings of the electron devices meeting (IEDM)*. IEEE International, Baltimore, MD.

70. Huh, W., Jang, S., Lee, J. Y., Lee, D., Lee, D., Lee, J. M., ... Lee, C. H. (2018). Synaptic barristor based on phase-engineered 2D heterostructures. *Advanced Materials, 30*(35), 1801447.

71. Sun, L., Zhang, Y., Hwang, G., Jiang, J., Kim, D., Eshete, Y. A., ... Yang, H. (2018). Synaptic computation enabled by joule heating of single-layered semiconductors for sound localization. *Nano Letters, 18*(5), 3229–3234.

72. Tian, H., Guo, Q., Xie, Y., Zhao, H., Li, C., Cha, J. J., ... Wang, H. (2016). Anisotropic black phosphorus synaptic device for neuromorphic applications. *Advanced Materials, 28*(25), 4991–4997.

73. Zhou, F., Chen, J., Tao, X., Wang, X., & Chai, Y. (2019). 2D materials based optoelectronic memory: Convergence of electronic memory and optical sensor. *Research, 2019*, 9490413.

Synaptic Devices Based on 2D Semiconductors

74. Jiang, J., Hu, W., Xie, D., Yang, J., He, J., Gao, Y., & Wan, Q. (2019). 2D electric-double-layer phototransistor for photoelectronic and spatiotemporal hybrid neuromorphic integration. *Nanoscale*, *11*(3), 1360–1369.

75. Wang, S., Chen, C., Yu, Z., He, Y., Chen, X., Wan, Q., ... Zhou, P. (2018). A MoS_2/PTCDA hybrid heterojunction synapse with efficient photoelectric dual modulation and versatility. *Advanced Materials*, *31*(3), e1806227.

76. Ni, Z., Wang, Y., Liu, L., Zhao, S., Xu, Y., Pi, X., & Yang, D. (2018). Hybrid structure of silicon nanocrystals and 2D WSe_2 for broadband optoelectronic synaptic devices. *IEEE International Electron Devices Meeting (IEDM)*, 38.5.1–38.5.4, San Francisco, CA.

77. He, H.-K., Yang, R., Zhou, W., Huang, H. M., Xiong, J., Gan, L., ... Guo, X. (2018). Photonic potentiation and electric habituation in ultrathin memristive synapses based on monolayer MoS_2. *Small*, *14*(15), 1800079.

78. Wang, S., Chen, C., Yu, Z., He, Y., Chen, X., Wan, Q., Shi, Y., Zhang, D. W., Zhou, H., Wang, X., & Zhou, P. (2018). MoS_2/PTCDA hybrid heterojunction synapse with efficient photoelectric dual modulation and versatility. *Advanced Materials*, *31*, 1806227.

79. Ahmed, T., Kuriakose, S., Mayes, E. L. H., Ramanathan, R., Bansal, V., Bhaskaran, M., ... Walia, S. (2019). Optically stimulated artificial synapse based on layered black phosphorus. *Small*, *15*(22), 1900966.

8 Sensorimotor Devices Based on Two-Dimensional Semiconductor Materials

8.1 INTRODUCTION TO BIOINSPIRED SENSORIMOTOR DEVICES

The emulation of human sensorimotor system and its functionalities is a fundamental step toward the technological advancement for the next generation of bioinspired electronic devices for humanoid and robotic applications [1–9]. The performance of human body is not only based on naturally and intricately designed networks of sensors, receptors, and nociceptors, but also based on the functionalities of neural signal processors and sensorimotors [10–12]. The external stimuli including the mechanical, optical and chemical exciters are detected and received by the biological receptors of our body [13–15]. The biological synapses are the fundamental bricks of neural system, facilitating the signal transfer by bioneurotransmitters to adjacent neuron to produce informative signals. Informative signals either convey the messages to brain or deliver the orders to sensorimotor system[16–18]. Furthermore, brain functionalities are also executed by intricately connected network of synapses [19]. Therefore, the development of artificial synaptic devices represents a viable strategy to expand our capability for fabrication of bioinspired electronic devices. Although considerable attention has been paid to the research and development of organic and inorganic synaptic imitations, the current artificial instruments are still in their rudimentary stages [20–27] because of the lack of commitments toward the development of materials and devices emulating the synaptic performance of the human sensorimotor system. However, from the engineering point of view mimicking the sophisticated biological sensorimotor functionalities of human body is still a hurdle to overcome.

The Internet of Things (IoT) is one of the main technological advancements by which the interconnected network of software and hardware facilitate the fabrication of artificial sensorimotor devices in which a sensory component plays the role of receptors in the human body [1, 28]. Ultrathin 2D materials with their incredible sensory and memristor functionalities are among the main candidates to play the role of sensors, receptors and nociceptors in artificial sensorimotor system [29–32]. Considering the type of stimuli, various sensing mechanisms of 2D materials were employed to ignite the interconnected network of sensors and motors and then to mimic the behavior of fabricated human sensorimotor system. Considering the sensory role of 2D materials, optical stimulus is one of the most exciting sources to

275

trigger several kinds of reactions in sensorimotor systems. For instance, the light or visual cognition, is among the most important sensory functions with tremendous applications in optoelectronics, artificial visualization systems, communication, and information transition and/or processing technology [33–36]. Visual light cognition aside, the light-driven operation of an artificial optical sensory system accompanied by a motor device can facilitate the development of bioinspired soft robotic device and humanoid systems [1, 33]. Particularly, the self-powered, self-biased wireless sensors and synapses with low power consumption are highly desired for bioinspired neurological optoelectronics [36–39]. It is worthwhile to mention that the excitation sources are not merely restricted into the light and optical sensory mechanisms. In this concept, the memristive characteristic is another unique feature of ultrathin 2D materials that can be utilized to fabricate the memory component of sensorimotor device [31, 40–42]. In this view, the synaptic or memristor device performs like a memory sensor in which the controlled transfer of optical or voltaic input signals excite the motor component of sensorimotor system. In this chapter the applications of ultrathin films of 2D materials as the receptor, nociceptor and memristor components of the sensorimotor system are introduced. In this view, several practical cases of employment of 2D materials and ultrathin films in sensorimotor systems are discussed and their working mechanisms are explained.

8.2 OPTOELECTRONIC DEVICES

In optogenetics the muscular motion including the contraction and relief of biological muscular fibers can be controlled by optical triggering and stimulation of the sensorimotor neurons. These motoneurons are naturally and genetically modified to be responsive to the light and optical stimuli [43–45]. This concept has attracted the researcher's attention since the approach is highly promising to retrieve the defective muscles [46, 47]. An optically stimulated biological sensorimotor system consists of the opto-genetically neural (Figure 8.1a) junction that transfers presynaptic pulses (Figure 8.1b) through axon to muscles (Figure 8.1c). At neuromuscular junction (Figure 8.1d), the presynaptic action potentials are transmitted to the muscle fibers by biological chemical synapses which release the biological neurotransmitters in neuromuscular junctions (Figure 8.1e) [1, 33, 48, 49]. Following the delivery of action potentials by neurotransmitters, the excitatory postsynaptic potentials are generated in the muscle fibers which facilitate the contraction of fiber muscles (Figure 8.1f). The same mechanism is duplicated to design the artificial sensorimotor systems which emulate the behavior of optogenetically stimulated sensorimotor systems. The optically stimulated neurons and synapses are the fundamental components of optogenetically stimulated natural sensorimotor system [50, 51]. The ultrasensitive light stimulated organic and inorganic 2D materials are the main candidate to be employed as the optical receptors in the interconnected sensorimotor systems. The following lines provide detailed descriptions on the recently developed artificial sensorimotor systems in which an optically stimulated or voltaic sensors play the role of receptors in system. In this regard, the devices are divided based on the type of ultrathin 2D materials employed as sensorimotor components and based on the sensory mechanisms.

Sensorimotor Devices Based on 2D Materials

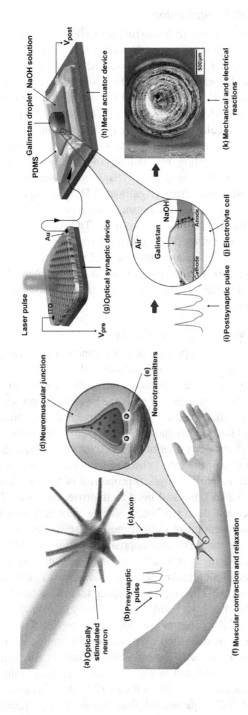

FIGURE 8.1 The graphical scheme of optogenetically engineered neuron system in human body and the designed artificial optoelectronic sensorimotor system. (a to f) In a biological system (a) the specific kind of neurons which contain photosensitive proteins can be stimulated optically. (b) The presynaptic pulses will be transmitted through (c) axons to (d) neuromuscular junctions. (e) Presynaptic pulses are transmitted to muscle fibers by neurotransmitters to cause the (f) muscular contraction and relaxation. Which are the postsynaptic reactions. Analogously, (g to k) the artificial sensorimotor system consists of (g) an optical synaptic device integrated into a (h) liquid metal actuator system. The pulsed lights trigger the optical synaptic device to generate (i) postsynaptic pulses. Then the postsynaptic potential pulses are transmitted into the liquid metal actuator device to oscillate the galinstan droplet in NaOH solution (j) the galinstan bath droplet in NaOH solution is in fact an electrolyte cell. The device output can be recorded either as (k) mechanical or electrical reactions (Reprinted with permission from Karbalaei Akbari, M., & Zhuiykov, S. (2019). A bioinspired optoelectronically engineered artificial neurorobotics device with sensorimotor functionalities. *Nature Communications*, *10*(1), 3873).

8.2.1 Sensorimotor Device Based on Ultrathin TiO_2 Film

8.2.1.1 Working Mechanism of Sensorimotors

Nanostructured metal oxide semiconductors (NMOSs) are overwhelmingly acknowledged to be reliable materials for employment in environmental sensors and receptors [52–55]. An artificial sensorimotor system was introduced in which an atomic layer deposited (ALD) TiO_2 optical synaptic device (Figure 8.1g) is integrated into to a metal actuator system (Figure 8.1h) that facilitated the artificial emulation of sensorimotor system [1]. The reason why a visible light sensitive optical synapse is used returns to its resemblance to optically stimulated synapses in motoneurons. The ability of output control is another advantage of the optical synaptic device, which is facilitated by the control of resistive switching mechanisms of device by employing patterned light pulses. The optically stimulated TiO_2 synapse generates the output potentials transmitted to the liquid metal actuator (galinstan) [1]. The liquid metal droplet in a bath of NaOH solution constitutes an electrochemical cell (EC) [56–60], which receives the postsynaptic pulses (Figure 8.1i) from the TiO_2 optical synaptic device. By imposing the postsynaptic pulses (Figure 8.1j) the reconfiguration of the charge distribution on the surface of galinstan droplet in NaOH solution is facilitated. This in turn causes the mechanical oscillation of liquid metal in NaOH bath (Figure 8.1k). This interconnected process resembles to an artificial neuromuscular bioelectronic system in neurorobotic devices.

8.2.1.2 Ion Intercalation for Visible Light Sensitivity

The visible light reception is the most important sensory function in a receptor device designed to mimic the human sensory functionalities and its cognition capabilities. Ultrathin 2D NMOSs are attractive candidates for application in synaptic devices since they benefit from the advantages of resistive switching, low energy consumption and also facile integration in complementary metal oxide semiconductor (CMOS) technology [61]. However, as most of them considered to be high bandgap semiconductors, the high energy radiation of the light can stimulate and trigger their responsivity. TiO_2 is one of the well-known semiconductor oxides with such tunable optical properties. To employ the optical properties of the well-known high bandgap TiO_2 oxide films (3.2 eV), we still need to improve the visible light sensitivity of this metal oxide semiconductor [62–65]. Visible light receptors play a remarkable role in visible light reception of light stimuli [66–68]. The bandgap engineering of 2D materials is the most possible approach to attain visible light absorption [69]. To fulfill the requirements of visible light sensory functions in the sensorimotor device, the ion intercalation technique was innovatively employed to alter the bandgap of 2D TiO_2 films [1]. The employed doping technique has broadened the optical sensitivity of ultrathin high bandgap TiO_2 film to the visible light region. A three-electrode design of Au conector was fabricated on Si/SiO_2 wafers, in which ~7.0 nm thick TiO_2 film was deposited over source and drain electrodes, while the gate electrode remained untouched. The uniform coverage of Au granular surface by ultrasmooth TiO_2 film is depicted in Figure 8.2b & c, facilitating the conformal charge transfer between ionic solid solution and TiO_2. A solid electrolyte (SE) containing In-atoms (0.1 M $InCl_4$ aqueous solution dissolved in polyethylene

Sensorimotor Devices Based on 2D Materials 279

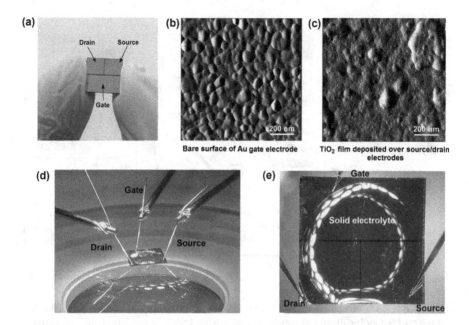

FIGURE 8.2 (a) The three electrodes set-up. (b) The bare surface of Au gate electrode is shown in the AFM image. (c) The ultra-uniform film of ALD TiO$_2$ is deposited on source and drain electrodes shown in the AFM image. (d) The three electrodes set-up for In-ion doping in TiO$_2$ film and (e) corresponding enlarged image that demonstrates the gate, source, drain and solid electrolyte (Reproduced with permission from Karbalaei Akbari, M., & Zhuiykov, S. (2019). A bioinspired optoelectronically engineered artificial neurorobotics device with sensorimotor functionalities. *Nature Communications*, *10*(1), 3873).

glycol diacrylate) was dropped cast over the three Au electrodes (Figure 8.2a, b, c) to intercalate In-ions into 7.0 nm thick TiO$_2$ film. Then a direct bias voltage (V_G) of 2.0 V was applied on SE/TiO$_2$ interface (Figure 8.2d, e) to generate the driving force for intercalation of In-ions into TiO$_2$ sublayers.

The intercalation mechanism of In-ions into 2D TiO$_2$ sublayer requires that the In-ions overcome the surface diffusion barrier and bulk diffusion barrier at TiO$_2$ interface [70, 71]. To electrochemically characterize the In-ion intercalation process in 2D TiO$_2$ sublayer, the variation of drain-source current (I_{DS}) vs the gate voltage (V_G) was monitored and depicted in Figure 8.3a. The graphical scheme shows the device structure in which the TiO$_2$ film is deposited over source/drain electrodes, while the SE containing In-ions is deposited over Au gate/TiO$_2$ film [1]. By applying an external V_G, In-ions are pushed to accumulate at the surface of 7.0 nm think TiO$_2$ film. It finally results in an increase in channel conductance of TiO$_2$ film. By further increase of V_G, In-ions would overcome the surface diffusion barrier and the bulk diffusion barrier to inject into TiO$_2$ sublayer (Figure 8.3b). The voltage sweep from upper limit (2.0 V) to lower voltages creates an internal field (E_{int}) accompanied by ion concentration gradient. The non-equilibrium distribution of In-ions causes concentration gradient at the TiO$_2$/SE interface. It acts as the driving force to extract

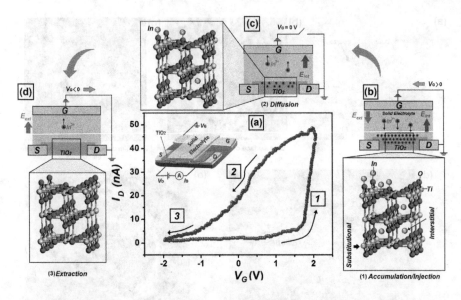

FIGURE 8.3 The graphical scheme of In-ion intercalation into ultrathin ALD TiO$_2$ film and the corresponding electrical measurements during In-ion intercalation. (a) The variation of channel current (I_D) vs. the gate voltage (V_G). (b) By applying positive gate voltage, the In-ions are pushed to accumulate at the interface between TiO$_2$ and solid electrolyte. Thus, a concentration gradient field (E$_{int}$) is caused, which results in the increase of channel conductance in ultrathin 2D TiO$_2$ film. Finally, the bulk diffusion occurs. (c) The extraction of intercalated In-ions from the TiO$_2$ film by sweeping the V_G from upper limit to lower voltage. (d) The extraction of In-ions from the ultrathin TiO$_2$ channel (Reprinted with permission from Karbalaei Akbari, M., & Zhuiykov, S. (2019). A bioinspired optoelectronically engineered artificial neurorobotics device with sensorimotor functionalities. *Nature Communications*, *10*(1), 3873).

In-ions from oxide layer toward the SE/TiO$_2$ interface. When the $V_G = 0$, the chemical potential gradient will cause an E_{int}, and consequently, the In-ions are driven back into SE (Figure 8.3c). A negative V_G will extract the accumulated In-ions from TiO$_2$ film (Figure 8.3d). The occurrence of hysteresis in I_{DS}-V_G curves is fully representative of the intercalation/de-intercalation mechanism, which is called SE based doping or ion functionalization method [1, 70, 71]. It also confirms that the reversible extraction of In-ions is possible only by imposing reverse driving force (-V_G) for detrapping of In-ions.

The ion intercalation mechanism in ultrathin 2D TiO$_2$ film was also studied by capacitance-frequency measurements of device (Figure 8.4a), where three individual steps were observed. The insignificant value of specific capacitance in high frequencies (≈ 0.020 μF cm^{-2} at higher frequencies > 10^5 Hz) refers to the capacitance of bulk electrolyte (Region I). The formation of electrolyte double layer (EDL) at SE/TiO$_2$ interface at higher frequencies is the result of frequency decline. The rapid increase of capacitance in region II (1 ~ 10^5 Hz) is the manifestation of the ion migration and accumulation at the interface. As the capacitance increases at the frequencies lower than 1 Hz, the pseudo-capacitance stage starts (Stage III), accompanied by

Sensorimotor Devices Based on 2D Materials 281

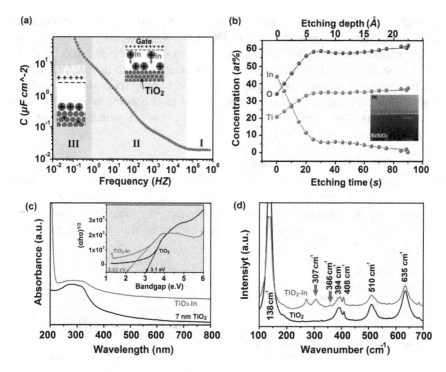

FIGURE 8.4 (a) The capacitance-frequency (*c-f*) graph of In-ion intercalation process in ultrathin ALD TiO$_2$ film. (b) The XPS surface profile of atomic concentration of In-ion doped TiO$_2$ film. The measurements depict how the concentration of In, O and Ti atoms are changing vs. the thickness of TiO$_2$ film. (c) The absorption spectra and bandgap values of TiO$_2$ and the In-doped TiO$_2$ films. (d) The Raman characteristic vibration of TiO$_2$ and the In-doped TiO$_2$ (Reproduced with permission from Karbalaei Akbari, M., & Zhuiykov, S. (2019). A bioinspired optoelectronically engineered artificial neurorobotics device with sensorimotor functionalities. *Nature Communications*, *10*(1), 3873).

electrochemical doping [1]. The XPS analysis shows the chemical composition of In-ion intercalated TiO$_2$ film which depicts the successful incorporation of In-ions in 2D TiO$_2$ film (8.4b). The concentration of In-atoms in top layer of ultrathin TiO$_2$ film is 45 at. % which sharply declined down to ~4 at. % at the etching depth of 30 Å. It means that the In-ion intercalation is limited to the surface layer of ultrathin 2D TiO$_2$ film. However, this small amount of intercalated ions can considerably affect the conductance and resistive behavior of ultrathin TiO$_2$ films [1]. Specifically, one of the main consequences of In-ion intercalation is the modulation of the optical bandgap of ultrathin TiO$_2$ film (Figure 8.4c), where the bandgap has shifted from 3.1 eV toward the visible region of light (2.03 eV) [1]. The proposed mechanism is substitution of In-atoms in the TiO$_2$ microstructure, which results in the formation of Ti$_{15}$In$_1$O$_{32}$ phase [72, 73]. The acquired visible light sensitivity is the main outcome of In-ion intercalation facilitates the visible light assisted sensitivity of ultrathin TiO$_2$ synaptic device. Considering the ionic size of In and Ti (81 pm for In^{3+} and 53 pm for Ti^{4+}) the substitutional In-ion intercalation in TiO$_2$ structure is the thermodynamically preferable condition even though both kind of interstitial and substitutional incorporations of In-ions in TiO_2 films were reported [74]. Due to ionic size mismatch between In and Ti, a structural distortion in the In-ion doped TiO_2 film is expected [75] upon the substantial replacement of ionic In into TiO_2 lattice. This process ultimately affects the Raman characteristics of intercalated TiO_2 films (Figure 8.4d). The characteristic vibration peak of In-O bonding is observed in the Raman spectra of In-ion intercalated TiO_2 film at 307 cm^{-1} (Figure 8.4d) confirming the incorporation of In-ions in 2D TiO_2 film.

8.2.1.3 The Sensorimotor System

It was found that the two-terminal sandwiched ITO/In-doped TiO_2/Au device shows optical synaptic characteristics [1], in which the conductance of memristor device is changed by the illumination of sequential pulses of visible light. The simplified graphical scheme of sensorimotor system is presented in Figure 8.5a. In this

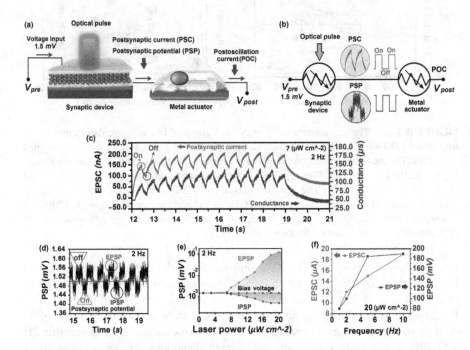

FIGURE 8.5 The graph of sensorimotor system and its optoelectronic properties of optical 2D synaptic device (a) The graphical scheme depicts the sensorimotor system and (b) its corresponding electrical circuit. (c) The typical EPSC and conductance variation for ITO/2D In-doped TiO_2/Au synaptic device when it was illuminated by 7 μW cm^{-2} pulsed light with 530 nm wavelength and 2 Hz frequency (d) the graph shows the corresponding variation of PSP of the same device. (e) The variation of PSP vs. the power intensity of laser light. (f) The variation of EPSC and EPSP vs. the frequency of the pulsed light. (Reproduced with permission from Karbalaei Akbari, M., & Zhuiykov, S. (2019). A bioinspired optoelectronically engineered artificial neurorobotics device with sensorimotor functionalities. *Nature Communications*, *10*(1), 3873).

Sensorimotor Devices Based on 2D Materials

scheme, the mechanical oscillation of liquid metal actuator component is facilitated by the transfer of patterned post synaptic potential (PSP) pulses of optical synaptic device into the liquid metal galinstan utilized as the mechanical component of sensorimotor system [1]. The schematic graph of electrical circuit of sensorimotor device is presented in Figure 8.5b. The sequential *on/off* cyclic pulsed lights alter the conductance of ultrathin TiO_2 synaptic device. The outputs of optical synapse are the postsynaptic potential (PSP) and current (PSC) signals. By applying an optical pulse, the device conductance increased, which was accompanied by the simultaneous enhancement of excitatory postsynaptic current (EPSC) (Figure 8.5c) and the decrease of inhibitory postsynaptic potentials (IPSP). The sequential illumination of pulsed light caused PSP oscillation where the PSP either jumps to higher voltage (light pulse-*off*) or suddenly falls to lower voltage (light pulse-*on*) than that of the imposed bias potential (1.5 mV) (Figure 8.5d). This behavior is very similar to the EPSP and IPSP phenomena in biological neurons [76, 77]. The level of conductance and output potential were consistent with the power intensity of illumination source (Figure 8.5e). The light frequency has also the accumulative impact on potential and current output of synaptic device (Figure 8.5f). Thus, the transmitted PSP to the liquid metal actuator works as the electrocapillarity force [60, 78] which alters the surface energy of liquid metal droplet in NaOH bath and at the same time changes the physical configuration of liquid metal.

The actual top view of liquid metal actuator accompanied by the light contrast developed images, the 3D model and the graphical scheme of galinstan droplet in NaOH solution demonstrate the mechanism of liquid metal oscillation (Figure 8.6a). The concave interface is formed at the air/solution/galinstan junction point because of the difference in surface energy of galinstan/solution and galinstan/air [60, 79]. The charge distribution on the surface of galinstan droplet in NaOH solution can be altered by imposing the sequential PSP pulses of optical synapse when the galinstan/NaOH system acts as the electrolyte cell. The mechanical oscillation of galinstan droplet is caused by the sequential alteration of charge distribution. The accurately captured image in Figure 8.6a depicts the exact moment of deformation of the conical head of the galinstan droplet, which is inclined to the cathodic pole. The mechanical deformation continues until the pressure difference drops on both sides of galinstan and system returns to its equilibrium condition. This mechanical deformation is called continuous electrowetting (CEW)/de-electrowetting (DEW), which is famous for its low power consumption [80]. Figure 8.6b shows the vibration of a galinstan droplet after receiving sequential PSP pulses. The controlled oscillation of galinstan droplet was achieved by using opto-electronically generated potential pulses of TiO_2 synaptic device (Figure 8.7c) which ultimately transmitted to the liquid metal actuator. Series of systematic experiments were carried out and confirmed the capability of artificial sensorimotor system to mimic the functionalities of neural system. For example, the mechanical oscillation of a galinstan droplet was monitored when the typical 20 μW cm^{-2} pulsed visible light at 10 Hz frequency was used to stimulate the optical synapse to later generate PSP pulses. After pulse transfer, the galinstan droplet started to oscillate. The mechanical oscillation was monitored by the optical microscope facilitated by the high-speed camera [1]. By comparing the sequential captured photos, the border displacement of $\sim$ 1280 μm diameter galinstan droplet

284 Ultrathin 2D Semiconductors

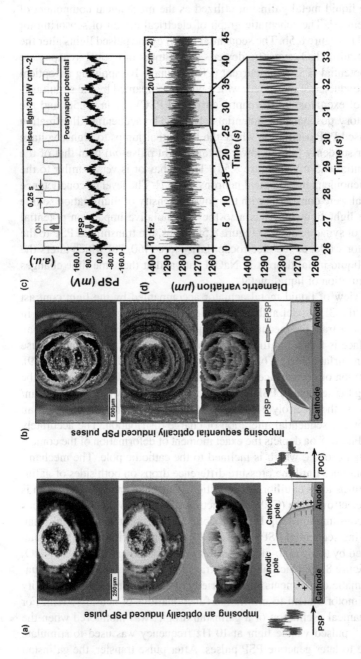

FIGURE 8.6 The monitoring of liquid metal galinstan oscillator by high-speed imaging optical camera. (a) The actual top view, the optical contrast image and the generated three-dimensional (3D) model of the galinstan droplet are shown when a singular postsynaptic pulse is transmitted from optical synaptic device to galinstan actuator. (b) The consecutive laser pulse caused periodic oscillation of galinstan droplet. (c) From top to down, the graph of patterned pulsed light, and corresponding variation of PSP for optical synaptic device. (d) The typical transitional motion of a 1280 μm diameter galinstan droplet by applying sequential PSPs originated from the optical synapse (Reproduced with permission from Karbalaei Akbari, M., & Zhuiykov, S. (2019). A bioinspired optoelectronically engineered artificial neurorobotics device with sensorimotor functionalities. *Nature Communications*, *10*(1), 3873).

Sensorimotor Devices Based on 2D Materials

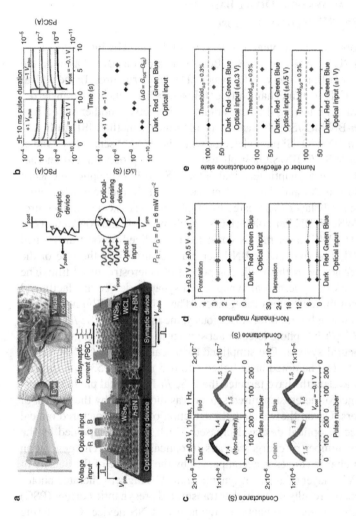

FIGURE 8.7 (a) The schematic of human optical nerve system. The optical h-BN/WSe$_2$ photodetector is integrated to h-BN/WSe$_2$ 2D heterostructures synaptic device. The simplified electrical circuit of the ONS device is plotted. The pulsed laser dot source illuminated the photodetector with three different wavelengths 655 nm (red), 532 nm (green), and 405 nm (blue) with a fixed power density of 6 mW.cm^{-2}. (b) EPSC and IPSP and (c) the conductance variation of ONS device under various light illumination, that is, dark versus red (R), green (G), and blue (B) or RGB lights. The input voltages were ±1 V. The LTP and LTD curves for different illumination condition. The input voltage pulses were ± 0.3 V. (d) The nonlinearity magnitude and the (e) number of effective conductance which are extracted from different illumination wavelengths (Reprinted with permission from Seo, S., Jo, S. H., Kim, S., Shim, J., Oh, S., Kim, J. H., ... Park, J. H. (2018). Artificial optic-neural synapse for colored and color-mixed pattern recognition. *Nature Communications*, 9(1), 5106).

was recorded during oscillation as it is presented in Figure 8.1d. The sinusoidal PSPs are caused by the stable periodic expansions and contractions of galinstan droplet. All of the above mentioned observations confirm the capabilities of developed artificial sensorimotor system in which the 2D TiO_2 oxide plays the fundamental role of receptor of the external optical stimulus to generate postsynaptic informative pulses [1].

8.2.2 OPTIC-NEURAL SYNAPTIC DEVICE BASED ON h-BN/WSe$_2$ 2D HETEROSTRUCTURE

A heterostructured optical neural synaptic (ONS) was fabricated by integrating a 2D synaptic device and 2D optical sensors [81]. Both units are fabricated based on the visible light responsive h-BN/WSe$_2$ van der Waals 2D heterostructures. In the optical sensing component of device, mechanically exfoliated h-BN flake was transferred to the SiO$_2$/Si substrate. Following the deposition of h-BN flake, another WSe$_2$ flake was transferred to h-BN/SiO$_2$/Si substrate by reside transfer method [81] to finally develop a 2D WSe$_2$/h-BN/SiO$_2$/Si photodetector with visible light sensitivity. One of the Au electrodes of WSe$_2$/h-BN/SiO$_2$/Si photodetector was grounded and another one was connected to the presynaptic terminal of the WSe$_2$/h-BN synaptic unit, when the bottom electrode was Au and the top-electrode was Pt. In the heterostructured synaptic devices, the h-BN layer was treated by O$_2$ plasma to create a weight control layer (WCL) at the WSe$_2$/h-BN heterointerfaces which modulate the conductivity of 2D WSe$_2$ channel. The WCl layer mostly consisted of oxidized boron transformed from h-BN layer [81]. The oxidized h-BN layer plays significantly the role of the charge trapping layer acting as the WCL component of heterostructured film. The interconnected network of optic-neural synapse and synaptic device has facilitated the colored- and color-mixed pattern recognition. The schematic of human optical nerve system is depicted in Figure 8.7a. Analogously, the integrated WSe$_2$/h-BN synaptic device with WSe$_2$/h-BN photodetector act similar to the optical cognition system in the human neural system. The simplified electrical circuit of ONS device is presented in Figure 8.7b. The doted laser lights with 655 nm (red), 532 nm (green), and 405 nm (blue) wavelengths have targeted the WSe$_2$/h-BN optical photodetectors in which the resistance of heterostructured sensor was modulated by the control of the density of carriers. Up on the visible light detection of heterostructured optical sensor, the resistance of optical device decreased which was accompanied by the generation of higher number of charge carriers in optical sensors. The generation of higher number of charge carries is accompanied by the increase of the density of trapped charges in WCL layer of adjacent synaptic unit. The higher generated photocurrent in optical sensors results in the higher transmitted presynaptic current (PSC) to the adjacent heterostructured synaptic component of ONS device. Considering the wavelength and intensity of optical pulses, the density of trapped charges in WCL layer of synaptic device can be altered and controlled. The light with shorter wavelength caused the generation of higher number of carriers at the constant light intensity. For example, the postsynaptic current increased from 1.55 nA to 2.29 µA by changing the light source from red to blue lasers, which was accompanied by the increase of conductance from 0.78 nS to 0.74 µS [81]. The nonlinearity of conductance was observed for the devices in all of the light wavelength, since the

Sensorimotor Devices Based on 2D Materials 287

conductance graphs kept the curved shape in all of the wavelengths of illumination. However, the levels of conductance values were different for various sources of light illuminations (Figure 8.7c). The nonlinear conductance is not preferable to achieve the sufficient number of usable conductance states since the applied voltage in synaptic component of ONS disables the induction of a sufficiently large conductance difference in the saturated regions [82]. To reduce the power consumption in ONS, a threshold ΔG was defined. In order to fulfill the energy consumption requirements, the ΔG value should exceed a certain percentage of G_{max}/G_{min}. In doing so, the pulsed signals, which induce small ΔG, are excluded since they are not appropriate for the neuromorphic computing and thus the amount of power consumption decreased considerably [81]. It was observed that the long-term potentiation (LTP) and long-term depression (LTD) characteristics in ONS device were the function of wavelength of illumination source. Furthermore, the linearity and the number of effective conductance states were almost independent of the wavelength of illumination source, which are highly important parameters for rates of pattern recognition (Figure 8.7 d and e). In general, the device demonstrated a close linear weight update trajectory while providing a large number of conductance states with less than 1% variation per state. The energy consumption of device is only 66 fj for a 0.3 V voltage spike. The credibility of the heterostructured 2D ONS unit was confirmed when it was working under the ambient conditions. Therefore, the outputs of 2D WSe_2/h-BN ONS device were successfully used to develop an optic-neural network (ONN) accompanied by a simple perception method. By applying the ONN, the colored and the color-mixed pattern recognition was facilitated emulating the human visionary system. Thus, the recognition rate of more than 90% was achieved for the color recognition by using the above mentioned 2D WSe_2/h-BN/SiO_2/Si heterostructured device.

8.2.3 Ultrathin ZnO Photodetectors for Stretchable Sensorimotors

An artificial organic optoelectronic sensorimotor system was developed based on a stretchable organic nanowire synaptic transistor in which an ultrathin ZnO photodetector film receives and transfers the informative optical signals to a stretchable organic nanowire synaptic transistor (ONWST) [33]. The transmission of postsynaptic current pulses to a polymer actuator caused mechanical deformation of polymeric component employed as the motor-responsive unit. The fabricated sensorimotor device facilitated the wireless optical cognition of environmental stimuli and assisted the transition of informative signals into the organic soft actuator, which later can emulate the performance of the artificial neuromuscular component of sensorimotor device. The main characteristics of the fabricated sensorimotor device are the ability to be self-powered and also the low-energy operation of neurologically bioinspired instrument [33]. The scheme of such device is depicted in Figure 8.8a and b. The photodetector is based on ultrathin 30 nm thick ZnO film deposited on the ITO substrates. A 40 nm thick hole extraction layer (CLEVIOS AI4083) was fabricated on the ZnO film and then was heat treated at 150°°C. The optical stimulation of photodetector by the patterned optical signals led to generation of excitatory presynaptic spikes (voltage pulses) transmitted into the ONWST synaptic transistor component. The stretchable synaptic transistor consists of a single wall carbon

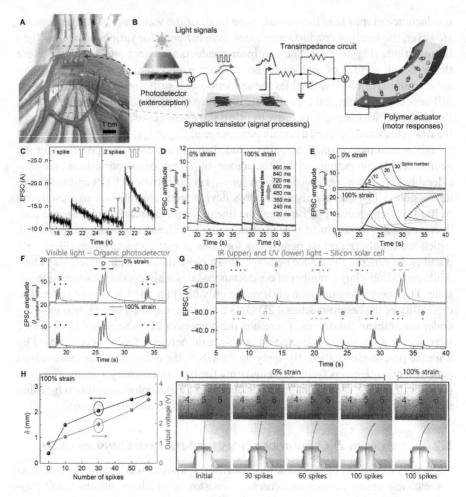

FIGURE 8.8 The schematic graph of organic optoelectronic synapse and neuromuscular electronic device. (a) Depicts the schematic of an organic optoelectronic synapse integrated into the human body structure. (b) The image shows the configuration of organic optoelectronic synapse with its photodetector and artificial synaptic component integrated into the neuromuscular artificial electronic device which composed of the artificial synapse, transimpedance circuit, and artificial muscle actuator. (c) The EPSC of device triggered by singular and doubled light spikes. The calculated PPF index is $A_2/A_1 = 1.42$. (d) and (e) depict the variation of EPSC amplitude of ONWST device for different values of strain from 0% to 100%. (f) and (g) demonstrates the amplitudes of EPSC of OWNST when the International Morse code of "*SOS*" and "Hello UNIVERSE" are used to reproduce by ONS device. (h) The maximum of polymer actuator strain (δ) and the corresponding generated output voltage by ONWST device when the spike number is increasing between 0 and 60. (i) The mechanical deformation of polymer actuator when the number of spikes is between 0 and 100 and the strain is from 0% to 100% (Reprinted with permission from Lee, Y., Oh, J. Y., Xu, W., Kim, O., Kim, T. R., Kang, J., ... Lee, T. W. (2018). Stretchable organic optoelectronic sensorimotor synapse. *Science Advances*, *4*(11), eaat7387).

nanotubes (SWCNTs) which were spray-coated on the SiO_2/Si substrate and transferred to styrene ethylene butylene substrate (SEBS). The carbon nanotube (CNT) source/drain electrodes were subsequently patterned on the SEBS substrate.

In the sensorimotor device, the visible light photodetector generated voltaic pulses (−1.1 V) after illumination. The induced output voltage stimulates the synaptic component to generate a nano-metric EPSC (−16.7 nA). The paired pulse facilitation (PPF) is observed after employing two consecutive imposed pulses with PPF index of (A_2/A_1) 1.42 (Figure 8.8c). It was observed that the spike-duration dependence plasticity (SDDP) and spike-number dependence plasticity characteristics of the device were similar at 0% and 100% strain, respectively (Figure 8.8d). The amplitude of EPSC showed a linear increasing relation with the spike number, that is, as the number of spikes increased, the EPSC values also were enhanced. However, when the spike number was higher than 10, the saturation behavior was observed (Figure 8.8e). This behavior was analogous to the biological muscle tension response during the twitch contraction, summation and incomplete tetanus [83, 84]. The optoelectronic sensorimotor device already demonstrated unique capability of interpretation of optical pulses, which represented the International Morse code [33]. In doing so, every letter in the English alphabet can induce a distinct EPSC amplitude response value. To this end, every letter was linearly correlated to the summation amplitude of EPSC peak. For example, the *SOS* emergency signal was expressed by three individual EPSC groups of signals. For instance, the letter "*S*" possesses the EPSC amplitude of around 24 and for the letter of "*O*" the EPSC amplitude was around 38. Figure 8.8f represents the EPSC amplitude for the SOS Morse signals in 0 strain and 100% strain. In fact, the stretching of device did not change the sensorimotor response. To further confirm the capability of the optical synaptic device, the IR and UV lights were also employed to create the words of "*hello*" and "*universe*" (Figure 8.8g). The developed optical synaptic device had the sufficient capacity to be employed as the telecommunication method for remote control of bioinspired artificial muscular systems. On the other hand, the capability of neuromuscular sensorimotor system was confirmed by using a polymer actuator [33, 85]. The polymer actuator was assembled into the ONWST synaptic device through a transimpedance circuit. The optical pulses generated the presynaptic impulse. The generated EPSCs from the organic synapse were converted to voltage signals and then transmitted to the polymer actuator to emulate the mechanical motion of a sensorimotor system. The artificial synaptic device generated a small voltage before illumination caused by its synaptic resting current. It resulted in small contraction of the polymer actuator. After light illumination, the generated EPSCs were converted to the voltage pulses and transmitted to the polymeric actuator. Here the amplitude of the output voltage increased by the increase of light frequencies which was accompanied by higher level of strain. Thus, there was a logical relationship between the number of spike illumination and the strain values of polymeric actuator. Nevertheless, the fabricated stretchable organic optoelectronic sensorimotor synapse with its ultrathin ZnO photovoltaic components has clearly demonstrated its capability for optical wireless communication for application in the human-machine interface technology. Furthermore, the Internet of Things represents a promising strategy for development of biomimetic soft robotics, neurorobotics, and sensorimotor devices.

8.3 ULTRATHIN OXIDE FILMS NOCICEPTOR DEVICES

8.3.1 ARTIFICIAL NOCICEPTORS BASED ON 2D SiO_2 DIFFUSIVE MEMRISTORS

Emulation of human brain functionalities using neuromorphic-based artificial instruments is a substantial achievement in the development of artificially engineered bio-inspired electronic materials and devices to finally fabricate the humanoid robots based on the artificial intelligence technology [86–90]. Sensory functions of sensorimotors are critically important, especially for those systems which collect the informative data and signals from surrounding environment. Natural nociceptors are the key sensory receptors in the human body which take the responsibility of rapid recognition of external stimuli before they exceeds the hazardous threshold level [90–96]. The nociceptors react to mechanical forces, pressure, chemical molecules, and extreme cold or hot temperatures by generation of warning signals. These signals later transmitted and delivered to the central nervous system to initiate a sensorimotor reaction and diminish the destructive impacts of potential physical or chemical damages to the human organs [97]. Considering that the nociceptors are one of the main members of neural system of human body, the nociceptor networks are consequently expanded throughout the human body. Specifically, it has been demonstrated that the cornea has one the highest number of nociceptors in human body. The corneal nociceptors are mostly polymodal, which means they are activated by the mechanical, heat and cold stimuli and also by a large variety of exogenous irritant chemicals and endogenous inflammatory substances released by the damaged corneal tissues [98–100]. Nociceptors work in two individual modes. In no-adaptation mode, the nociceptors do not adapt to the intensity of noxious stimuli while the other receptors adapt and show reaction to the intensity of external stimuli by gradual or rapid increase of response signals [96]. In the sensitization mode, when the intensity of stimulus is higher than a specific threshold, the sensitivity and reaction of nociceptors are much more intense which may result in tissue damage [96]. In the damaged tissue the threshold value for nociceptor reaction is decreased, which means that nociceptor reacts much sooner to lower intensity of internal or external stimuli. In addition, the nociceptor reaction is much intense and vivid. These typical behavior of nociceptors in the damaged stage is called *allodynia* and *hyperalgesia*, respectively [101]. The nociceptive behaviors are similar to the natural reactions of burned skin to weak heat sources, or close to the reaction of the damaged eyes to sunlight.

The current humanoid technology and robotics employ the exteroceptive sensors or so-called tactile sensors to detect the environmental stimuli [102]. This sensory technology is mostly built on complementary metal oxide semiconductor (CMOS) devices [103, 104]. While the CMOS sensors have tangibly and distinguishably developed the functionalities of novel sensory devices, the technological concerns have limited further development of CMOS systems, especially in the size and scalability issues, which in turn have limited the capability of CMOS sensors for increasing the sensitivity of receptors [105, 106]. Technically, a sensory system consists of a receptor device accompanied by a signal processing module, both of them can work as the individual nano-instrument in integrated Internet of Things system. This configuration has decreased the operational energy consumption of devices to the lowest possible level and increased the sensitivity of receptor. While the use of nanodevices has tangibly improved the performance of CMOS sensory systems, their signal processing module still remains bulky. It is because of the complexity of the sensory and processing functions. Furthermore, the destroying effects of environmental radiation on CMOS units have restricted the credibility and sustainability of sensory devices for applications in harsh environments. Therefore, it is highly desirable to realize the sensory functions of nociceptors by using a low-energy single electronic unit. Diffusive memristors are among the main candidates to fulfill requirements of nociceptor devices, including no-adaptation, relaxation, and sensitization modes [107]. So far, memristors have established their unreplaceable position in a wide range of technological applications including the nonvolatile memory [108–115], analog neuromorphic computing [116–119], biosystem emulator devices and radiofrequency systematic instruments [120]. Figure 8.9 demonstrates the graphical design corresponding to a natural nociceptor system of the human body and artificial nociceptor circuit consisting of a diffusive memristor component [107].

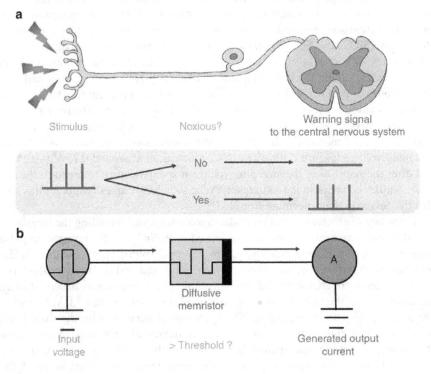

FIGURE 8.9 (a) The graphical scheme of natural nociceptor system in human body. When the noxious stimulus signal arrives at nerve ending, the nociceptor evaluates the amplitude of received signal with its pre-designed threshold value and then make the decision whether it should generate the warning signals to the spinal cord or brain system or not. (b) Analogously, in the artificial nociceptor device, when the amplitude of imposed pulse is higher than the threshold of diffusive memristor device, the artificial device is turned on and current pulses are generated (Reprinted with permission from Yoon, J. H., Wang, Z., Kim, K. M., Wu, H., Ravichandran, V., Xia, Q., … Yang, J. J. (2018). An artificial nociceptor based on a diffusive memristor. *Nature Communications*, 9(1), 417).

A diffusive memristor is basically a volatile switching device with the threshold dynamic [107]. The properties of diffusive memristor have facilitated the nociceptive characteristics in which the external stimuli are mostly the voltage pulses working as the action potentials. The threshold switching dynamic of memristor is quiet similar to the performance of a nociceptor, where the charge trapping/detrapping phenomenon has facilitated the nociceptive characteristics of artificial devices. If the amplitude of input electrical pulse in diffusive memristor is not strong enough to change the resistive states of memristor from high resistance state (HRS) to the low resistance state (LRS), the device will stay in its original HRS mode, which means the instrument adjusted itself to the threshold limit [107]. However, once the intensity of input voltage signal is higher than the threshold values, the output electrical characteristics of the device proportionally change the amplitude of the input voltaic pulses. This behavior is equivalent to the firing rates of natural nociceptors in nervous system [121, 122].

Recently developed ultrathin 2D nanostructures are the most desirable functional materials with resistive switching characteristics for memristor application. For example, a solid-state nociceptor based on diffusive memristor was fabricated using Pt (30 nm)/(SiO_2-11% Ag) 10 nm/Ag (1.0 nm)/Pt 15 nm sandwich structure [107]. 10 nm thick SiO_2 film is the switching layer positioned between Pt bottom and Pt top electrodes. 1.0 nm thick Ag film was laid between 2D switching layer and over Pt electrodes to perform as the reservoir of Ag atoms and prevent the Ag depletion during switching process. The cross-sectional and top view of diffusive memristor nociceptor, which has 55 μm^2 junction area are presented in Figure 8.10a. The *I-V* characteristics of memristor showed the bidirectional switching behavior under the sweeping voltage (Figure 8.10b). The device turned *on* at around 0.25 V (–0.25 V) and after the removal of the sweeping voltage it spontaneously returned to the *off-state*. Similar behavior in the subsequent cyclic switching curves reinforces this idea that the device is electroforming free [107].

Three key characteristics of an undamaged nociceptor including the threshold, no-adaptation, and relaxation, were observed in Pt/SiO_2-Ag/Pt through the pulse measurement of diffusive memristor. The external voltaic stimuli with different amplitudes, pulse widths, and numbers were employed and the main characteristics of device were characterized. The threshold phenomenon was observed when the amplitude of the pulsed voltaic signals increased to more than 1.0 V. However, the longer pulses have resulted in the higher output currents with the fixed pulse amplitude. The characteristics of nociceptive device are not only depended on the stimuli amplitude and frequencies but also on the number of external stimuli. Figure 8.11 shows the pulse response of the memristive nociceptors in which the amplitude of the external voltaic pulses and their corresponding output current are depicted. The threshold property was observed since the initiation of current jump was delayed. Furthermore, by increase of amplitude of the voltaic stimuli, the threshold time is subsequently decreased. Considering these observations, the formation of conducting path (CP) of Ag clusters or filaments can be the most plausible explanations for this kind of resistive behavior [123–126]. The development of CP or filamentary charge transfer between the top and bottom electrodes

Sensorimotor Devices Based on 2D Materials 293

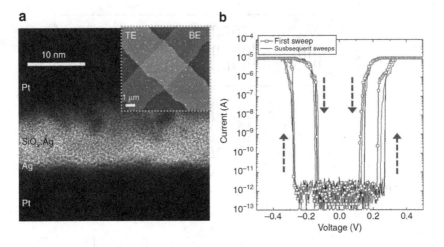

FIGURE 8.10 (a) The high-resolution cross-sectional TEM image of the diffusive memristor with Pt/SiO$_x$:Ag/Ag/Pt structure. The inset depicts the heterostructured cross-bar SEM image of diffusive memristor for 2.5 × 2.5 µm^2 junction area. (b) The typical cyclic *I-V* curve of device for threshold switching. The arrows indicate the direction of switching (Reprinted with permission from Yoon, J. H., Wang, Z., Kim, K. M., Wu, H., Ravichandran, V., Xia, Q., ... Yang, J. J. (2018). An artificial nociceptor based on a diffusive memristor. *Nature Communications*, 9(1), 417).

needs quite enough number of voltaic stimuli with enough amplitude of energy. When the bridge path was formed, imposing higher number of pulsed stimuli does not change the current level and current remains constant. This phenomenon is similar to the "*no-adaptation*" characteristics of nociceptor. The *relaxation*, which is the main characteristics of a nociceptor, was observed in Pt/SiO$_2$-Ag/Pt device [107]. When the consecutive voltage pulses with the different intensities and pulse time were imposed on the memristor, the device current did not return to its initial state. In nociceptors, when the intensity of the stimulus signal is stronger, it takes longer time until the response signal ultimately disappears after the stimulus signal is removed. Thus, the relaxation characteristic is similar to the behavior of nociceptor. When the intensity of stimulus signals is stronger, it takes longer time until the pain signal ultimately disappears (Figure 8.11b). The *sensitization* of nociceptors is recognized by the detection of *hyperalgesia* and *allodynia*. When a nociceptor is sensitized or harmed, it shows a nonexpected strong reaction to the external stimulus, which is similar to the reaction of burned skin to the heat source [127]. To introduce the injured or harmed state to the artificial nociceptor, a high amplitude of voltage stimulus was imposed on the memristor. As it is presented in Figure 8.11c, the threshold of the *on-state* of the nociceptor has shifted toward the lower voltages, while the current response of device has moved toward the higher intensities (Figure 8.11d). It was also confirmed that the sensitized Pt/SiO$_2$-Ag/Pt artificial nociceptor behaved similar to an injured natural nociceptor [107].

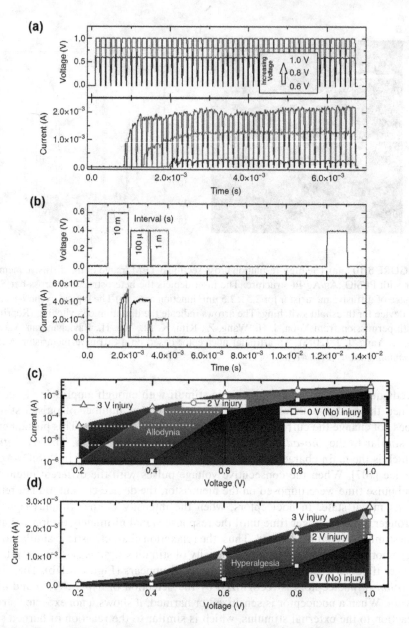

FIGURE 8.11 The electrical characteristics of memristor. (a) The response of memristive nociceptor device to the multiple number of voltage spikes (0.6, 0.8, and 1.0 V). The shorter incubation time and higher current were measured for voltage spikes with higher amplitude. (b) The upper panel of graph shows sequential voltage spikes (0.6 V and then 0.4 V) with different pulse intervals (100 μs, 1 ms and 10 ms). Lower panel depicts the output currents corresponding to the applied voltages. (c) The maximum of output current of devices for different voltage amplitudes which shows the *on-state* voltage toward lower threshold time (Allodynia) and (d) toward the higher current values (Hyperalgesia) (Reprinted with permission from Yoon, J. H., Wang, Z., Kim, K. M., Wu, H., Ravichandran, V., Xia, Q., … Yang, J. J. (2018). An artificial nociceptor based on a diffusive memristor. *Nature Communications*, 9(1), 417).

8.3.2 Thermal Nociceptor Based on 2D SiO$_2$ Diffusive Memristors

The nociceptive characteristics of Pt/SiO$_2$-Ag/Pt diffusive memristor were employed to fabricate an artificial thermal nociceptor [107]. A thermoelectric module was connected to a diffusive memristor. By heating the thermoelectric module, the voltage pulse was generated and transferred to the unit to trigger the nociceptor (Figure 8.12a). There is a linear relation between the generated voltage of thermoelectric module and the temperature of hotplate. As the input voltage exceeds the

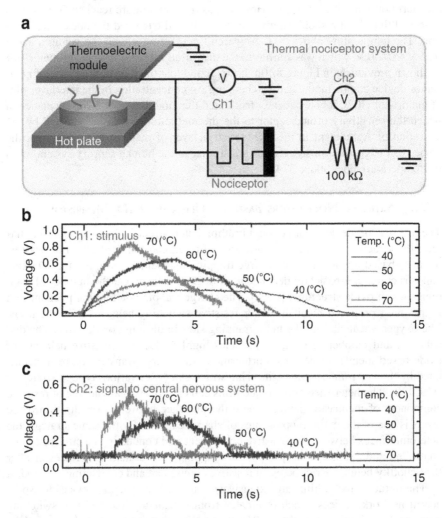

FIGURE 8.12 The thermal nociceptor device. (a) The simplified circuit of artificial nociceptor consists of a thermoelectric module and the diffusive memristor as nociceptor component. (b) The generated voltage of thermoelectric module at various temperatures. (c) The on and off switching of the threshold nociceptor device at various temperatures of thermal module (Reprinted with permission from Yoon, J. H., Wang, Z., Kim, K. M., Wu, H., Ravichandran, V., Xia, Q., ... Yang, J. J. (2018). An artificial nociceptor based on a diffusive memristor. *Nature Communications*, 9(1), 417).

threshold value of memristor, the generated current flows to the adjacent series resistor. The generated voltage pulses of thermoelectric device dropped ultimately on the memristor when it still at *off-state*. As the device was *turned on*, the voltage dropped ultimately on both memristor and resistor. When the thermoelectric module was attached to the hotplate, a voltage signals were generated and as the unit was subsequently detached from the hotplate, the voltage dropped (Figure 8.12b). Thus, the hotplate temperature acted as signal stimulus. It was observed that at 40°C the device did not react to the threshold switch, so the system was not triggered to generate the alarm outputs (Figure 8.12b). However, as the temperature reached 50°C, it was observed that the threshold external thermos-stimuli triggered the nociceptor reaction. The temperature above 50°C triggered the memristor with the voltages about 0.25 V to 0.30 V, which was similar to the measured threshold voltage of memristor as shown previously in Figure 8.10b. It was found that the threshold voltage of nociceptor device can be modulated by changing Ag concentration in the interlayer film of memristor [107]. It consequently facilitated the modulation of diffusive device in which the sensitivity of nociceptor to the thermal stimulus can be adjusted by the alteration of Ag content in the SiO_2 resistive layer of memristor. In doing so, the device can trigger alarm signals that are similar to the human sensory system when the temperature goes above 37°C.

8.3.3 ARTIFICIAL NOCICEPTORS BASED ON ULTRATHIN HfO_2 MEMRISTORS

The memristor devices based on ultrathin heterostructured oxide material has recently attracted the tremendous attention of researcher for application in resistive memory-based technology. The charge trapping mechanisms in ultrathin 2D oxide nanostructures are basically dependent on the availability and the number of trapping sites in oxide structure [1, 128]. The charge trapping-detrapping phenomenon is the main proposed mechanism for resistive switching behavior of oxide films. The oxygen vacancies are the main trapping sites in ultrathin oxide films. The distribution and number of oxygen vacancies highly affect the resistive behavior of oxide-based memristor devices. Furthermore, the charge transfer control is facilitated by the heterointerfaces engineering which can cover various type of strategies [129, 130]. The heterostructured oxide films can effectively play the role of resistive component of memristor device, where the interface between two ultrathin oxide layers is one of the main trapping sites of charge carriers. By the same strategy, the heterointerfaces between the resistive component and conducting electrodes can be controlled and engineered to provide the trapping sites at heterointerfaces or alter the Schottky height barrier between resistive component and conducting electrodes.

The ultrathin oxide films are extensively employed in synaptic devices for spike neural network devices which originates from the analog-like resistive switching functionality of oxide memristors. The threshold switching functionalities of ultrathin memristor devices has facilitated the nociceptive characteristics of these novel nanostructured based devices [128]. However, the synaptic and neural characteristics respond to the requirement of signal handling, transfer, and storage. The biocompatibility of heterostructured memristor devices is highly valuable from technical point of view for application in humanoid and robotic devices. Four critical nociceptive

Sensorimotor Devices Based on 2D Materials

characteristics, including, threshold, relaxation, allodynia, and hyperalgesia were reproduced by using a memristor device based on Pt/HfO$_2$/TiN heterostructured ultrathin 2D films [128]. The resistive switching mechanism in Pt/HfO$_2$/TiN device is examined by monitoring the *I-V* characteristics of device. The cyclic *I-V* curves of Pt/HfO$_2$/TiN devices at different compliance currents show that the sweeping curves followed the same line and overlapped (Figure 8.13a) [128]. It confirmed the threshold switching (TS) behavior of Pt/HfO$_2$/TiN memristor device when the compliance current was kept as low as Nano-ampere range [128]. By increasing the compliance current to micro ampere range (Figure 8.13b), the nonvolatile resistive switching behavior was observed confirming the switching mechanism was changed from TS to RS in Pt/HfO$_2$/TiN memristor device. At the negative bias voltages, the conductive states returned to the initial high resistance mode. The similar resistive switching behavior was already observed and reported during the study of resistive performance of all-metal oxide heterostructured Pt/Ta$_2$O$_5$/HfO$_2$/TiN device [131, 132] and Pt/NbO$_x$/TiO$_y$/NbO$_x$/TiN memristor device [133]. The switching mechanisms of the mentioned ultrathin heterostructured devices were explained by the trapping-induced electronic bipolar resistive switching mechanism [131–133]. The charge trapping mechanism and dual switching behavior of memristor device i.e.

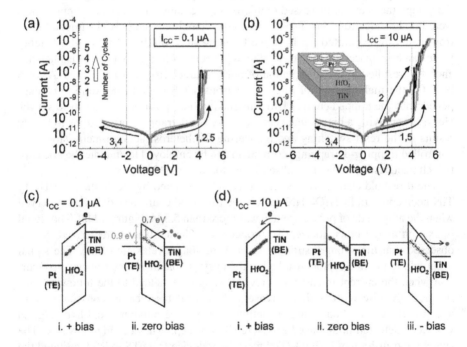

FIGURE 8.13 The resistance-switching behavior of PHT device. (a) The *I-V* curve obtained at the compliance current (I_{cc}) of 0.1 μA and (b) at 10 μA. (c) The graphical scheme of band alignment of devices at (c) 0.1 μA and (d) at 10 μA I_{cc} condition. Reprinted with permission from Kim, Y., Kwon, Y. J., Kwon, E. D., Yoon, K. J., Yoon, J. H., Yoo, S., Kim, H. J., Park, T. H., Han, J. W., Kim, K. M. & Hwang, C. S. (2018). Nociceptive memristor. *Advanced Materials*, 30, 1704320).

threshold and resistive switching in $Pt/HfO_2/TiN$ device can be explained by the electron trapping/detrapping phenomenon with respect to the external bias voltage [128]. The measured and calculated trap level in HfO_2 nanofilm was 0.7 eV below the conduction band edge [128]. The trap level corresponds to the energy level of the charged oxygen vacancies in the nanostructured 2D HfO_2 film. The switching mechanism of heterointerfaced structure is explained based on the energy level of heterostructured components. By imposing sufficiently enough applied voltage on Pt electrode in $Pt/HfO_2/TiN$ device, the trap level in HfO_2 film was pulled down to the below the Fermi energy level of TiN electrode (Figure 8.13c). Consequently, the trapping sites in HfO_2 can be filled by the injection of electrons from TiN electrode into ultrathin film of HfO_2. When the trap sites are completely filled, the electrons suddenly flow through the HfO_2 film by the trap-assisted tunneling mechanism, thus the resistance value decreased and device switched to low resistive switching state [128]. By removing the applied bias voltage on Pt electrode, the energy level of trapped sites moved upward and went higher than the Fermi energy level of TiN electrode. An internal electric field is induced by the difference between the work function of Pt and TiN, even under zero bias voltage. (Figure 8.13c). The combination of these parameters resulted in the detrapping of electrons from ultrathin 2D HfO_2 films, consequently the device returned to its initial high resistive switching state. When the compliance current increased to micro ampere range, not only the trapping sites were filled by electrons, but also due to higher compliance current, the trapped electrons can still be emitted out from the trapping sites in HfO_2 film and their empty place can be replaced by the other following electrons (Figure 8.13d). This phenomenon has resulted in the observation of noisy characteristics in I-V curves. At zero bias, the thermal emission of trapped electrons needs quite longer time to be happened, thus the trapped charges still remained in their positions and decayed slowly. However, applying a bias voltage can intensify the detrapping of electrons to finally results in the reset switching and recovering the heterostructured memristor. This controlled trapping-detrapping mechanism can be employed to emulate the nociceptive characteristics of the natural sensorimotor system [128].

The threshold characteristics were precisely emulated by the artificial $Pt/HfO_2/TiN$ nociceptor. In $Pt/HfO_2/TiN$ device, the threshold characteristics were observed when the amplitude of pulsed voltage was more than 5.5 V (Figure 8.14a). The signal relaxation as the other nociceptive properties of device was characterized. It was observed that the stronger stimulus caused the longer relaxation time. The signal accumulation in nociceptors resulted in the unstoppable increase of nociceptive current when the current intensity of previous signal was added to the following one. It especially observed that the decrease of delay time between the consecutive voltage stimuli has resulted in higher output current. The allodynia and hyperalgesia were also detected in damaged $Pt/HfO_2/TiN$ nociceptor device (Figure 8.14b). The switching transition of $Pt/HfO_2/TiN$ memristor device from TS to RS facilitated the reproduction of allodynia and hyperalgesia. To emulate the reflex action of the nervous system a circuit system was designed where the $Pt/HfO_2/TiN$ memristor works as the nociceptor component. Generally in the human body, when the nociceptors are subjected by noxious stimuli, the nociceptor responses are generated and transmitted to the spinal cord. After analyzing the informative hazardous signals, the motor

Sensorimotor Devices Based on 2D Materials

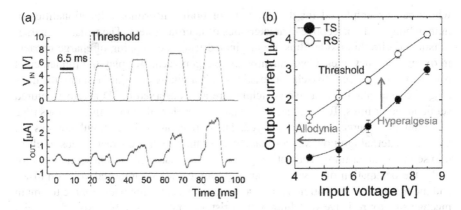

FIGURE 8.14 (a) The threshold characteristics for Pt/HfO$_2$/TiN device. (b) The allodynia and hyperalgesia properties of Pt/HfO$_2$/TiN device. Reprinted and reproduced with permission from Kim, Y., Kwon, Y. J., Kwon, E. D., Yoon, K. J., Yoon, J. H., Yoo, S., Kim, H. J., Park, T. H., Han, J. W., Kim, K. M. & Hwang, C. S. (2018). Nociceptive memristor. *Advanced Materials*, *30*, 1704320).

neurons trigger the effectors to alleviate the effect of hazardous stimulus. A p-type enhancement mode field effect transistor (P-FET) was employed in reflex device. This reflex device represented the spinal cord system to amplify the response of the nociceptors. In reflex device, at the zero-gate voltage the P-FET device stayed *on-state*. When the gate voltage increased positively the P-FET device turned off. The observation of electrical characteristics of reflex device gave valuable information. A weak current response was detected when the imposed gate voltage of 4.5 V was applied on device. As it was observed the 4.5 V measured voltage, could not strongly trigger the reaction of memristive nociceptor device since the memristor still stayed in the high resistance state. The corresponding declined current for the imposed 4.5 V stimulus voltage was about 10–50 µA. By the termination of imposed voltage signals, the response current of P-FET returned to its initial states. However, when the amplitude of voltage stimulus exceeded the threshold of memristor nociceptors (5.5 V), the nociceptor showed immediate reaction to the stimulus and the P-FET current fell down to 2 µA. The response of device was similar to the reflex reaction of neuron system to the hazardous external stimuli. These observations confirmed the applicability of the ultrathin 2D heterostructured oxide films for the successful development of artificial nociceptor and sensorimotor systems.

CONCLUSION

The chapter investigates the application of ultrathin 2D materials in the Internet of Things technology. Mimicking the functionalities of human nervous system with their complexities is highly challenging indeed. Recent progress on the development of artificial bioinspired systems reinforces the idea of fabrication of humanoid robotic sensory systems by the employment of ultrathin 2D nanostructures. The memristive characteristics of ultrathin films facilitated outstanding novel properties which can be employed for development of both sensorimotor functionalities of human body and nociceptive characteristics of neuron system. The resistive characteristics of ultrathin oxide films are originated from the trapping of charge carriers in oxygen vacancy sites. Consequently, the oxygen vacancies play the fundamental role in evoking the resistive characteristics of ultrathin oxides. Furthermore, the trapping phenomenon and resistive switching are also observed in 2D heterostructured semiconductor films. To modulate the resistance level in ultrathin films, various type of stimulation sources can be employed. The optical and voltaic stimuli sources are the main external ignition sources that can alter the resistive characteristics of ultrathin semiconductor films. The optical generation of charge carriers in semiconductor body, the occupation of oxygen vacancy trapping sites by imposed voltaic pulses, and the filamentary formation of conduction channels in semiconductor are the main mechanisms for resistive switching of memristor components of bioinspired artificial receptors, nociceptors, and sensorimotors devices. The sensorimotor functionalities are facilitated by the integration of ultrathin 2D oxide or semiconductor films into the mechanical components of accelerators, where the generated informative signals of memristor and synaptic devices are employed to mimic the mechanical contraction and expansion of human muscle, or are used for optical pattern recognition and emulation of human optical visionary system. Artificial thermal, voltaic and optical nociceptors are the other main members of bioinspired 2D nanodevices that are designed and developed to modulate the nociceptive functionalities of human sensory system. Several successful cases of the development of sensorimotor devices were reported in this chapter, in which the main receptor and nociceptor components are ultrathin 2D semiconductor films. Consequently, it was demonstrated that the ultrathin 2D films of metal oxide and semiconductors have an undeniable potential to be employed as the main semiconductor component in various bioinspired devices.

REFERENCES

1. Karbalaei Akbari, M., & Zhuiykov, S. (2019). A bioinspired optoelectronically engineered artificial neurorobotics device with sensorimotor functionalities. *Nature Communications*, *10*(1), 3873.
2. Kaltenbrunner, M., Sekitani, T., Reeder, J., Yokota, T., Kuribara, K., Tokuhara, T., ... Someya, T. (2013). An ultra-lightweight design for imperceptible plastic electronics. *Nature*, *499*(7459), 458–463.
3. Tee, B. C. K., Chortos, A., Berndt, A., Nguyen, A. K., Tom, A., McGuire, A., ... Bao, Z. (2000). A skin-inspired organic digital mechanoreceptor. *Science*, *350*(6258), 313–316.
4. Castro, M. C. F., & Cliquet, A. (2000). Artificial sensorimotor integration in spinal cord injured subjects through neuromuscular and electrotactile stimulation. *Artificial Organs*, *24*(9), 710–717.
5. Desmurget, M., & Sirigu, A. (2015). Revealing humans' sensorimotor functions with electrical cortical stimulation. *Philosophical Transactions of the Royal Society of London Series B*, *370*(1677), 20140207.
6. Lipomi, D. J., Vosgueritchian, M., Tee, B. C. K., Hellstrom, S. L., Lee, J. A., Fox, C. H., & Bao, Z. (2011). Skin-like pressure and strain sensors based on transparent elastic films of carbon nanotubes. *Nature Nanotechnology*, *6*(12), 788–792.
7. Someya, T., Bao, Z., & Malliaras, G. G. (2016). The rise of plastic bioelectronics. *Nature*, *540*(7633), 379–385.

Sensorimotor Devices Based on 2D Materials

8. Sekitani, T., Yokota, T., Kuribara, K., Kaltenbrunner, M., Fukushima, T., Inoue, Y., ... Someya, T. (2016). Ultra-flexible organic amplifier with biocompatible gel electrodes. *Nature Communications, 7*, 11425.

9. Kim, Y., Chortos, A., Xu, W., Liu, Y., Oh, J. Y., Son, D., ... Lee, T. W. (2018). A bioinspired flexible organic artificial afferent nerve. *Science, 360*(6392), 998–1003.

10. Bouchard, K. E., Mesgarani, N., Johnson, K., & Chang, E. F. (2013). Functional organization of human sensorimotor cortex for speech articulation. *Nature, 495*(7441), 327–332.

11. White, L. E., Andrews, T. J., Hulette, C., Richards, A., Groelle, M., Paydarfar, J., & Purves, D. (1997). Structure of the human sensorimotor system. II: Lateral symmetry. *Cerebral Cortex, 7*(1), 31–47.

12. Jens Volkmann, M. D. (1998). Oscillations of the human sensorimotor system as revealed by magnetoencephalography. *Movement Disorders, 13*, 73–76.

13. Miller, L. E., Montroni, L., Koun, E., Salemme, R., Hayward, V., & Farnè, A. (2018). Sensing with tools extends somatosensory processing beyond the body. *Nature, 561*(7722), 239–242.

14. Khojasteh, B., Janko, M., & Visell, Y. (2018). Complexity, rate, and scale in sliding friction dynamics between a finger and textured surface. *Scientific Reports, 8*(1), 13710.

15. Petrini, F. M., Bumbasirevic, M., Valle, G., Ilic, V., Mijović, P., Čvančara, P., ... Raspopovic, S. (2019). Sensory feedback restoration in leg amputees improves walking speed, metabolic cost and phantom pain. *Nature Medicine, 25*(9), 1356–1363.

16. Susman, L., Brenner, N., & Barak, O. (2019). Stable memory with unstable synapses. *Nature Communications, 10*(1), 4441.

17. Alcamí, P., & Pereda, A. E. (2019). Beyond plasticity: The dynamic impact of electrical synapses on neural circuits. *Nature Reviews: Neuroscience, 20*(5), 253–271.

18. Zador, A. M. (2000). The basic unit of computation. *Nature Neuroscience, 3*, 1167.

19. Ohtaka-Maruyama, C., Okamoto, M., Endo, K., Oshima, M., Kaneko, N., Yura, K., ... Maeda, N. (2019). Synaptic transmission from subplate neurons controls radial migration of neocortical neurons. *Science, 360*(6386), 313–317.

20. Xu, W., Min, S. Y., Hwang, H., & Lee, T. W. (2016). Organic core-sheath nanowire artificial synapses with femtojoule energy consumption. *Science Advances, 2*(6), e1501326.

21. Zhu, Z., & Lu, W. D. (2018). Optogenetics-inspired tunable synaptic functions in memristors. *ACS Nano, 12*(2), 1242–1249.

22. Qin, S., Wang, F., Liu, Y., Wan, Q., Wang, X., Xu, Y., ... Zhang, R. (2017). A lightstimulated synaptic device based on graphene hybrid phototransistor. *2D Materials, 4*(3), 035022.

23. Walters, B. J., Hallengren, J. J., Theile, C. S., Ploegh, H. L., Wilson, S. M., & Dobrunz, L. E. (2014). A catalytic independent function of the deubiquitinating enzyme USP14 regulates hippocampal synaptic short-term plasticity and vesicle number. *Journal of Physiology, 592*(4), 571–586.

24. Kumar, M., Kim, J., & Wong, C. P. (2019). Transparent and flexible photonic artificial synapse with piezo-phototronic modulator: Versatile memory capability and higher order learning algorithm. *Nano Energy, 63*, 103843.

25. Zhu, X., & Lu, W. D. (2018). Optogenetics-inspired tunable synaptic functions in memristors. *ACS Nano, 12*(2), 1242–1249.

26. Yu, S., Wu, Y., Jeyasingh, Kuzum, D., & Wong, H. S. (2011). An electronic synapse device based on metal oxide resistive switching memory for neuromorphic computation. *IEEE Transactions, 8*, 2729–2737.

27. Wang, Z., Joshi, S., Savelev, S. E., Jiang, H., Midya, R., Lin, P., ... Yang, J. J. (2017). Memristors with diffusive dynamics as synaptic emulators for neuromorphic computing. *Nature Materials, 16*(1), 101–108.

28. Saha, H. N., Auddy, S., Pal, S., Kumar, S., Subhadeep, J., Singh, R., Singh, R., Banerjee, S., Sharan, P., & Maity, A. (2017). Internet of things (IoT) on bio-technology. *8th Annual Industrial Automation and Electromechanical Engineering Conference (IEMECON)*. 16-18 Aug. 2017. Bangkok, Thailand. doi: 10.1109/IEMECON.2017.8079624

29. Tian, H., Zhao, L., Wang, X., Yeh, Y. W., Yao, N., Rand, B. P., & Ren, T. L. (2017). Extremely low operating current resistive memory based on exfoliated 2D perovskite single crystals for neuromorphic computing. *ACS Nano, 11*(12), 12247–12256.

30. Zhang, H., Yoo, S., Menzel, S., Funck, C., Cüppers, F., Wouters, D. J., ... Hoffmann-Eifert, S. (2018). Understanding the coexistence of two bipolar resistive switching modes with opposite polarity in Pt/TiO$_2$/Ti/Pt nanosized ReRam devices. *ACS Applied Materials and Interfaces, 10*(35), 29766–29778.

31. Yuan, J., & Lou, J. (2015). Memristor goes two-dimensional. *Nature Nanotechnology, 10*(5), 389–390.

32. Zhang, L., Gong, T., Wang, H., Guo, Z., & Zhang, H. (2019). Memristive devices based on emerging two-dimensional materials beyond graphene. *Nanoscale, 11*(26), 12413.

33. Lee, Y., Oh, J. Y., Xu, W., Kim, O., Kim, T. R., Kang, J., ... Lee, T. W. (2018). Stretchable organic optoelectronic sensorimotor synapse. *Science Advances, 4*(11), eaat7387.

34. Bryson, J. B., Machado, C. B., Crossley, M. D., Stevenson, D., Bros-Facer, V., Burrone, J., ... Lieberam, I. (2014). Optical control of muscle function by transplantation of stem cell–derived motor neurons in mice. *Science, 344*(6179), 94–97.

35. Llewellyn, M. E., Thompson, K. R., Deisseroth, K., & Delp, S. L. (2010). Orderly recruitment of motor units under optical control in vivo. *Nature Medicine, 16*(10), 1161–1165.

36. Magown, P., Shettar, B., Zhang, Y., & Rafuse, V. F. (2015). Direct optical activation of skeletal muscle fibres efficiently controls muscle contraction and attenuates denervation atrophy. *Nature Communications, 6*, 8506.

37. Paul, A. K., & Sato, T. (2017). Localization in wireless sensor networks: A survey on algorithms, measurement techniques, applications and challenges. *Journal of Sensor and Actuator Networks, 6*(4), 24.

38. Vossiek, M., Wiebking, L., Gulden, P., Wieghardt, J., Hoffmann, C., & Heide, P. (2003). Wireless local positioning. *IEEE Microwave Magazine, 4*(4), 77–86.

39. Al-Karaki, J. N., & Kamal, A. E. (2004). Routing techniques in wireless sensor networks: A survey. *IEEE Wireless Communications, 11*(6), 6–28.

40. Sangwan, V. K., Jariwala, D., Kim, I. S., Chen, K. S., Marks, T. J., Lauhon, L. J., & Hersam, M. C. (2015). Gate-tunable memristive phenomena mediated by grain boundaries in single-layer MoS$_2$. *Nature Nanotechnology, 10*(5), 403–406.

41. Zhou, F., Chen, J., Tao, X., Wang, X., & Chai, Y. (2019). 2D Materials based optoelectronic memory: Convergence of electronic memory and optical sensor. *Research, 2019*, 9490413.

42. Yan, X., Wang, K., Zhao, J., Zhou, Z., Wang, H., Wang, J., ... Zhou, P. (2019). A new memristor with 2D Ti$_3$C$_2$T$_x$ MXene flakes as an artificial bio-synapse. *Small, 15*(25), 1900107.

43. Bruegmann, T., Van Bremen, T., Christoph, T., Vogt, C., Fleischmann, B. K., & Sasse, P. (2015). Optogenetic control of contractile function in skeletal muscle. *Nature Communications, 6*, 7153.

44. Magown, P., Shettar, B., Zhang, Y., & Rafuse, V. F. (2015). Direct optical activation of skeletal muscle fibres efficiently controls muscle contraction and attenuates denervation atrophy. *Nature Communications, 6*, 8506.

45. Srinivasan, S. S., Maimon, B. E., Diaz, M., Song, H., & Herr, H. M. (2018). Closed-loop functional optogenetic stimulation. *Nature Communications, 9*(1), 5303.

46. Vajtay, T. J., Bandi, A., Upadhyay, A., Swerdel, M. R., Hart, R. P., Lee, C. R., & Margolis, D. J. (2019). Optogenetic and transcriptomic interrogation of enhanced muscle function in the paralyzed mouse whisker pad. *Journal of Neurophysiology, 121*(4), 1491–1500.

Sensorimotor Devices Based on 2D Materials 303

47. Bryson, J. B., Machado, C. B., Crossley, M., Stevenson, D., Bros-Facer, V., Burrone, J., ... Lieberam, I. (2014). Optical control of muscle function by transplantation of stem cell-derived motor neurons in mice. *Science, 344*(6179), 94–97.
48. Llewellyn, M. E., Thompson, K. R., Deisseroth, K., & Delp, S. L. (2010). Orderly recruitment of motor units under optical control in vivo. *Nature Medicine, 16*(10), 1161–1165.
49. Magown, P., Shettar, B., Zhang, Y., & Rafuse, V. F. (2015). Direct optical activation of skeletal muscle fibres efficiently controls muscle contraction and attenuates denervation atrophy. *Nature Communications, 6*, 8506.
50. Thompson, A. C., Stoddart, P. R., & Jansen, E. D. (2014). Optical stimulation of neurons. *Journal of Neuroscience, 34*, 7704–7714.
51. Go, M. A., & Daria, V. R. (2017). Light-neuron interactions: Key to understanding the brain. *Journal of Optics, 19*(2), 023002.
52. Lee, E., Yoon, Y. S., & Kim, D. J. (2018). Two-dimensional transition metal dichalcogenides and metal oxide hybrids for gas sensing. *ACS Sensors, 310*(10), 2045–2060.
53. Dral, A. P., and Elshof, J. E. (2018). 2D metal oxide nanoflakes for sensing applications: Review and perspective. *Sensors and Actuators. Part B, 272*, 369–392.
54. Karbalaei Akbari, M., Hai, Z., Depuydt, S., Kats, E., Hu, J., & Zhuiykov, S. (2017). Highly sensitive, fast-responding, and stable photodetector based on ALD-developed monolayer TiO_2. *IEEE Transactions on Nanotechnology, 16*(5), 880–887.
55. Shavanova, K., Bakakina, Y., Burkova, I., Shtepliuk, I., Viter, R., Ubelis, A., ... Khranovskyy, V. (2016). Application of 2D non-graphene materials and 2D oxide nanostructures for biosensing technology. *Sensors, 16*(2), 223.
56. Kalantar-Zadeh, K., Tang, J., Daeneke, T., O'Mullane, A. P., Stewart, L. A., Liu, J., ... Dickey, M. D. (2019). Emergence of liquid metals in nanotechnology. *ACS Nano, 137*(7), 7388–7395.
57. Majidi, L., Gritsenko, D., & Xu, J. (2017). Gallium-based room-temperature liquid metals: Actuation and manipulation of droplets and flows. *Frontiers in Mechanical Engineering, 3*, 9.
58. Tang, S. Y., Khoshmanesh, K., Sivan, V., Petersen, P., Mullane, A. P. O., Abbott, D., Mitchell, A., & Kalantar-zadeh, K. (2014). Liquid metal enables pump, *111*, 3304–3309.
59. Tang, S. Y., Sivan, V., Petersen, P., Zhang, W., Morrison, P. D., Kalantar-zadeh, K., ... Khoshmanesh, K. (2014). Liquid metal actuator for inducing chaotic advection. *Advanced Functional Materials, 24*(37), 5851–5858.
60. Khan, M. R., Trlica, C., & Dickey, M. D. (2015). Recapillarity: Electrochemically controlled capillary withdrawal pf a liquid metal alloy from microchannel. *Advanced Functional Materials, 25*(5), 671–678.
61. Sawa, A. (2008). Resistive switching in transition metal oxides. *Materials Today, 11*(6), 28–36.
62. Nowotny, J. (2012). Basic properties of TiO_2. In: Nowotny, J. *Oxide semiconductors for solar energy conversion* (pp. 145–163). Boca Raton, FL: CRC Press.
63. Nowotny, J. (2012). Applications. In: Nowotny, J. *Oxide semiconductors for solar energy conversion* (pp. 323–385). Boca Raton, FL: CRC Press.
64. Scanlon, D. O., Dunnill, C. W., Buckeridge, J., Shevlin, S. A., Logsdail, A. J., Woodley, S. M., ... Sokol, A. A. (2013). Band alignment of rutile and anatase TiO_2. *Nature Materials, 12*(9), 798–801.
65. Evtushenko, Y. M., Romashkin, S. V., Trofimov, N. S., & Chekhlova, T. K. (2015). Optical properties of TiO_2 thin films. *Physics Procedia, 73*, 100–107.
66. Shang, L., Zhang, W., Xu, K., & Zhao, Y. (2019). Bio-inspired intelligent structural color materials. *Materials Horizons, 6*(5), 945–958.
67. Tadepalli, S., Slocik, J. M., Gupta, M. K., Naik, R. R., & Singamaneni, S. (2017). Bio-optics and bioinspired optical materials. *Chemical Reviews, 117*(20), 12705–12763.
68. Liu, Z., Leow, W. R., & Chen, X. (2018). Bio-inspired plasmonic photocatalysts. *Small, 3*, 1800295.

69. Ge, C. H., Li, H. L., Zhu, X. L., & Pan, A. L. (2017). Band gap engineering of atomically thin two-dimensional semiconductors. *Chinese Physics. Part B, 26*(3), 034208.
70. Yang, C. S., Shang, D. H., Liu, N., Fuller, E. J., Agrawal, S., Talin, A. A., ... Sun, Y. (2018). All-solid state synaptic transistor with ultralow conductance for neuromorphic computing. *Advanced Functional Materials, 28*(42), 1804170.
71. Yang, C. S., Shang, D. H., Liu, N., Shi, G., Shen, X., Yu, R. C., Li, Y. Q., & Sun, Y. (2017). A synaptic transistor based on quasi-2D molybdenum oxide. *Advanced Materials, 29*, 1700906.
72. Khan, M., Lan, Z., & Zeng, Y. (2018). Analysis of indium oxidation state on the electronic structure and optical properties of TiO_2. *Materials, 11*(6), 952.
73. Berengue, O. M., Rodrigues, A. D., Dalmaschio, C. J., Lanfredi, A. J. C., Leite, E. R., & Chiquito, A. J. (2010). Structural characterization of indium oxide nanostructures: A Raman analysis. *Journal of Physics. Part D, 43*(4), 045401.
74. Nowotny, J., Bak, T., & Alim, M. A. (2015). Dual mechanism of indium incorporation into TiO_2 (Rutile). *Journal of Physical Chemistry C, 119*(2), 1146–1154.
75. Qian, L., Du, Z. L., Yang, S. Y., & Jin, Z. S. (2005). Raman study of titania nanotube bysoft chemical process. *Journal of Molecular Structure, 749*(1–3), 103–107.
76. Abbott, L. F., & Nelson, S. B. (2000). Synaptic plasticity: Taming the beast. *Nature Neuroscience, 3*, 1178–1183.
77. Citri, A., & Malenka, R. C. (2008). Synaptic plasticity: Multiple forms, functions, and mechanisms. *Neuropsychopharmacology, 33*(1), 18–41.
78. Goush, R. C., Dang, J. H., Moorefield, M. R., Zhang, G. B., Hihara, L. H., Shiroma, W. A., & Ohta, A. T. (2015). Self-actuation of liquid metal via redox reaction. *ACS Applied Materials and Interfaces, 8*(1), 6–10.
79. Yi, L., Ding, Y., Yuan, B., Wang, L., Tian, L., Chen, C., ... Liu, J. (2016). Breathing to harvest energy as a mechanism towards making a liquid metal beating heart. *RSC Advances, 6*(97), 94692–94698.
80. Tang, S. Y., Sivan, V., Khoshmanesh, K., O'Mullane, A. P., Tang, X., Gol, B., ... Kalantar-zadeh, K. (2013). Electrochemically induced actuation of liquid metal marbles. *Nanoscale, 5*(13), 5949–5957.
81. Seo, S., Jo, S. H., Kim, S., Shim, J., Oh, S., Kim, J. H., ... Park, J. H. (2018). Artificial optic-neural synapse for colored and color-mixed pattern recognition. *Nature Communications, 9*(1), 5106.
82. Querlioz, D., Bichler, O., Dollfus, P., & Gamrat, C. (2013). Immunity to device variations in a spiking neural network with memristive nanodevices *IEEE Trans., 12*(3), 288–295.
83. Head, S. I., & Arber, M. B. (2013). An active learning mammalian skeletal muscle lab demonstrating contractile and kinetic properties of fast- and slow-twitch muscle. *Advances in Physiology Education, 37*(4), 405–414.
84. Cleworth, D., & Edman, K. A. P. (1969). Laser diffraction studies on single skeletal muscle fibers. *Science, 163*(3864), 296–298.
85. Kim, O., Kim, H., Choi, U. H., & Park, M. J. (2016). One-volt-driven superfast polymer actuators based on single-ion conductors. *Nature Communications, 7*, 13576.
86. Yang, X., Zhou, T., Zwang, T. J., Hong, G., Zhao, Y., Viveros, R. D., ... Lieber, C. M. (2019). Bioinspired neuron-like electronics. *Nature Materials, 18*(5), 510–517.
87. Le Feuvre, R. A., & Scrutton, N. S. (2018). A living foundry for synthetic biological materials: A synthetic biology roadmap to new advanced materials. *Synthetic and Systems Biotechnology, 3*(2), 105–112.
88. Jung, Y. H., Park, B., Kim, J. U., & Kim, T. I. (2018). Bioinspired electronics for artificial sensory systems. *Advanced Materials, 31*, 1803637.

Sensorimotor Devices Based on 2D Materials

89. Fujii, H., Setiadi, A., Kuwahara, Y., & Akai-Kasaya, M. (2017). Single walled carbon nanotube-based stochastic resonance device with molecular self-noise source. *Applied Physics Letters, 111*(13), 133501.

90. Yu, J., Horsley, J. R., & Abell, A. D. (2018). Peptides as bioinspired electronic materials: An electrochemical and first-principles perspective. *Accounts of Chemical Research, 519*(9), 2237–2246.

91. Gold, M. S., & Gebhart, G. F. (2010). Nociceptor sensitization in pain pathogenesis. *Nature Medicine, 16*(11), 1248–1257.

92. Holmes, D. (2016). The pain drain. *Nature, 535*(7611), S2–S3.

93. Holmes, D. (2018). Reconstructing the retina. *Nature, 561*(7721), S2–S3.

94. Gilbert, C. D., & Li, W. (2013). Top-down influences on visual processing. *Nature Reviews: Neuroscience, 14*(5), 350–363.

95. Dubin, A. E., & Patapoutian, A. (2010). Nociceptors: The sensors of the pain pathway. *Journal of Clinical Investigation, 120*(11), 3760–3772.

96. Handwerker, H. O. (2009). Nociceptors and characteristics. In M. D. Binder, N. Hirokawa & U. Windhorst (Eds.), *Encyclopedia of neuroscience.* Berlin, Heidelberg: Springer.

97. Gold, M. S., & Gebhart, G. F. (2010). Nociceptor sensitization in pain pathogenesis. *Nature Medicine, 16*(11), 1248–1257.

98. Belmonte, C. (2013). Ocular nociceptors. In G. F. Gebhart & R. F. Schmidt (Eds.), *Encyclopedia of pain.* Berlin, Heidelberg: Springer.

99. Belmonte, C. (1996). Signal transduction in nociceptors: General principles. In C. Belmonte & F. Cervero (Eds.), *Neurobiology of nociceptors* (pp. 243–257). Oxford: Oxford University Press.

100. Gilbert, C. D., & Li, W. (2013). Top-down influences on visual processing. *Nature Reviews: Neuroscience, 14*(5), 350–363.

101. Jensen, T. S., & Finnerup, N. B. (2014). Allodynia and hyperalgesia in neuropathic pain: Clinical manifestations and mechanisms. *Lancet Neurology, 13*(9), 924–935.

102. Wang, Z., Dong, C., Wang, X., Li, M., Nan, T., Liang, X., ... Sun, N. (2018). Highly sensitive integrated flexible tactile sensors with piezoresistive $Ge_2 Sb_2 Te_5$ thin films. *npj Flexible Electronics, 2*(1), 17.

103. Dahiya, R. S., Metta, G., Valle, M., & Sandini, G. (2010). Tactile sensing-from humans to humanoids. *IEEE Transactions on Robotics, 26*(1), 1–20.

104. Dahiya, R. S., Cattin, D., Adami, A., Collini, C., Barboni, L., Valle, M., ... Brunetti, F. (2011). Towards tactile sensing system on chip for robotic applications. *IEEE Sensors Journal, 11*(12), 3216–3226.

105. Hossain, M. S., Al-Dirini, F., Hossain, F. M., & Skafidas, E. (2015). High performance graphene nano-ribbon thermoelectric devices by incorporation and dimensional tuning of nanopores. *Scientific Reports, 5*, 11297.

106. Sadek, A. S., Karabalin, R. B., Du, J., Roukes, M. L., Koch, C., & Masmanidis, S. C. (2010). Wiring nanoscale biosensors with piezoelectric nanomechanical resonators. *Nano Letters, 10*(5), 1769–1773.

107. Yoon, J. H., Wang, Z., Kim, K. M., Wu, H., Ravichandran, V., Xia, Q., ... Yang, J. J. (2018). An artificial nociceptor based on a diffusive memristor. *Nature Communications, 9*(1), 417.

108. Strukov, D. B., Snider, G. S., Stewart, D. R., & Williams, R. S. (2008). The missing memristor found. *Nature, 453*(7191), 80–83.

109. Wedig, A., Luebben, M., Cho, D. Y., Moors, M., Skaja, K., Rana, V., ... Valov, I. (2006). Nanoscale cation motion in TaO_x, HfO_x and TiO_x memristive systems. *Nature Nanotechnology, 11*(1), 67–74.

110. Yang, J. J., Pickett, M. D., Li, X., Ohlberg, D. A., Stewart, D. R., & Williams, R. S. (2008). Memristive switching mechanism for metal/oxide/metal nanodevices. *Nature Nanotechnology*, *3*(7), 429–433.

111. Lee, M. J., Lee, C. B., Lee, D., Lee, S. R., Chang, M., Hur, J. H., … Kim, K. (2011). A fast, high-endurance and scalable non-volatile memory device made from asymmetric Ta_2O_{5-x}/TaO_{2-x} bilayer structures. *Nature Materials*, *10*(8), 625–630.

112. Waser, R., Dittmann, R., Staikov, G., & Szot, K. (2009). Redox-based resistive switching memories–nanoionic mechanisms, prospects, and challenges. *Advanced Materials*, *21*(25–26), 2632–2663.

113. Kwon, D. H., Kim, K. M., Jang, J. H., Jeon, J. M., Lee, M. H., Kim, G. H., … Hwang, C. S. (2010). Atomic structure of conducting nanofilaments in TiO_2 resistive switching memory. *Nature Nanotechnology*, *5*(2), 148–153.

114. Yoon, J. H., Song, S. J., Yoo, I. H., Seok, J. Y., Yoon, K. J., Kwon, D. E., … Hwang, C. S. (2014). Highly uniform, electroforming-free, and self-rectifying resistive memory in the $Pt/Ta_2O_5/HfO_{2-x}/TiN$ structure. *Advanced Functional Materials*, *24*(32), 5086–5095.

115. Valov, I., Waser, R., Jameson, J. R., & Kozicki, M. N. (2011). Electrochemical metallization memories fundamentals, applications, prospects. *Nanotechnology*, *22*(25), 254003.

116. Prezioso, M., Bayat, F. M., Hoskins, B. D., Adam, G. C., Likharev, K. K., & Strukov, D. B. (2015). Training and operation of an integrated neuromorphic network based on metal-oxide memristors. *Nature*, *521*(7550), 61–64.

117. Chakrabarti, B., Lastras-Montaño, M. A., Adam, G., Prezioso, M., Hoskins, B., Payvand, M., … Strukov, D. B. (2017). A multiply-add engine with monolithically integrated 3D memristor crossbar/CMOS hybrid circuit. *Scientific Reports*, *7*, 42429.

118. Gaba, S., Sheridan, P., Zhou, J., Choi, S., & Lu, W. (2013). Stochastic memristive devices for computing and neuromorphic applications. *Nanoscale*, *5*(13), 5872–5878.

119. Yu, S., Wu, Y., Jeyasingh, R., Kuzum, D., & Wong, H. S. P. (2011). An electronic synapse device based on metal oxide resistive switching memory for neuromorphic computation. *IEEE Transactions on Electron Devices*, *58*(8), 2729–2737.

120. Pi, S., Ghadiri-Sadrabadi, M., Bardin, J. C., & Xia, Q. (2015). Nanoscale memristive radiofrequency switches. *Nature Communications*, *6*, 7519.

121. Cain, D. M., Khasabov, S. G., & Simone, D. A. (2001). Response properties of mechanoreceptors and nociceptors in mouse glabrous skin: An in vivo study. *Journal of Neurophysiology*, *85*(4), 1561–1574.

122. Neugebauer, V., & Li, W. (2002). Processing of nociceptive mechanical and thermal information in central amygdala neurons with knee-joint input. *Journal of Neurophysiology*, *87*(1), 103–112.

123. Yang, Y., Gao, P., Gaba, S., Chang, T., Pan, X., & Lu, W. (2012). Observation of conducting filament growth in nanoscale resistive memories. *Nature Communications*, *3*, 732.

124. Yang, Y., Gao, P., Li, L., Pan, X., Tappertzhofen, S., Choi, S., … Lu, W. D. (2014). Electrochemical dynamics of nanoscale metallic inclusions in dielectrics. *Nature Communications*, *5*, 4232.

125. Hsiung, C., Liao, H. W., Gan, J. Y., Wu, T. B., Hwang, J. C., Chen, F., & Tsai, M. J. (2010). Formation and instability of silver nanofilaments in Ag-based programmable metallization cells. *ACS Nano*, *4*(9), 5414–5420.

126. Song, M., Kwon, K., & Park, J. (2017). Electro-forming and electro-breaking of nanoscale Ag filaments for conductive-bridging random-access memory cell using Ag-doped polymer-electrolyte between Pt electrodes. *Scientific Reports*, *7*(1), 3065.

127. Kumar, M., Kim, H. S., & Kim, J. (2019). A highly transparent artificial nociceptor. *Advanced Materials*, *31*, 1900021.

128. Kim, Y., Kwon, Y. J., Kwon, E. D., Yoon, K. J., Yoon, J. H., Yoo, S., Kim, H. J., Park, T. H., Han, J. W., Kim, K. M. & Hwang, C. S. (2018). Nociceptive memristor. *Advanced Materials*, *30*, 1704320.
129. Wang, M., Cai, S., Pan, C., Wang, C., Lian, X., Zhuo, Y., ... Miao, F. (2018). Robust memristors based on layered two-dimensional materials. *Nature Electronics*, *1*(2), 130–136.
130. Ren, Y., Hu, L., Mao, J. Y., Yuan, J., Zeng, Y., Ruan, S., ... Han, S. (2018). Phosphorene nano-heterostructure based memristors with broadband response synaptic plasticity. *Journal of Materials Chemistry C*, *6*(35), 9383–9393.
131. Yoon, J. H., Song, S. J., Yoo, I. H., Seok, J. Y., Yoon, K. J., Kwon, D. E., ... Hwang, C. S. (2014). Highly uniform, electroforming-free, and self-rectifying resistive memory in the $Pt/Ta_2O_5/H_fO_{2-x}/TiN$ Structure. *Advanced Functional Materials*, *24*(32), 5086.
132. Yoon, J. H., Kim, K. M., Song, S. J., Seok, J. Y., Yoon, K. J., Kwon, D. E., ... Hwang, C. S. (2015). $Pt/Ta_2O_5/HfO_2-x/Ti$ resistive switching memory competing with multilevel NAND flash. *Advanced Materials*, *27*(25), 3811.
133. Kim, K. M., Zhang, J., Graves, V., Yang, J. J., Choi, B. J., Hwang, C. S., ... Williams, R. S. (2016). Low-power, self-rectifying, and forming-free memristor with an asymmetric programing voltage for a high-density crossbar application. *Nano Letters*, *16*(11), 6724.

Index

γ-AlOOH nanosheets, 92

1,2 ethanedithiol, 52, 53
1-ethyl-3-methylimidazolium bis-
    (trifluoromethanesulfonyl)-imide, 252
1H MoS_2/1T′$MoTe_2$ hetero-interfaces, 28
1T phase, 13, 155–156
2D/3D nano-array based devices, 147
2D black phosphorous, 266
2D composite materials, 183, 184
2D electronic characteristics, 145
2D graphene films, 8, 59, 115, 121, 236
2D graphene nanosheets, 152
2D graphene/TMDCs based heterostructures, 22
2D hetero-interfaces, 111–138
   band alignment, 113, 115
   charge transport, 116, 118
   devices based on, 131–138
      electronic, 131–133
      magnetic, 133–136
      spintronic and valleytronic, 137–138
   electrical contact, 127–131
      charge-injection mechanism, 130–131
      geometry of interfaces, 128–130
   homogenous junctions, 119–124
      doping and passivation, 122
      strain and dielectric modulation, 122–124
      structural properties, 119–121
   interlayer excitons generation, 118–119
   metal/semiconductor (MS) heterogeneous
      junctions, 125, 127
   overview, 111–112
   semiconductor/semiconductor, 124–125
2D heterostructured-based synapses, 260,
    262–263
2D heterostructured EuO/graphene films, 134
2D heterostructured MoS_2/PTCDA
    heterojunction, 264
2D heterostructured Sb_2Te_3/Bi_2Te_3 metal
    chalcogenides, 59
2D heterostructured WSe_2/EuS, 138
2D HfO_2 nanosheets, 90
2D layered crystals, 79, 101
2D magnetic materials, 135
2D materials, *see* Two-dimensional materials
2D metal chalcogenide films, ALD of, 49–60
   layered heterostructures, 59–60
   layered SnS and SnS_2 films, 56–59
   MoS_2, 50–53
   WS_2, 54–56

WSe_2, 56
2D metal oxide films, ALD of, 60–68
   Al_2O_3, 67–68
   MoO_3, 63–65
   TiO_2, 67
   WO_3, 65, 67
2D monolayer films, 121–123
2D MoS_2
   ALD of, 50–53
   doping of, 154–156
   film, 16, 20, 50, 52, 53, 118, 129, 183, 206,
      209, 210, 242, 244, 260, 265
   memristor based on, 204–210
      CVD MoS_2 2D film, 206–208
      for efficient-energy radio frequency,
         208–210
      oxidized MoS_2 nanosheets, 205–206
2D nano-materials bandgap, 121
2D nanostructures, 3, 43–45, 49, 71, 90, 92, 94,
    99, 101, 107, 214, 292, 299
2D oxide films, 7, 8, 63, 72, 80, 82, 90, 95, 99
2D perovskite synaptic devices, 245–247
2D plasmonic nanostructures, 166
2D post-transition metal chalcogenide, 80
2D semiconductor/metal hetero-interfaces, 129,
    131, 138
2D SiO_2 diffusive memristors
   artificial nociceptors based on, 290–293
   thermal nociceptors based on, 295–296
2D surface oxides of liquid metals, 79–107
   characteristics, 82–83
   monophasic and biphasic, 80–81
   overview, 79–80
   synthesis and applications, 83–107
      green synthesis method of ultrathin flux
         membranes, 92, 94–95
      reactive environment for, 88–90
      screen printing of Ga_2O_3 and GaS, 84
      semiconducting 2D gallium phosphate
         nanosheets, 97–99
      semiconducting GaN and InN nanosheets,
         99–101
      semiconducting SnO monolayers, 95, 97
      SnO/In_2O_3 2D van der Waals
         heterostructure, 101–102
      sonochemical-assisted functionalization
         of galinstan, 102–103, 105, 107
      wafer-scale screen printing of GaS, 85,
         87–88
2D TiO_2 semiconductor, 161

Index

2D transition metal dichalcogenides (TMDCs)
films, 4, 6, 9, 11–15, 22, 24, 28, 30, 31, 50, 59, 130, 138, 146, 163
on flexible substrates, 18, 20–21
on rigid substrates, 16–18
conventional techniques, 16, 18
metal-organic chemical vapor deposition (MOCVD), 16
post annealing process, 18
predeposition methods, 18
pulsed laser deposited (PLD) technique, 18
sulfurization/selenization of Mo film, 18
2D vertical memristor materials, 183
2D WO_3 film, 65, 67, 161
2D WS_2 single-crystal film, 10
2D WSe_2/h-BN/SiO_2/Si photodetector, 286, 287
2H phase, 13, 50, 156

AAO, *see* Anodic aluminum oxide
Acoustic cavitation, 156
AFM, *see* Antiferromagnetic states; Atomic force microscope
aGNR, *see* Armchair graphene nanoribbon
Al_2O_3 2D nanosheets, 94
Al_2O_3 gate electrode, 265
ALD, *see* Atomic layer deposition
Allodynia, 290, 293, 297, 298
AlN-GaN nanoribbons, 124
Ammonolysis method, 99–101
Ammonolysis reaction, 101
ANN, *see* Artificial neural network
Anodic aluminum oxide (AAO), 160
Antiferromagnetic (AFM) states, 135
Anti-solvent vapor assisted crystallization method, 202
Area-selective ALD, 45
Armchair graphene nanoribbon (aGNR), 120, 121
Artificial neural network (ANN), 233, 234
Artificial neurons, 230
Artificial neurotransmitter, 256
Artificial nociceptors
based on 2D SiO_2 diffusive memristors, 290–293
based on ultrathin HfO_2 memristors, 296–299
Artificial synaptic devices, 229–269
2D heterostructured-based, 260, 262–263
biological synapse *vs.*, 230–236
challenges of data processing, 229–230
design and structure, 236
electrochemical metallization/conductive bridge (ECM/CB), 249–250, 252
electro-iono-photoactive 2D MoS_2, 242, 244–245
In-ion doped ultrathin TiO_2 optical, 247–249
ionic transport in 2D perovskite, 245–247

Li-ions intercalated quasi-2D α-MoO_2, 255–257
optical RRAM synaptic devices, 264–268
optical synaptic devices *vs.* photodetectors, 263
phase-change, 257–259
polycrystalline 2D MoS_2, 241–242
proton intercalated quasi-2D α-MoO_2, 252–253, 255
thermochemical/joule heating, 259–260
As_2P_2 heterojunction, 124
Atomic force microscope (AFM), 51, 52, 95, 99, 100–102, 124, 151, 158, 196
Atomic layer deposition (ALD), 18, 43–73, 102, 184, 194, 196, 247
2D metal chalcogenide films, 49–60
layered heterostructures, 59–60
layered SnS and SnS_2 films, 56–59
MoS_2, 50–53
WS_2, 54–56
WSe_2, 56
2D metal oxide films, 60–68
Al_2O_3, 67–68
MoO_3, 63–65
TiO_2, 67
WO_3, 65, 67
overview, 43–45, 47
parameters, 47–49
precursor, 48–49
window, 47–48
Au–M–O interfaces, 7
Au/MoS_2/Au memristor device, 210
Au nanoparticles, 161, 217
Au/Ti/5–7-layer h-BN/Cu device, 250
Axon-multi-synaptic network, 262, 263

Band alignment, 22, 24, 102, 113–117, 121, 124, 125, 216, 242
Band bending, 113, 116, 216
Band gap, 1, 2, 11, 15, 22, 85, 95, 99, 101, 102, 113, 124, 127, 163, 235
Band-to-band alignment, 132
Band-to-band transition, 244
Band-to-band tunneling (BTBT), 113, 116
Beam lithography technique, 158
Bioinspired sensorimotor devices, 275–276, 300
Bio-microelectronic systems (bio-MEMS), 72
Bioneurotransmitters, 275
Bipolar RS mechanism, 175, 178, 191, 194, 199–201, 208–210, 212, 242, 259, 297
Bis(N,N'-diisopropylacetamidinato) tin (II) (Sn(amd)$_2$), 56
Bis(ter-butylimido) bis(dimethylamino) tungsten(VI) (^{t}BuN)$_2$W(NMe$_2$)$_2$, 65
Black–blue phosphorene lateral heterojunction, 121

Index

311

Black phosphene, 1
Black phosphorus (BP)-based synaptic
device, 266
Block polymer lithography method, 121
Boehmite, 92, 94
BP-based synaptic device, *see* Black phosphorus-
based synaptic device
Bridge-grain boundary memristor, 207
Broken gap, 113
BTBT, *see* Band-to-band tunneling

$C_{12}H_{30}N_4Mo$ precursor, 65
Carbon nanotube–based synapses, 234
Carbon nanotubes (CNTs), 129, 264, 289
Catalytic-assisted growth, 10
Catching, 230
CB, *see* Conduction band
CBM, *see* Conduction band minimum
Central processing unit (CPU), 229
CEW, *see* Continuous electrowetting
CFs, *see* Conductive filaments
Charge carrier mechanism, 163
Charge carrier mobility, 1, 9, 14, 15, 21, 32,
145, 175
Charge-injection mechanisms, 127, 130–131
Charge trapping/detrapping mechanisms, 116,
184, 214–217, 222, 236, 242, 263,
264, 268, 296–298
Chemical bonding, 7, 111, 131
Chemical doping, 116, 156, 253
Chemical reactions, 5, 44, 47, 49, 156
Chemical solution–based techniques, 184
Chemical vapor deposition (CVD), 1–33, 43, 102
development, 31
exploration of novel 2D materials, 31–32
growth, 9–15
doping, 13
grain size, 9–10
layer number, 10–12
morphology, 12
orientation, 12
phase, 12–13
quality and defects, 14–15
heterostructured growth, 21–28, 32
lateral 2D, 26–28
vertical 2D, 22–26
high-quality 2D films, 31
low-temperature 2D films, 32
overview, 1–4
parameters, 4–9
induced coupled plasma chemical vapor
deposition (ICP-CVD), 9
plasma-enhanced, 8
precursor, 5
pressure, 6
substrate, 6–8

temperature, 5–6
wafer-scale continuous growth, 15–21, 31
2D TMDCs films on flexible substrates,
18, 20–21
2D TMDCs films on rigid substrates,
16–18
Chemisorption, 47, 63
Classical conditioning, 244
CMOS, *see* Complementary metal oxide
semiconductor
CNTs, *see* Carbon nanotubes
Complementary metal oxide semiconductor
(CMOS), 131, 171, 179, 255, 256,
278, 290, 291
Complementary resistive switching (CRS),
187–194, 196–199
in heterostructured devices, 191–194
in In-doped TiO_2 ultrathin film, 196–199
in TaO_x, HFO_x, and TiO_x devices,
187–189, 191
Conduction band (CB), 24, 102, 113, 161, 162,
178, 179
Conduction band minimum (CBM), 22, 24, 113,
116, 118, 121, 125
Conductive bridge synapses, 234, 236, 250
Conductive channels, 173–175, 177, 178, 189,
194, 206, 207, 250, 252
Conductive electrodes, 174, 177, 182, 185, 212,
214, 233, 234
Conductive filaments (CFs), 183, 187, 194, 196,
199, 201–203, 207, 210, 246, 250, 259
Continuous electrowetting (CEW), 283
Corneal nociceptors, 290
Corrugated substrates, 123
Counter-eight- wise-switching mechanism, 189
CPU, *see* Central processing unit
CRS, *see* Complementary resistive switching
Cu/double-layer MoS_2/Ag device, 210
Cu/Ta_2O_5/Pt system, 181
CVD, *see* Chemical vapor deposition

Data processing, 171, 222, 229–230, 258
Data storage, 62, 172, 208, 215, 216, 220, 222,
229, 230
Data transfer traffic, 229
De-electrowetting (DEW), 283
DEMETFSI, *see* N,N-diethyl-N-(2-methoxy ethyl)
-N-methyl ammonium-bis-(triflu
oromethylsulfonyl)-imide
Dendritic growth, 9
Density functional theory (DFT), 67, 81
Deposition process, 6, 10, 13, 47, 65
DESe, *see* DiEthyl selenide
DEW, *see* De-electrowetting
DEZ, *see* DiEthylZinc
DFT, *see* Density functional theory

312 Index

DiEthyl selenide (DESe), 56
DiEthylZinc (DEZ), 54
Diffusive memristor, 291–295
DIGS, *see* Disorder induced gap states
Dip coating, 18, 159
Dip-pen lithography (DPN), 158
Direct printing technique, 158
Disk arrays, 151
Disorder induced gap states (DIGS), 115
Double-layer MoS_2 memristor, 210
DPN, *see* Dip-pen lithography
Drop casting, 185

e-beam lithography techniques, 165
EBL, *see* Electron-beam lithography
ECM, *see* Electrochemical metallization
Edge-contact configuration, 128
Edge magnetism, 122
Edge modification, 122
EDL, *see* Electrolyte double layer
EELS, *see* Electron energy loss spectrum
Elastic modulus, 122
Electrical properties, 7, 67, 90, 95, 138, 268
Electrical-pulse-triggered memristor
      devices, 186–215
  based on 2D insulating *h*-BN, 212–215
  based on 2D MoS_2, 204–210
    CVD MoS_2 2d film, 206–208
    for efficient-energy radio frequency,
      208–210
    oxidized MoS_2 nanosheets, 205–206
  based on 2D MoS_2, $MoSe_2$, WS_2, and WSe_2,
    210–212
  bipolar RS in Pt/TiO_2/Ti/Pt 2D memristors,
    199–201
  complementary RS in In-doped TiO_2
    ultrathin film, 196–199
  CRS in heterostructured devices, 191–194
  CRS in TaO_x, HfO_x, and TiO_x devices,
    187–189, 191
  filamentary RS in TiO_2 ultrathin film,
    194–196
  RS in exfoliated 2D perovskite single crystal,
    201–204
Electric field assisted migration, 246
Electric field–assisted switching
    mechanisms, 181
Electric field–induced mechanism, 175
Electrochemical doping process, 253, 255, 281
Electrochemical metallization (ECM), 177,
    179, 183, 187, 189, 191, 197, 209,
    210, 214, 236
Electrochemical metallization/conductive
    bridge (ECM/CB) artificial synapses,
    249–250, 252
Electrochemical transistors, 256
Electroforming process, 174

Electro-iono-photoactive 2D MoS_2 synaptic
    device, 242, 244–245
Electrolyte double layer (EDL), 280
Electromagnetic waves, 146, 147
Electron-based probe techniques, 7
Electron beam deposition, 18
Electron-beam lithography (EBL), 158
Electron diffraction techniques, 7
Electron effects, 214
Electron-electron coupling, 21
Electron energy loss spectrum (EELS), 68,
    90, 97, 99
Electronic devices and 2D heterostructured
    films, 131–133
Electronic mode, 242, 244
Electronic properties, 2, 12, 13, 15, 22, 24, 119,
    121–125, 127, 235, 236
Electronic structure, 3, 7, 122, 145, 147
Electron magnetic field (EM), 146
Electron mobility, 15, 163
Electron–phonon coupling, 21
Electron tunneling, 82, 116
Electron wind mechanism, 178
Electrostatic doping, 151, 163, 259
Electrostatic/electron ion synapses, 236
Electrostatic gating, 116
Electrostatic potential, 82
Electrostatic tuning, 121
Elemental doping, 13
EM, *see* Electron magnetic field
Energy band system, 131
Epitaxial effects, 7
Epitaxial growth, 24, 31
EPSC, *see* Excitatory postsynaptic current
Evaporation-based techniques, 43
Excitatory postsynaptic current (EPSC), 231,
    232, 244, 247, 249, 252, 253, 255,
    266–267, 283, 289
Exteroceptive sensors, 290

Fabrication techniques, 3, 43, 84, 160, 166, 172,
    183, 184, 186, 187, 221, 250, 268
Fermi energy, 145, 148, 151, 298
Fermi-level pining effect, 236, 258
Ferro electric materials-based synapses, 234
Ferromagnetic (FM) states, 135
FET, *see* Field effect transistors
FIB, *see* Focused ion beam lithography
Field effect transistors (FET), 16, 43, 56, 71, 95,
    97, 127, 131–132, 152, 163, 215, 234
Filamentary resistive switching, 177,
    194–197, 202
Floating-gate memories, 132
FM, *see* Ferromagnetic states
FM/Gr/FM sandwiched structures, 137
F-N tunneling, *see* Fowler–Nordheim tunneling
Focused ion beam (FIB) lithography, 158

Index

313

Fourier transform infrared spectroscopy (FTIR)
measurements, 67, 151
Fowler–Nordheim (F–N) tunneling, 178, 217
FTIR, *see* Fourier transform infrared
spectroscopy measurements
Fuse-antifuse mechanism, 194

Ga–In alloy, 81
Galinstan (Ga–In–Sn), 81, 83, 88–90, 92,
102–103, 105, 107, 283
Gallium (Ga) crystals, 80
Gallium hydroxide 2D nanosheets, 105
Gallium oxide, 83, 105
Gaseous precursors, 5, 31
Gas injection method, 90
Gate-assisted modulation, 258
Gate voltage, 115, 116, 154, 183, 192, 216, 217,
220, 242, 252, 258, 279, 299
Gibbs free energy, 83, 89
GNM, *see* Graphene nanomesh
GNR, *see* Graphene nanoribbon
Gra/BP, *see* Graphene/black phosphorus 2D
heterostructured device
Grain boundaries, 3, 6, 9, 10, 12, 184, 201,
206–207, 210, 212, 214, 247
Graphene/black phosphorus (Gra/BP) 2D
heterostructured device, 136
Graphene electrode, 115, 202, 206, 245
Graphene/*h*-BN hetero-interfaces, 28, 115
Graphene hybrid nanostructures, 166
Graphene monolayer, 121
Graphene/MoS$_2$ heterojunction, 127
Graphene/MoS$_2$ heterostructures, 22
Graphene nanodisk, 151
Graphene nanomesh (GNM), 121
Graphene nanoribbon (GNR), 120–122, 129,
149, 151, 152, 166
Graphene nanostructure, 121, 151
Graphene plasmons, 148
Green synthesis method, 92, 94–95
Growth mechanism, 4, 5, 9, 15, 30, 31, 44, 52,
65, 71, 82

Hardware implementation, 234
h-BN, *see* Hexagonal-boron nitride
h-BN/graphene heterostructure, 25, 26
hcp, *see* Hexagonal closed pack monolayer
Hebbian theory, 232, 244
Heptamer antennas, 165
Heterostructured 2D-based nonvolatile optical
memory devices, 217, 220–221
Heterostructured 2D films, 3, 21, 22, 32, 44,
135, 187
electronic devices based on, 131–133
magnetic devices based on, 133–136
spintronic and valleytronic, 137–138
Heterostructured artificial synapse, 257

Heterostructured CVD growth, 21–28
lateral 2D, 26–28
metal/insulator, 28
semiconductor/semiconductor, 26, 28
vertical 2D, 22–26
metal/insulator, 25–26
metal/semiconductor, 22
semiconductor/insulator, 24–25
semiconductor/semiconductor, 22, 24
Hexagonal-boron nitride (*h*-BN), 1, 24–26, 28,
97, 124–127, 131, 220, 235, 250, 286
Hexagonal closed pack (hcp) monolayer, 159
Hexatomicring interface, 121
High-k materials, 201
High-quality 2D films, 4, 9, 31, 84
High-quality single crystal films, 43
High resistance state (HRS), 172, 173, 191–194,
196–202, 204, 206, 210, 215, 259,
292, 299
High-resolution transmission electron
microscope (HRTEM), 59, 92, 97,
99, 102, 194
High temperature CVD process, 5
High temperature deposition factors, 44
High-vacuum deposition-based methods, 79, 101
h-MoS$_2$ film, 52
Hofstadter's butterfly effect, 25
Homeostatic synaptic plasticity, 244
Hot spot, 161
Hot-wire assisted ALD (HWALD), 72
HRS, *see* High resistance state
HRTEM, *see* High-resolution transmission
electron microscope
Human brain operation, 230
Humanoid robotics, 275, 276, 290, 296, 299
Human optical visionary system, 300
Human sensorimotor system, 275
HWALD, *see* Hot-wire assisted ALD
Hybridization, 101, 129, 131, 152
Hydrographical printing, 105
Hyperalgesia, 290, 293, 297, 298

ICP-CVD, *see* Induced coupled plasma
chemical vapor deposition
ILD, *see* Internal level difference
Imprint fluid material, 160
Indium tin oxide (ITO), 83
Induced coupled plasma chemical vapor
deposition (ICP-CVD), 9, 16, 31
Infrared (IR) pattern, 149
Inhibitory postsynaptic current (IPSC), 231,
232, 264
Inhibitory postsynaptic potentials (IPSP), 283
In-In$_2$O$_3$/BaF$_3$ substrate, 151
In-ion doped ultrathin TiO$_2$ optical synaptic
devices, 247–249
Inorganic precursors, 48

314 Index

In-situ ellipsometry map, 64, 65
In–Sn alloys, 83
Intercalation/de-intercalation mechanism, 280
Interfacial magnetic exchange, 138
Interlamination coupling, 152
Interlayer tunneling effect, 116
Internal level difference (ILD), 260
Internal time difference (ITD), 260
Internet of Things (IoT), 275, 289, 290, 299
Interneural signaling, 231
Ion doping, 196, 247
Ion gel, 151
Ionic transfer, 201, 214
Ion intercalation, 185, 196, 252, 256, 278–282
Ionotronic mode, 242, 244
IoT, *see* Internet of Things
IPSC, *see* Inhibitory postsynaptic current
IPSP, *see* Inhibitory postsynaptic potentials
IR, *see* Infrared pattern
IR light sources, 220
ITD, *see* Internal time difference
ITO, *see* Indium tin oxide

Joule heating, 174, 175, 177, 178, 194, 202, 207, 214, 259–260

Laser beams interference, 159
Laser diode, 11
Laser interface lithography (LIL), 159
Lateral 2D heterostructures, 21, 26, 28, 32, 111, 112, 124
Lateral heterojunctions, 111, 112, 121, 122
Lateral memristors, 183
Lattice constant, 28, 100, 115
Lattice disorders (LD), 250
Lattice growth, 12
Layer-by-layer (LBL) growth, 6, 9
Layered double hydroxides (LDH), 95
Layer-over-layer (LOL) morphology, 9
LBL, *see* Layer-by-layer growth
LBL deposition mechanism, 49
LBL interactions, 21
LD, *see* Lattice disorders;
    Low-dimensional materials
LDH, *see* Layered double hydroxides
LEDs, *see* Light-emitting diode
Light absorption, 124, 151, 161, 163, 278
Light-assisted resistive switching, 172
Light-emitting diode (LEDs), 2, 11, 113
Light-matter interaction phenomena, 145–147, 163, 221
Light-plasmon coupling, 151
Li-ions intercalated quasi-2D α-MoO_2 artificial synaptic transistor, 255–257
LIL, *see* Laser interface lithography
Liquid metal galinstan, 156, 283
Liquid metal printing approach, 89, 102

Liquid metal vdW transfer method, 102
Liquid phase exfoliation, 3, 185
Lithium ion intercalation technique, 185
Localized surface plasmons resonance (LSPR), 146–147
LOL, *see* Layer-over-layer morphology
Long-term depression/inhibition, 232, 233, 256–258, 264, 287
Long-term plasticity (LTP), 232
Long-term potentiation (LTP), 232, 233, 242, 244, 247, 255, 287
Low-dimensional (LD) materials, 1
Low energy charge transfer phenomenon, 116
Low pressure chemical vapor deposition (LPCVD), 16
Low resistance state (LRS), 172, 173, 177, 189–194, 196–202, 204, 206, 208, 210, 215, 292
Low-temperature chalcogenization, 84
LPCVD, *see* Low pressure chemical vapor deposition
LRS, *see* Low resistance state
LSPR, *see* Localized surface plasmons resonance
LTP, *see* Long-term plasticity;
    Long-term potentiation

Magnéli nanofilaments, 194
Magnetic devices and 2D heterostructured films, 133–136
Magnetic exchange field (MEF), 133, 135, 138
Magnetoresistance (MR) effect, 136
Magnetron sputtering, 18
Mass spectroscopy analysis, 52
MBE, *see* Molecular beam epitaxy
Mechanical cleavage, 184
Mechanical delamination, 90, 95, 99, 101
Mechanical exfoliation, 3, 13, 21, 79, 84, 90, 98, 101, 111, 184, 185
MEF, *see* Magnetic exchange field
Memory-transistors (memtransistors), 241
Memory unit, 116, 132, 215, 217, 220, 222, 229
Memristive characteristics, 171, 173, 184, 187, 196, 201, 205–207, 214, 215, 276
Memristive switching, 172, 175, 257
Memristor devices, 3, 171–222
    device structure, 182–185
        atomic-layered 2D materials, 184
        electrodes, 185
        fabrication techniques, 184–185
        lateral memristors, 183
        tip-based memristors, 183–184
        vertical memristors, 182–183
    electrical-pulse-triggered, 186–215
        based on 2D insulating h-BN, 212–215
        based on 2D MoS_2, 204–210
        based on 2D MoS_2, $MoSe_2$, WS_2, and WSe_2, 210–212

Index

bipolar RS in Pt/TiO$_2$/Ti/Pt 2D memristors, 199–201

CRS in heterostructured devices, 191–194

CRS in In-doped TiO$_2$ ultrathin film, 196–199

CRS in TaO$_x$, HfO$_x$, and TiO$_x$ devices, 187–189, 191

filamentary RS in TiO$_2$ ultrathin film, 194–196

RS in exfoliated 2D perovskite single crystal, 201–204

heterostructured 2D-based nonvolatile optical memory, 217, 220–221

materials and their RS mechanisms, 174–182
  anion-based, 174–175, 177–179
  cation-based, 179–182

nonvolatile optical resistive memories *vs.* optical sensors, 215–216

optical memories, 216–217

overview, 171–172

resistive switching characteristics, 172–174

Metal/2D semiconductor heterostructures, 163

Metal-induced gap states (MIGS), 115

Metal-insulator-metal (MIM) structures, 206, 212

Metal-insulator-semiconductor (MIS), 161

Metallic tungsten, 56

Metal nitride semiconductors, 80

Metal-organic chemical vapor deposition (MOCVD), 16

Metal-organic precursors, 31, 48

Metal oxide nanopores, 160

Metal oxide semiconductor field effect transistors (MOSFET), 62

Metal/semiconductor (MS) heterogeneous junctions, 125, 127

Metal-semiconductor (MS) interfaces, 147, 161

Metaplasticity properties, 244

Metastable phases, 13

MIGS, *see* Metal-induced gap states

MIM, *see* Metal-insulator-metal structures

MIS, *see* Metal-insulator-semiconductor

MoCl$_5$ precursors, 50

Mo(CO)$_6$, 52, 63

MOCVD, *see* Metal-organic chemical vapor deposition

Mo film sulfurization/selenization, 18

Moiré patterns, 8, 115

Molecular beam epitaxy (MBE), 102, 185

Molybdenum bronze (H$_x$MoO$_3$), 253, 255

Momentum transfer mechanism, 178

Monolayer 2D materials, 122

Monophasic and biphasic liquid metals and alloys, 80–81

MoO$_3$ film, 18, 50, 63, 64

MoS$_2$/BN/graphene van der Waals heterostructured unit, 220

MoS$_2$ film, 6, 11, 13–18, 20, 50–53, 118, 129, 163, 165, 210, 260, 264

MoS$_2$/graphene heterostructured 2D films, 116

MoS$_2$/*h*-BN heterostructures, 24

MoS$_2$ plasmonic photodetector, 165

MoS$_2$/SiO$_2$ optical memristor, 216

MoS$_2$/VS$_2$ 2D heterostructure, 136

MoS$_2$/WS$_2$ lateral 2D heterostructures, 26

MOSFET, *see* Metal oxide semiconductor field effect transistors

Motoneurons, 276, 278

Mott potential, 82

MR, *see* Magnetoresistance effect

MR-based synapses, 234

MS heterogeneous junctions, *see* Metal/semiconductor heterogeneous junctions

MS interfaces, *see* Metal-semiconductor interfaces

Multiple-pen DPN methods, 159

Multiple stage synthesis approach, 99

Multithreading, 230

N^2,N^3-di-tert-butylbutane-2,3-diamine tin (II), 56

Nanodisk antenna, 165

Nanofabrication techniques, 159, 160

Nanogap length, 183

Nanoimprint lithography (NIL), 160, 172

Nanoparticles, 52, 103, 105, 147, 148, 152, 156, 161, 165, 217

Nanopillar structures, 123

Nanoscale dimensionality, 1

Nanoscaled nonvolatile memristors, 171

Nanosphere lithography (NSL), 159

Nanostructured hybrid plasmonic photonic instruments, 166

Nanostructured metal oxide semiconductors (NMOSs), 278

Natural nociceptors, 290–293

Natural oxide layer, 82, 83

NDC, *see* Negative differential conductance

N-doped graphene nanoribbon, 122

NDR, *see* Negative differential resistance

Negative differential conductance (NDC), 121

Negative differential resistance (NDR), 122, 125, 127, 178, 207, 208

Neumann technology-based computers, 171

Neural synaptic electronics, 234

Neuromorphic computation, 268

Neuromorphic devices, 222

Neuromorphic properties, 235, 242

Neuromorphic system, 268

Neuromuscular junction, 276

Neuromuscular sensorimotor system, 289

Neurotransmitters, 231, 232, 253, 256, 276

Nickelocene decomposition, 26

NIL, *see* Nanoimprint lithography

Index

NM, *see* Nonmagnetic states
NMOSs, *see* Nanostructured metal oxide semiconductors
N,N-diethyl-N-(2-methoxy ethyl)-N-methyl ammonium-bis-(trifluoromethylsulfon yl)-imide (DEMETFSI), 242
"No-adaptation" characteristics, 293
Nonmagnetic (NM) states, 135
Nonvolatile optical resistive memories *vs.* optical sensors, 215–216
Nonvolatile resistive switching, 177, 210, 250, 252, 297
NSL, *see* Nanosphere lithography
Nucleation, 4–6, 9, 10, 12, 28, 31, 44, 50, 52, 54, 59, 67, 71, 79, 105, 156

O_2 plasma, 63
Octatomicring interface, 121
Ohmic losses, 148
ONN, *see* Optic-neural network
ONS, *see* Optical neural synaptic device
ONWST, *see* Organic nanowire synaptic transistor
Optical assisted resistive switching, 172
Optical Fono-resonance unit, 154
Optical microphotograph, 252
Optical near-field enhancement, 146
Optical neural synaptic (ONS) device, 286, 287
Optical photodetectors, 9, 85
Optical pulses, 172, 215, 216, 220, 244, 247, 248, 283, 286, 289
Optical RAM, 263
Optical resistive random access memories (RRAM) synaptic devices, 264–268
Optical sensing, 172, 215, 222, 263, 286
Optical sensors, 127, 215–216, 222, 286
Optical sensory functions, 215
Optical sensory system, 276
Optical stimulation, 172, 215, 242, 247, 287
Optical synaptic devices *vs.* photodetectors, 263
Optic-neural network (ONN), 287
Optoelectronic memories, 215
Optoelectronic memristor, 187, 222
Optoelectronic properties, 9, 12, 15, 103, 107, 138, 266
Optogenetics, 276
Organic-inorganic hybrid perovskites, 201
Organic nanostructured-based synapses, 234
Organic nanowire synaptic transistor (ONWST), 287, 289
Organic optoelectronic sensorimotor synapse, 288, 289
Organometallic precursors, 48
Oxidation process, 64, 82, 83
Oxidation/reduction chemical valence synapse, 236
Oxygen Bridge bonding, 67

Oxygen vacancies, 173, 174, 187, 189, 191, 193, 194, 196–198, 200, 201, 296, 298, 300

Paired pulsed depression (PPD), 232, 260
Paired pulsed facilitation (PPF), 232, 233, 247, 250, 253, 255, 256, 260, 264, 265, 289
Pavlov's dog experiment, 244
P-doped graphene nanoribbon, 122
PEA, *see* Phenethylammonium
Peak force tunneling microscope, 90
PE-ALD, *see* Plasma-enhanced atomic layer deposition
PECVD method, 8, 9, 31
Persistent conductivity, 244
Persistent photoconduction (PPC) effect, 264
Perylene-3,4,9,10-tetracarboxylic acid tetrapotassium salt (PTAS), 6, 20, 264
Perylene-3,4,9,10-tetracarboxylic dianhydride (PTCDA), 6, 20, 264
P-FET, *see* P-type enhancement mode field effect transistor
Phase alteration, 253, 257
Phase-change materials-based synapses, 234
Phase change memories, 214, 234, 257
Phase-change synapses, 236, 257–259
Phase-change transition, 257
Phase transformation, 13, 50, 56, 156, 174, 177, 178, 235
Phenethylammonium (PEA), 202
Phosphorene/graphene heterostructures (PNR/GNR), 127
Photoactive mode, 242, 244
Photolithography (PL), 16, 84, 85, 158–160
Photoluminescence (PL), 15, 28, 50, 53, 85, 105, 115, 119
Photoresponsivity, 22, 88, 102, 127, 161, 163, 165, 216, 217, 220, 221
Phototransistor devices, 88
Photovoltaic devices, 11
Photovoltaic effects, 28
Physical adsorption, 47, 56
Physisorption, 51
PI, *see* Polyimide plastic substrate
Piezoelectricity, 97–99
PL, *see* Photolithography; Photoluminescence
Plasma-enhanced atomic layer deposition (PE-ALD), 47, 48, 52, 72
Plasmonic characteristics and photonic devices, 145–166
    Au-WO_3-TiO_2 heterojunction, 161
    doping of 2D MoS_2, 154–156
    fabrication of nanostructured arrays, 158–161
        anodic aluminum oxide (AAO), 160
        dip-pen lithography (DPN), 158–159
        electron-beam lithography (EBL), 158
        focused ion beam (FIB) lithography, 158
        laser interface lithography (LIL), 159

Index

nanoimprint lithography (NIL), 160
nanosphere lithography (NSL), 159–160
overview, 145–146
photodetector devices, 163, 165
plasmonic phenomena in 2D graphene,
148–154
hybrid devices, 152, 154
structural design, 149, 151–152
principles, 146–148
surface functionalization of galinstan alloy,
156–157
Plasmonic Fano resonance devices, 165
Plasmonic modulation, 156
Plasmonic phenomena in 2D graphene, 148–154
hybrid devices, 152, 154
structural design, 149, 151–152
Plasmonic properties, 105, 148, 154, 166
Plasmonic structures, 145–148, 152
Plasmon resonance, 149, 151–154, 157, 161, 166
Plasmon resonance enhanced multicolor
photodetector, 165
PLD, *see* Pulsed laser deposited technique
PMMA, *see* Poly methyl methacrylate
based substrates
p-MSB/WSe$_2$ electrogating switchable
photodetector, 116
PN, *see* Polyethylene naphthalate substrate
PNR/GNR, *see* Phosphorene/graphene
heterostructures
Polycrystalline 2D MoS$_2$ synaptic device,
241–242
Polycrystalline granular structure, 9
Polyethylene naphthalate (PN) substrate, 266
Polyimide plastic (PI) substrate, 20
Polymer actuator, 287, 289
Polymeric substrates, 20
Poly methyl methacrylate (PMMA) based
substrates, 24
Poole-Frenkel emission, 179
Postammonolysis treatment, 99
Postannealing process, 50, 52, 53, 57, 58, 65,
67, 71, 72
Postdeposition annealing process, 63
Postsynaptic current (PSC), 231, 232, 242, 247,
258, 283, 286, 287
Postsynaptic outputs, 231
Postsynaptic potential (PSP) pulses, 276, 283
Post-treatment process, 84, 99, 107
PO$_x$/BP 2D heterostructured device, 260, 262
PPC, *see* Persistent photoconduction effect
PPD, *see* Paired pulsed depression
PPF, *see* Paired pulsed facilitation
Precursor condensation, 47
Predeposition methods, 18
Prefetching, 230
Presynaptic signal, 231, 233, 244
Programming voltage, 206, 220–222

Projection printing technique, 158
Proton intercalated quasi-2D α-MoO$_2$ artificial
synaptic transistor, 252–253, 255
Proton ionic transmitters, 252
PSC, *see* Post synaptic current
PSP, *see* Post synaptic potential pulses
PTAS, *see* Perylene-3,4,9,10-tetracarboxylic acid
tetrapotassium salt
PTCDA, *see* Perylene-3,4,9,10-tetracarboxylic
dianhydride
Pt/HfO$_2$/TiN memristor device, 297, 298
Pt/TiO$_2$/Pt thin-film devices, 173, 195, 200, 201
Pt/TiO$_2$/Ti/Pt memristor device, 199–201
Pt/TiO$_{2-x}$/TiN$_x$O$_y$/TiN memristor, 191
P-type enhancement mode field effect
transistor (P-FET), 299
Pulsed laser deposited (PLD) technique, 18, 102
Pump–probe technique, 118

QDs, *see* Quantum dots
Quantum confinement, 1, 147
Quantum dots (QDs), 183

Radio-frequency switches (RFS), 208
RAM, *see* Random access memories
Raman characterization technique, 102
Raman spectra, 13, 105, 107, 282
Raman spectroscopy, 13, 50, 85, 99, 156
Raman studies, 52, 90
RAMBUS, 230
Random access memories (RAM), 230
Random local potential fluctuation (RLFP), 244
Rapid thermal annealing (RTA), 52
Rectification, 120–122, 125, 127, 173
Redox reaction, 175, 187, 189, 194
Redox/valence type artificial synapses, 252–257
Li-ions intercalated quasi-2D α-MoO$_2$,
255–257
proton intercalated quasi-2D α-MoO$_2$,
252–253, 255
Reduced graphene oxide (rGO), 6
Reflection magnetic circular dichroism
(RMCD), 135
Relaxation characteristics, 293
Resistive random access memories (RRAMs),
187, 199, 234, 259, 263
Resistive switching (RS), 10, 171, 233, 236, 246,
264, 278, 296
bipolar in Pt/TiO$_2$/Ti/Pt 2D memristors,
199–201
characteristics, 171–172
in exfoliated 2D perovskite single crystal,
201–204
filamentary in TiO$_2$ ultrathin film, 194–196
memristive materials and, 174–182
anion-based, 174–175, 177–179
cation-based, 179–182

RFS, *see* Radio-frequency switches
rGO, *see* Reduced graphene oxide
RLFP, *see* Random local potential fluctuation
RMCD, *see* Reflection magnetic circular dichroism
Room-temperature liquid metals, 75
RRAMs, *see* Resistive random access memories
RS, *see* Resistive switching
RTA, *see* Rapid thermal annealing

SAED, *see* Selected area electron diffraction
Sb_2Te_3 layer, 59
SBH, *see* Schottky barrier height
Scanning transmission electron microscopy (STEM), 31, 65
Scanning tunneling electron microscope (STEM), 202
Scanning tunneling microscopy (STM), 7, 22, 68, 115, 187, 189, 196, 210
Scanning tunneling spectroscopy (STS), 115
Schottky barrier, 87, 121, 128–131, 161, 165, 178, 214, 216, 217, 258
Schottky barrier height (SBH), 115, 127, 129, 222
Scotch-tape-based method, 21
Screen printing technique, 84
    2D Ga_2O_3 and GaS, 84
    wafer-scale, 85, 87–88
Screw-dislocation driven (SDD) growth, 9
SEBS, *see* Styrene ethylene butylene substrate
Selected area electron diffraction (SAED), 99
Selective area deposition technique, 84
Selenium nanoparticles, 156
Self-limiting growth mechanism, 44, 82, 84
Self-limiting layer synthesis (SLS), 50, 52
Self-limiting metal oxidation reaction, 82
Self-saturation mechanism, 44, 47
Sensitization of nociceptors, 293
Sensorimotor devices, 275–299
    2D SiO_2 diffusive memristors
        artificial nociceptors based on, 290–293
        thermal nociceptors based on, 295–296
    artificial nociceptors
        based on 2D SiO_2 diffusive memristors, 290–293
        based on ultrathin HfO_2 memristors, 296–299
    bioinspired, 275–276
    optoelectronic devices, 276–289
        based on *h*-BN/WSe_2 2D heterostructure, 286–287
        based on ultrathin TiO_2 film, 278–283, 286
        ultrathin ZnO photodetectors, 287, 289
Sensorimotor neurons, 276
Sensory functions, 215, 276, 278, 290, 291

SERS, *see* Surface enhanced Raman scattering
Short-term plasticity (STP), 232, 249, 250, 253, 258, 260, 263, 264
Silicon micro-fabrication techniques, 15
Silver nanoparticles, 105
Single-crystal 2D films, 10, 31
Single particle transportation, 116
Single wall carbon nanotubes (SWCNTs), 287, 289
Si/SiO_2 substrate, 6, 16, 52, 59, 65, 67, 84–87, 95, 99, 138, 165, 242
SLS, *see* Self-limiting layer synthesis
Software implementation, 234
Solid electrolyte/MoO_3 interface, 255
Solidification, 5
Solid precursors, 5
Solid-state electrochemical interaction, 174
Solid-state nociceptor, 292
Solid substrate, 7
Solution-based synthesis, 185
Sonochemical-assisted reactions, 103
Sonochemical reactions, 156
Spatiotemporal learning rule (STLR), 264
Spike-timing-dependent plasticity (STDP), 232, 233, 260, 267
Spin coating, 18, 159, 185
Spin filter effect, 122
Spin polarization, 122, 135, 137
Spintronic and valleytronic 2D heterostructured films, 137–138
Spin valve effect, 137
SP mode, *see* Surface plasmon mode
SPP, *see* Surface plasmon polariton
Sputtering techniques, 18, 43
Stacking order, 11, 12
Stack orientations, 71, 115
Staggered gap, 113
Static RAM, 234
STDP, *see* Spike-timing-dependent plasticity
STEM, *see* Scanning transmission electron microscopy
STLR, *see* Spatiotemporal learning rule
STM, *see* Scanning tunneling microscopy
STM redox-based lithography techniques, 184, 189
Stoichiometric 2D oxide nanofilms, 49
STP, *see* Short-term plasticity
Straddling gap, 113
Straddling heterostructures, *see* Type I heterostructure
Structural phase, 3, 4, 12, 13, 257
STS, *see* Scanning tunneling spectroscopy
Styrene ethylene butylene substrate (SEBS), 289
Sulfurization, 10, 18, 84, 85, 87
Sulfur vacancies, 13, 14, 206, 210

Index

Surface activation mechanisms, 71
Surface enhanced Raman scattering (SERS), 105, 157
Surface plasmon (SP) mode, 148
Surface plasmon polariton (SPP), 146–148
SWCNTs, *see* Single wall carbon nanotubes
Synaptic plasticity, 232, 235, 236, 241–245, 250, 255, 258, 266, 268

TCM, *see* Thermochemical memories; Thermochemical memristor
TDMASn, *see* Tetrakis (dimethylamino) tin
TDMAT, *see* Tetrakis (dimethylamino) titanium precursor
TEM, *see* Transmission electron microscopy
Tensile strength, 125
Tetrakis (dimethylamido) molybdenum $(Mo(NMe_2)_4$, 52
Tetrakis (dimethylamino) tin (TDMASn), 57
Tetrakis (dimethylamino) titanium (TDMAT) precursor, 67
TF, *see* Thickness fluctuations
TFET, *see* Tunnel field effect transistor
Thermal annealing, 59
Thermal decomposition, 5, 20–22
Thermal deposition, 18
Thermal-diffusion process, 177, 182
Thermal evaporation, 9
Thermally assisted RS mechanism, 177, 178
Thermal nociceptors and 2D SiO_2 diffusive memristors, 295–296
Thermionic emission, 130, 131, 161
Thermochemical memories (TCM), 214
Thermochemical memristor (TCM), 177
Thermochemical synapses, 236
Thermodynamic-activated deposition mechanism, 5
Thermoelectric module, 295, 296
Thermoplastic polymer, 160
Thickness fluctuations (TF), 250
Thin-film depositions, 22, 43, 45, 59
Threefold lattice symmetry, 12
Threshold switching (TS), 175, 177, 178, 259, 292, 296, 297
Tin acetate $(Sn(OAc)_4)$, 57
Tin precursors, 56
TiO_2 optical synaptic device, 247–249
Tip-based memristors, 183–184
TMA, *see* Trimethylaluminum
TMDC/mica vertical heterostructures, 24
TMDCs, *see* Transition metal dichalcogenides
TMPS, *see* Transition metal phosphorus trichalcogenides
TMR, *see* Tunneling magnetoresistance
Top contacts configuration, 128, 129
Top-down and bottom-up nanofabrication, 160

Transition metal dichalcogenides (TMDCs), 1, 5, 6, 10, 24, 32, 51, 56, 97, 124, 145, 163
defects in, 14
monolayer, 50, 113, 124, 137, 138
structural characteristics of, 13
Transition metal oxides, 1, 5, 65, 97, 174, 175, 187, 234
Transition metal phosphorus trichalcogenides (TMPS), 135
Transmission electron microscopy (TEM), 24, 52, 85, 101, 202, 210, 212
Trap-assisted tunneling mechanism, 298
Trap-to-trap tunneling, 179
Trimethylaluminum (TMA), 67
Trions, 119
TS, *see* Threshold switching
Tungsten electrode, 85, 87
Tungsten hexacarbonyl $W(CO)_6$, 65
Tunnel field effect transistor (TFET), 113, 132
Tunneling magnetoresistance (TMR), 137
Turing, Alan, 230
Two-dimensional (2D) materials, 1–4
Two-step annealing approach, 87
Two-terminal sandwiched ITO/In-doped TiO_2/Au device, 282
Type-I band alignment, *see* Straddling gap
Type I heterostructure, 22, 116
Type-II band alignment, *see* Staggered gap
Type II heterostructures, 22
Type III hetero-interfaces, 24, 113

Ultrafast electron/holes separation, 118
Ultrasonic-assisted technique, 156
Ultra-sonication process, 105, 159
Ultrathin oxide film, 8, 60, 90, 171, 296, 300
Ultrathin TiO_2 film, sensorimotor device based on, 278–286
ion intercalation for visible light sensitivity, 278–282
system, 282–283, 286
working mechanism, 278
Unipolar RS mechanism, 175, 177, 181, 211
UV-Nil technique, 160

Valance change mechanism, 194
Valence band maximum (VBM), 22, 24, 113, 116, 118, 121, 125
Valence change memory (VCM), 173, 177, 183, 187, 189, 196–197, 199–201, 210, 214
Valleytronic characteristics, 137
Van der Waals (vdW) interaction, 12, 21, 50, 111, 115
Van der Waals 2D graphene heterostructures, 236
Van der Waals exfoliation, *see* Mechanical delamination
Van der Waals (vdWs) force, 84, 185

Index

Vapor phase–based direct growth, 4
VBM, *see* Valence band maximum
VCM, *see* Valence change memory
vdW, *see* Van der Waals
Vertical 2D heterostructures, CVD growth
    of, 21–26
    metal/insulator, 25–26
    metal/semiconductor, 21–23
    semiconductor/insulator, 24–25
    semiconductor/semiconductor, 23–24
Vertical memristors, 182–183
Visible light-assisted technologies, 161
Visible light reception, 278
Visible light sensitivity, 278–282, 286
Visual light cognition, 276
Volatile RS, 250
Voltaic driven memristors, 171–172
Von Neumann computers, 229, 230, 234

Wafer-scale continuous CVD growth, 15–21, 31
    TMDCs films on flexible substrates, 18,
        20–21
    TMDCs films on rigid substrates, 16–18
        conventional techniques, 16, 18
        metal-organic chemical vapor deposition
          (MOCVD), 16

post annealing process, 18
predeposition methods, 18
pulsed laser deposited (PLD)
    technique, 18
sulfurization/selenization of Mo film, 18
WCL, *see* Weight control layer
WCl_6, 56
Weight control layer (WCL), 286
WS_2/MoS_2 heterostructure, 115, 118, 119
WS_2/SnS layered semiconductor
    heterojunctions, 59
WS_2/VS_2 2D heterostructure, 135
WSe_2/CrI_3 heterostructure, 135
$WSe_2/$graphene barristor, 257
WSe_2/h-BN heterostructured 2D device, 217, 220
WSe_2/MoS_2 photodetectors, 115

XPS analysis, 50, 67, 68, 101, 113, 198, 281
X-ray photoelectron spectroscopy, 101
XRD, 85, 92, 94, 99, 101, 105

Young's modulus, 95

Zeeman splitting, 138
Zigzag graphene nanoribbon (zGNR), 120, 122
ZnS film, 55